Progress in Mathematical Physics
Volume 30

Poincaré Seminar 2002

Vacuum Energy – Renormalization

Bertrand Duplantier
Vincent Rivasseau
Editors

Birkhäuser Verlag
Basel · Boston · Berlin

Editors:

Bertrand Duplantier
Service de Physique Théorique
Orme des Merisiers
CEA - Saclay
F-91191 Gif-sur-Yvette Cedex
e-mail: bertrand.duplantier@spht.saclay.cea.fr

Vincent Rivasseau
Laboratoire de Physique Théorique
Université Paris XI
F-91405 Orsay Cedex
e-mail: Vincent.Rivasseau@th.u-psud.fr

2000 Mathematics Subject Classification 81Txx, 83F05, 85A40

This book is also available as a softcover edition (ISBN 3-7643-0527-4).

A CIP catalogue record for this book is available from the Library of Congress, Washington D.C., USA

Bibliographic information published by Die Deutsche Bibliothek
Die Deutsche Bibliothek lists this publication in the Deutsche Nationalbibliografie; detailed bibliographic data is available in the Internet at <http://dnb.ddb.de>.

ISBN 3-7643-0579-7 Birkhäuser Verlag, Basel – Boston – Berlin

Member of the BertelsmannSpringer Publishing Group
Printed on acid-free paper produced of chlorine-free pulp. TCF ∞
Printed in Germany
ISBN 3-7643-0579-7 (hardcover edition)
ISBN 3-7643-0527-4 (softcover edition)

9 8 7 6 5 4 3 2 1

www.birkhauser.ch

Foreword

This book is the first in a series of lectures that have been held in a new kind of seminar, the *Séminaire Poincaré*, which is directed towards a large audience of physicists and of mathematicians.

The goal of this seminar is to provide up-to-date information about general topics of great interest in physics. Both the theoretical and experimental aspects are covered, with some historical background. Inspired by the Bourbaki seminar in mathematics in its organization, hence nicknamed 'Bourbaphy, the Poincaré Seminar is held twice a year at the Institut Henri Poincaré in Paris, with contributions prepared in advance. Particular care is given to the pedagogical nature of the presentation so as to fulfill the goal of being readable by a large audience of scientists.

This volume contains the lectures of the first two such seminars, held in 2002. The first one is devoted to the vacuum energy, in particular the Casimir effect and the nature of the cosmological constant. The second one is devoted to renormalization, giving a comprehensive account of its mathematical structure and applications to high energy physics, statistical mechanics and classical mechanics.

We hope that the publication of this series will serve the community of physicists and mathematicians at professional or graduate student level.

We thank the Centre National de la Recherche Scientifique and the Commissariat à l'Énergie Atomique for sponsoring the Seminar. Special thanks are due to Chantal Delongeas for the preparation of the manuscript.

Bertrand Duplantier
Vincent Rivasseau

Contents

Part I

The Vacuum Energy

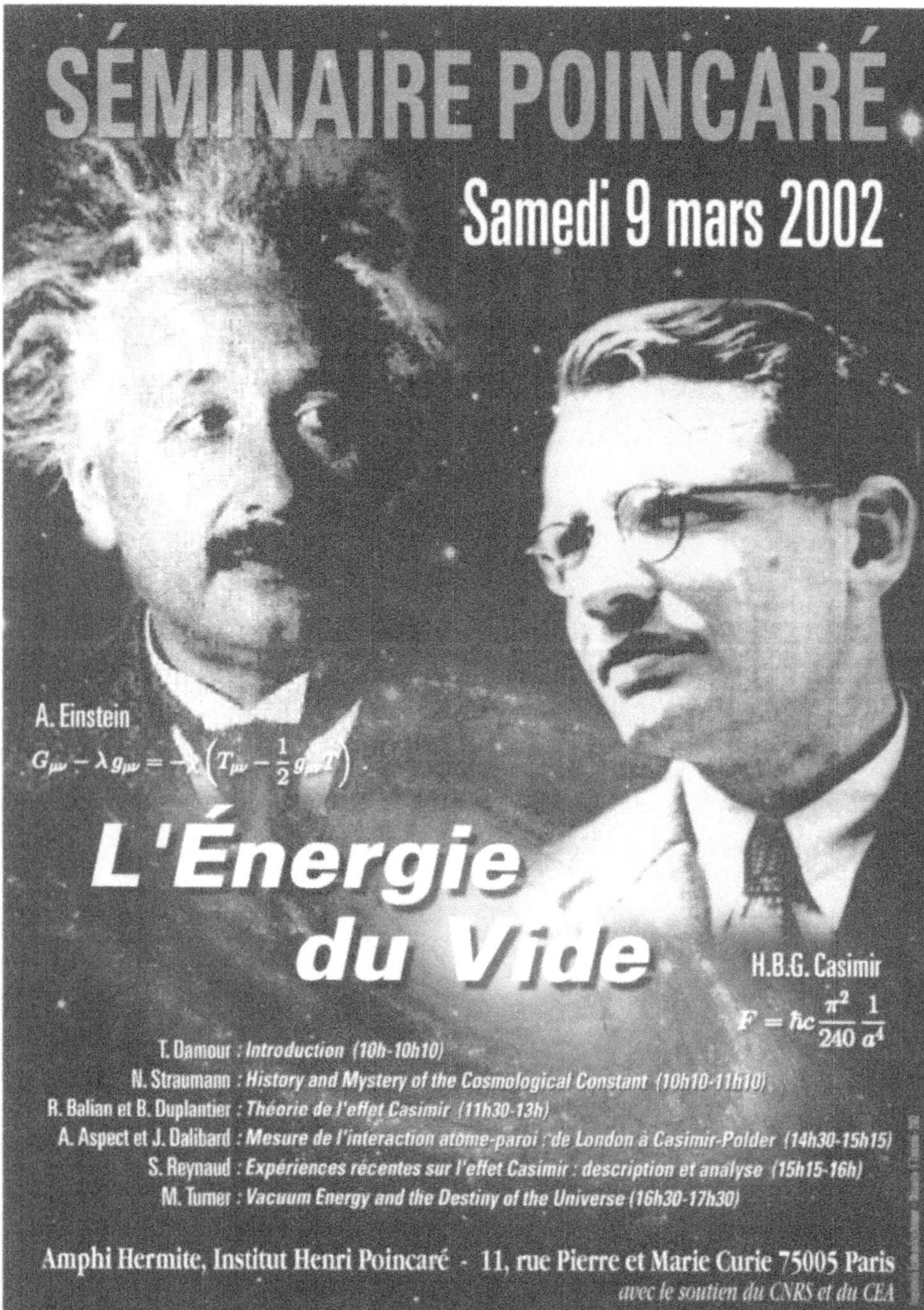
SÉMINAIRE POINCARÉ
Samedi 9 mars 2002
A. Einstein
$G_{\mu\nu} - \lambda g_{\mu\nu} = -\kappa\left(T_{\mu\nu} - \frac{1}{2} g_{\mu\nu} T\right)$
L'Énergie du Vide
H.B.G. Casimir
$F = \hbar c \frac{\pi^2}{240} \frac{1}{a^4}$
T. Damour : Introduction (10h-10h10)
N. Straumann : History and Mystery of the Cosmological Constant (10h10-11h10)
R. Balian et B. Duplantier : Théorie de l'effet Casimir (11h30-13h)
A. Aspect et J. Dalibard : Mesure de l'interaction atome-paroi : de London à Casimir-Polder (14h30-15h15)
S. Reynaud : Expériences récentes sur l'effet Casimir : description et analyse (15h15-16h)
M. Turner : Vacuum Energy and the Destiny of the Universe (16h30-17h30)
Amphi Hermite, Institut Henri Poincaré - 11, rue Pierre et Marie Curie 75005 Paris
avec le soutien du CNRS et du CEA

Poincaré Seminar 2002, 3 – 5

Introduction

Thibault Damour

Ce document rassemble les contributions écrites des conférenciers du premier Séminaire Poincaré de physique théorique (Paris, 9 mars 2002). Cette nouvelle série de séminaires a l'ambition de jouer, pour la physique théorique, un rôle analogue au rôle joué par les séminaires Bourbaki en mathématique: à la fois introduire à, et faire le point sur, un problème important sur lequel beaucoup est déjà connu, mais qui est encore en plein développement.

Le problème retenu pour ce premier séminaire "Bourbaphy" est l'*énergie du vide*. Il s'agit là d'une des questions centrales de la physique, qui est intimement liée à deux des grandes révolutions scientifiques du vingtième siècle: la Théorie Quantique (née des méditations de Planck concernant l'énergie qu'il faut attribuer à l'infinité des modes d'oscillations du champ électromagnétique dans une cavité dont les parois sont maintenues à une certaine température), et la Théorie de la Relativité Générale (dont l'application à la Cosmologie conduisit Einstein à proposer l'existence possible d'une "constante cosmologique", équivalente à attribuer une densité d'énergie et une pression au "vide"). Ces deux aspects de ce que l'on rassemble aujourd'hui sous le vocable "énergie du vide" sont longtemps restés conceptuellement séparés l'un de l'autre. De plus, pendant de longues années, tous deux sont restés très loin de toute vérification expérimentale. La situation actuelle est radicalement différente. Le "problème de la constante cosmologique" est aujourd'hui conceptuellement identifié au problème de la définition de l'énergie moyenne, dans le vide, de l'infinité d'oscillateurs quantiques associés à tous les champs physiques.

Sur le front expérimental, après l'importante réalisation de Casimir (que l'énergie quantique du vide en présence d'une cavité dépendait de la taille et de la forme de cette cavité et conduisait, en particulier, à une force attractive entre deux plaques conductrices), de nombreux expérimentateurs ont cherché à mettre en évidence cet "effet Casimir". Le succès avec lequel ils ont finalement réussi est discuté en détail dans ce document. Quant à la mesure expérimentale de la "constante cosmologique", c'est-à-dire de la densité moyenne de l'énergie du vide à l'échelle cosmologique, elle est longtemps restée le Monstre du Loch Ness de la cosmologie: elle ressortait de temps en temps comme une possibilité que l'on ne pouvait pas exclure, mais le "consensus" tendait bientôt à "sauver les phénomènes" sans introduire une telle constante supplémentaire dont rien ne motivait la présence (au niveau $\rho_{\rm vide} \sim 10^{-29}$ g cm^{-3} pertinent pour jouer un rôle dans l'univers cosmologique actuel). La situation a complètement changé il y a très peu d'années, et le consensus actuel est que les deux tiers de l'énergie moyenne remplissant l'espace cosmologique doit être attribué à une sorte d'"énergie du vide" (appelée, dans ce contexte, l'"énergie sombre", et qu'il faut distinguer de la "matière sombre", maintenant reléguée à ne rendre compte que du tiers de la densité cosmologique moyenne). L'expérience nous oblige donc à prendre au sérieux les prédictions combinées des scientifiques qui, après Planck et Einstein, suivis de Casimir, Pauli, Zel'dovich et beaucoup d'autres, ont conçu le Vide Quantique comme un milieu complexe, fourmillant d'une incessante activité interne donnant lieu à une densité moyenne d'énergie et de pression. Mais ceci nous conduit à l'un des plus profonds paradoxes de la physique fondamentale actuelle: pourquoi la valeur "résiduelle" de l'énergie du vide $\rho_{\rm vide}$ est d'ordre 10^{-29} g cm^{-3}, c'est-à-dire, en unités $\hbar = c = 1$, de l'ordre de 10^{-46} GeV^4? Une telle valeur est si éloignée des échelles de densité d'énergie associées à la physique que l'on connait (échelles du vide de la chromodynamique quantique $\sim (0.1 \text{ GeV})^4$, du secteur électrofaible $\sim (100 \text{ GeV})^4$, ou bien échelle "de Planck" $\sim (10^{19} \text{ GeV})^4$) qu'il s'est révélé, jusqu'à présent, impossible d'inventer un mécanisme pouvant naturellement l'"expliquer".

L'ensemble des contributions rassemblées dans ce document constitue une introduction quasi exhaustive aux aspects essentiels du problème de l'énergie du vide. Le texte de Norbert Straumann ouvre la discussion et brosse un tableau très complet de la position conceptuelle du problème, nourrie par une grande connaissance de son histoire, tout en donnant aussi un exposé très détaillé (et d'une clarté rare) de la physique des données cosmologiques suggérant l'existence d'une valeur non nulle de $\rho_{\rm vide} \sim 10^{-29}$ g cm^{-3} (c'est-à-dire $\Omega_\Lambda \sim 2/3$). La théorie de l'effet Casimir est ensuite discutée en détail par deux des théoriciens dont les travaux ont fait progresser de manière significative cette théorie: Bertrand Duplantier, ayant à l'esprit la vocation didactique des séminaires Bourbaphy, présente d'abord une introduction relativement élémentaire (mais assez complète) des aspects essentiels de l'effet Casimir dans les cas géométriques les plus simples, qui correspondent par ailleurs à la situation expérimentale. Roger Balian attaque ensuite des aspects plus généraux et plus fins de l'effet Casimir: comment l'énergie Casimir dépend-elle de la forme des parois et de la température? Dans quelle condition les diverses limites que l'on doit prendre conduisent-elles à un résultat final bien défini? Les vérifications expérimentales, en laboratoire, de l'effet Casimir sont ensuite discutées en grand détail dans les contributions suivantes. D'abord, Alain Aspect (qui fit l'exposé oral) et Jean Dalibard discutent les aspects théoriques et expérimentaux de l'interaction entre un atome et une paroi. Les effets dus au temps de retard dans cette interaction sont (théoriquement et expérimentalement) très importants. Ils donnent ensuite une discussion détaillée des résultats expérimentaux actuels pour l'interaction atome-paroi (effet Casimir–Polder). Ensuite, Astrid Lambrecht et Serge Reynaud (qui fit l'exposé oral) nous présentent une discussion détaillée des expériences récentes sur l'effet Casimir original (interaction paroi-paroi), et du cadre théorique nécessaire à leur interprétation. Il est fascinant de voir comment la précision des expériences les plus récentes (de l'ordre de 1 %) a permis non seulement de confirmer la réalité de la prédiction originale de Casimir, mais a aussi mis en évidence les effets auxiliaires liés à la réflectivité imparfaite des parois. La journée, et ce document, se terminent par le brillant exposé de Michael Turner qui tire sept leçons profondes de l'existence d'une "énergie sombre" cosmologique. Ce dernier exposé montre encore, s'il en était besoin, la place exceptionnelle que joue la problématique de l'énergie du vide dans la physique (et l'astrophysique) actuelle.

Mais il est temps de laisser au lecteur le plaisir de découvrir la richesse de ce problème. Remercions les auteurs des textes ici rassemblés d'avoir pris le temps de préparer des contributions claires, nourries par l'histoire, mises à jour et donnant accès à la littérature essentielle d'une des questions clefs de la physique actuelle.

Thibault Damour
Institut des Hautes Études Scientifiques
35, route de Chartres
F-91440 Bures-sur-Yvette, France

Poincaré Seminar 2002, 7 – 51

On the Cosmological Constant Problems and the Astronomical Evidence for a Homogeneous Energy Density with Negative Pressure

Norbert Straumann

Abstract. In this article the cosmological constant problems, as well as the astronomical evidence for a cosmologically significant homogeneous exotic energy density with negative pressure (quintessence), are reviewed for a broad audience of physicists and mathematicians. After a short history of the cosmological term it is explained why the (effective) cosmological constant is expected to obtain contributions from short-distance physics corresponding to an energy scale at least as large as the Fermi scale. The actual tiny value of the cosmological constant by particle physics standards represents, therefore, one of the deepest mysteries of present-day fundamental physics. In a second part I shall discuss recent astronomical evidence for a cosmologically significant vacuum energy density or an effective equivalent, called quintessence. Cosmological models, which attempt to avoid the disturbing cosmic coincidence problem, are also briefly reviewed.

1 Introduction

In recent years important observational advances have led quite convincingly to the astonishing conclusion that the present Universe is dominated by an exotic homogeneous energy density with *negative* pressure. I shall discuss the current evidence for this finding in detail later on, but let me already indicate in this introduction the most relevant astronomical data.

First, we now have quite accurate measurements of the anisotropies of the cosmic microwave background radiation (CMB). In particular, the position of the first acoustic peak in the angular power spectrum implies that the Universe is, on large scales, nearly flat (Sect. 6).

On the other hand, a number of observational results, for instance from clusters of galaxies, show consistently that the amount of "matter" (baryons and dark matter) which clumps in various structures is significantly *undercritical.* Hence, there must exist a *homogeneously* distributed exotic energy component.

Important additional constraints come from the Hubble diagram of type Ia supernovas at high redshifts. Although not yet as convincing, they support these conclusions (Sect. 5). More recently, the combination of CMB data and information provided by large scale galaxy redshift surveys have given additional confirmation.

Some of you may say that all this just shows that we have to keep the cosmological term in Einstein's field equations, a possibility has been considered during all the history of relativistic cosmology (see Sect. 2). From our present understanding we would indeed expect a non-vanishing cosmological constant, mainly on the basis of quantum theory, as will be discussed at length later on. However, if a cosmological term describes the astronomical observations, then we are confronted with two difficult problems, many of us worry about:

The first is the *old mystery*: Since all sorts of vacuum energies contribute to the effective cosmological constant (see Sect. 4), we wonder why the total vacuum energy density is so incredibly small by all particle physics standards. Theoreticians are aware of this profound problem since a long time, — at least those who think about the role of gravity among the fundamental interactions. Most probably, we will only have a satisfactory answer once we shall have a theory which successfully combines the concepts and laws of general relativity about gravity and spacetime structure with those of quantum theory.

Before the new astronomical findings one could at least hope that we may one day have a basic understanding for a vanishing cosmological constant, and there have been interesting attempts in this direction (see, e.g., Ref. [1]). But now we are also facing a *cosmic coincidence* problem: Since the vacuum energy density is constant in time — at least after the QCD phase transition —, while the matter energy density decreases as the Universe expands, it is more than surprising that the two are comparable just at the present time, while their ratio has been tiny in the early Universe and will become very large in the distant future.

This led to the idea that the effective cosmological constant we observe today is actually a *dynamical* quantity, varying with time. I want to emphasize already now that these so-called *quintessence* models do, however, not solve the first problem. (More on this in Sect.7.)

This paper is organized as follows. Section 2 is devoted to the instructive early history of the Λ-term, including some early remarks by Pauli on the quantum aspect connected with it. In Section 3 we recall important examples of vacuum energies in quantum electrodynamics and their physical significance under variable external conditions. We then shown in Section 4 that simple and less naive order of magnitude estimates of various contributions to the vacuum energy density of the Standard Model all lead to expectations which are in gigantic conflict with the facts. I then turn to the astronomical and astrophysical aspects of our theme. In Section 5 it be described what is known about the luminosity-redshift relation for type Ia supernovas. The remaining systematic uncertainties are discussed in some detail. Most space of Section 6 is devoted to the physics of the CMB, including of how the system of basic equations which govern its evolution before and after recombination is obtained. We then summarize the current observational results, and what has been learned from them about the cosmological parameters. We conclude in Section 7 with a few remarks about the goal of quintessence models and the main problems this scenario is facing.

2 On the history of the Λ-term

The cosmological term was introduced by Einstein when he applied general relativity for the first time to cosmology. In his paper of 1917 [2] he found the first cosmological solution of a consistent theory of gravity. In spite of its drawbacks this bold step can be regarded as the beginning of modern cosmology. It is still interesting to read this paper about which Einstein says: *"I shall conduct the reader over the road that I have myself travelled, rather a rough and winding road, because otherwise I cannot hope that he will take much interest in the result at the end of the journey."* In a letter to P. Ehrenfest on 4 February 1917 Einstein wrote about his attempt: *"I have again perpetrated something relating to the theory of gravitation that might endanger me of being committed to a madhouse. (Ich habe wieder etwas verbrochen in der Gravitationstheorie, was mich ein wenig in Gefahr bringt, in ein Tollhaus interniert zu werden.)"* [3].

In his attempt Einstein assumed — and this was completely novel — that space is globally *closed*, because he then believed that this was the only way to satisfy Mach's principle, in the sense that the metric field should be determined uniquely by the energy-momentum tensor. In addition, Einstein assumed that the Universe was *static*. This was not unreasonable at the time, because the relative velocities of the stars as observed were small. (Recall that astronomers only learned later that spiral nebulae are independent star systems outside the Milky Way. This was definitely established when in 1924 Hubble found that there were Cepheid variables in Andromeda and also in other galaxies. Five years later he announced the recession of galaxies.)

These two assumptions were, however, not compatible with Einstein's original field equations. For this reason, Einstein added the famous Λ-term, which is compatible with the principles of general relativity, in particular with the energy-momentum law $\nabla_\nu T^{\mu\nu} = 0$ for matter. The modified field equations in standard notation (see, e.g., [15]) and signature $(+ - - -)$ are

$$G_{\mu\nu} = 8\pi G T_{\mu\nu} + \Lambda g_{\mu\nu}. \tag{1}$$

For the static Einstein universe these equations imply the two relations

$$8\pi G\rho = \frac{1}{a^2} = \Lambda, \tag{2}$$

where ρ is the mass density of the dust filled universe (zero pressure) and a is the radius of curvature. (We remark, in passing, that the Einstein universe is the only static dust solution; one does not have to assume isotropy or homogeneity. Its instability was demonstrated by Lemaître in 1927.) Einstein was very pleased by this direct connection between the mass density and geometry, because he thought that this was in accord with Mach's philosophy. (His enthusiasm for what he called Mach's principle later decreased. In a letter to F. Pirani he wrote in 1954: *"As a matter of fact, one should no longer speak of Mach's principle at all. (Von dem Machschen Prinzip sollte man eigentlich überhaupt nicht mehr sprechen".)* [4])

In the same year, 1917, de Sitter discovered a completely different static cosmological model which also incorporated the cosmological constant, but was *anti-Machian*, because it contained no matter [5]. The model had one very interesting property: For light sources moving along static world lines there is a gravitational redshift, which became known as the *de Sitter effect*. This was thought to have some bearing on the redshift results obtained by Slipher. Because the fundamental (static) worldlines in this model are not geodesic, a freely-falling particle released by any static observer will be seen by him to accelerate away, generating also local velocity (Doppler) redshifts corresponding to *peculiar velocities*. In the second edition of his book [6], published in 1924, Eddington writes about this:

> *"de Sitter's theory gives a double explanation for this motion of recession; first there is a general tendency to scatter (...); second there is a general displacement of spectral lines to the red in distant objects owing to the slowing down of atomic vibrations (...), which would erroneously be interpreted as a motion of recession."*

I do not want to enter into all the confusion over the de Sitter universe. This has been described in detail elsewhere (see, e.g., [7]). An important discussion of the redshift of galaxies in de Sitter's model by H. Weyl [8] in 1923 should, however, be mentioned. Weyl introduced an expanding version of the de Sitter model[1]. For *small* distances his result reduced to what later became known as the Hubble law.

Until about 1930 almost everybody *knew* that the Universe was static, in spite of the two fundamental papers by Friedmann [9] in 1922 and 1924 and Lemaître's independent work [10] in 1927. These path breaking papers were in fact largely ignored. The history of this early period has — as is often the case — been distorted by some widely read documents. Einstein too accepted the idea of an expanding Universe only much later. After the first paper of Friedmann, he published a brief note claiming an error in Friedmann's work; when it was pointed out to him that it was his error, Einstein published a retraction of his comment, with a sentence that luckily was deleted before publication: *"[Friedmann's paper] while mathematically correct is of no physical significance"*. In comments to Lemaître during the Solvay meeting in 1927, Einstein again rejected the expanding universe solutions as physically unacceptable. According to Lemaître, Einstein was telling him: *"Vos calculs sont corrects, mais votre physique est abominable"*. On the other hand, I found in the archive of the ETH many years ago a postcard of Einstein to Weyl from 1923 with the following interesting sentence: *"If there is no quasi-static world, then away with the cosmological term"*. This shows once more that history is not as simple as it is often presented.

It also is not well-known that Hubble interpreted his famous results on the redshift of the radiation emitted by distant 'nebulae' in the framework of the de Sitter model. He wrote:

[1] I recall that the de Sitter model has many different interpretations, depending on the class of fundamental observers that is singled out.

> *"The outstanding feature however is that the velocity-distance relation may represent the de Sitter effect and hence that numerical data may be introduced into the discussion of the general curvature of space. In the de Sitter cosmology, displacements of the spectra arise from two sources, an apparent slowing down of atomic vibrations and a tendency to scatter. The latter involves a separation and hence introduces the element of time. The relative importance of the two effects should determine the form of the relation between distances and observed velocities."*

However, Lemaître's successful explanation of Hubble's discovery finally changed the viewpoint of the majority of workers in the field. At this point Einstein rejected the cosmological term as superfluous and no longer justified [11]. He published his new view in the *Sitzungsberichte der Preussischen Akademie der Wissenschaften.* The correct citation is:

Einstein. A. (1931). Sitzungsber. Preuss. Akad. Wiss. 235–37.

Many authors have quoted this paper but never read it. As a result, the quotations gradually changed in an interesting, quite systematic fashion. Some steps are shown in the following sequence:

- A. Einstein. 1931. Sitzsber. Preuss. Akad. Wiss. ...
- A. Einstein. Sitzber. Preuss. Akad. Wiss. ... (1931)
- A. Einstein (1931). Sber. preuss. Akad. Wiss. ...
- Einstein. A .. 1931. Sb. Preuss. Akad. Wiss. ...
- A. Einstein. S.-B. Preuss. Akad. Wis. ...1931
- A. Einstein. S.B. Preuss. Akad. Wiss. (1931) ...
- Einstein, A., and Preuss, S.B. (1931). Akad. Wiss. **235**

Presumably, one day some historian of science will try to find out what happened with the young physicist S.B. Preuss, who apparently wrote just one important paper and then disappeared from the scene.

Einstein repeated his new standpoint much later [12], and this was also adopted by many other influential workers, e.g., by Pauli [13]. Whether Einstein really considered the introduction of the Λ-term as "the biggest blunder of his life" appears doubtful to me. In his published work and letters I never found such a strong statement. Einstein discarded the cosmological term just for simplicity reasons. For a minority of cosmologists (O. Heckmann, for example [14]), this was not sufficient reason.

After the Λ-force was rejected by its inventor, other cosmologists, like Eddington, retained it. One major reason was that it solved the problem of the age of the Universe when the Hubble time scale was thought to be only 2 billion years (corresponding to the value $H_0 \sim 500\ km\ s^{-1} Mpc^{-1}$ of the Hubble constant). This was even shorter than the age of the Earth. In addition, Eddington and others overestimated the age of stars and stellar systems.

For this reason, the Λ-term was employed again and a model was revived which Lemaître had singled out from the many solutions of the Friedmann–Lemaître equations[2]. This so-called Lemaître hesitation universe is closed and has a repulsive Λ-force ($\Lambda > 0$), which is slightly greater than the value chosen by Einstein. It begins with a big bang and has the following two stages of expansion. In the first the Λ-force is not important, the expansion is decelerated due to gravity and slowly approaches the radius of the Einstein universe. At about the same time, the repulsion becomes stronger than gravity and a second stage of expansion begins which eventually inflates into a whimper. In this way a positive Λ was employed to reconcile the expansion of the Universe with the age of stars.

The repulsive effect of a positive cosmological constant can be seen from the following consequence of Einstein's field equations for the time-dependent scale factor $a(t)$:

$$\ddot{a} = -\frac{4\pi G}{3}(\rho + 3p)a + \frac{\Lambda}{3}a, \tag{3}$$

where p is the pressure of all forms of matter.

Historically, the Newtonian analog of the cosmological term was regarded by Einstein, Weyl, Pauli, and others as a *Yukawa term*. This is not correct, as I now show.

For a better understanding of the action of the Λ-term it may be helpful to consider a general static spacetime with the metric (in adapted coordinates)

$$ds^2 = \varphi^2 dt^2 + g_{ik} dx^i dx^k, \tag{4}$$

where φ and g_{ik} depend only on the spatial coordinate x^i. The component R_{00} of the Ricci tensor is given by $R_{00} = \bar{\Delta}\varphi/\varphi$, where $\bar{\Delta}$ is the three-dimensional Laplace operator for the spatial metric $-g_{ik}$ in (4) (see,e.g., [15]). Let us write Eq. (1) in the form

$$G_{\mu\nu} = \kappa(T_{\mu\nu} + T^{\Lambda)}_{\mu\nu} \qquad (\kappa = 8\pi G), \tag{5}$$

with

$$T^{\Lambda}_{\mu\nu} = \frac{\Lambda}{8\pi G} g_{\mu\nu}. \tag{6}$$

This has the form of the energy-momentum tensor of an ideal fluid, with energy density $\rho_\Lambda = \Lambda/8\pi G$ and pressure $p_\Lambda = -\rho_\Lambda$. For an ideal fluid at rest Einstein's field equation implies

$$\frac{1}{\varphi}\bar{\Delta}\varphi = 4\pi G\Big[(\rho + 3p) + \underbrace{(\rho_\Lambda + 3p_\Lambda)}_{-2\rho_\Lambda}\Big]. \tag{7}$$

[2] I recall that Friedmann included the Λ-term in his basic equations. I find it remarkable that for the negatively curved solutions he pointed out that these may be open or compact (but not simply connected).

Since the energy density and the pressure appear in the combination $\rho + 3p$, we understand that a positive ρ_Λ leads to a repulsion (as in (3)). In the Newtonian limit we have $\varphi \simeq 1 + \phi$ (ϕ : Newtonian potential) and $p \ll \rho$, hence we obtain the modified Poisson equation

$$\Delta\phi = 4\pi G(\rho - 2\rho_\Lambda). \tag{8}$$

This is the correct Newtonian limit.

As a result of revised values of the Hubble parameter and the development of the modern theory of stellar evolution in the 1950s, the controversy over ages was resolved and the Λ-term became again unnecessary. (Some tension remained for values of the Hubble parameter at the higher end of recent determinations.)

However, in 1967 it was revived again in order to explain why quasars appeared to have redshifts that concentrated near the value $z = 2$. The idea was that quasars were born in the hesitation era [16]. Then quasars at greatly different distances can have almost the same redshift, because the universe was almost static during that period. Other arguments in favor of this interpretation were based on the following peculiarity. When the redshifts of emission lines in quasar spectra exceed 1.95, then redshifts of absorption lines in the same spectra were, as a rule, equal to 1.95. This was then quite understandable, because quasar light would most likely have crossed intervening galaxies during the epoch of suspended expansion, which would result in almost identical redshifts of the absorption lines. However, with more observational data evidence for the Λ-term dispersed for the third time.

Let me conclude this historical review with a few remarks on the *quantum aspect* of the Λ-problem. Since quantum physicists had so many other problems, it is not astonishing that in the early years they did not worry about this subject. An exception was Pauli, who wondered in the early 1920s whether the zero-point energy of the radiation field could be gravitationally effective.

As background I recall that Planck had introduced the zero-point energy with somewhat strange arguments in 1911. The physical role of the zero-point energy was much discussed in the days of the old Bohr–Sommerfeld quantum theory. From Charly Enz and Armin Thellung — Pauli's last two assistants — I have learned that Pauli had discussed this issue extensively with O. Stern in Hamburg. Stern had calculated, but never published, the vapor pressure difference between the isotopes 20 and 22 of Neon (using Debye theory). He came to the conclusion that without zero-point energy this difference would be large enough for easy separation of the isotopes, which is not the case in reality. These considerations penetrated into Pauli's lectures on statistical mechanics [17] (which I attended). The theme was taken up in an article by Enz and Thellung [18]. This was originally written as a birthday gift for Pauli, but because of Pauli's early death, appeared in a memorial volume of Helv. Phys. Acta.

From Pauli's discussions with Enz and Thellung we know that Pauli estimated the influence of the zero-point energy of the radiation field — cut off at the classical electron radius — on the radius of the universe, and came to the conclusion that it "could not even reach to the moon".

When, as a student, I heard about this, I checked Pauli's unpublished[3] remark by doing the following little calculation:

In units with $\hbar = c = 1$ the vacuum energy density of the radiation field is

$$< \rho >_{vac} = \frac{8\pi}{(2\pi)^3} \int_0^{\omega_{max}} \frac{\omega}{2} \omega^2 d\omega = \frac{1}{8\pi^2} \omega_{max}^4,$$

with

$$\omega_{max} = \frac{2\pi}{\lambda_{max}} = \frac{2\pi m_e}{\alpha}.$$

The corresponding radius of the Einstein universe in eq. (2) would then be ($M_{pl} \equiv 1/\sqrt{G}$)

$$a = \frac{\alpha^2}{(2\pi)^{\frac{2}{3}}} \frac{M_{pl}}{m_e} \frac{1}{m_e} \sim 31 km.$$

This is indeed less than the distance to the moon. (It would be more consistent to use the curvature radius of the static de Sitter solution; the result is the same, up to the factor $\sqrt{3}$.)

For decades nobody else seems to have worried about contributions of quantum fluctuations to the cosmological constant. As far as I know, Zel'dovich was the first who came back to this issue in two papers [19] during the third renaissance period of the Λ-term, but before the advent of spontaneously broken gauge theories. The following remark by him is particularly interesting. Even if one assumes completely ad hoc that the zero-point contributions to the vacuum energy density are exactly cancelled by a bare term (see eq. (29) below), there still remain higher-order effects. In particular, *gravitational* interactions between the particles in the vacuum fluctuations are expected on dimensional grounds to lead to a gravitational self-energy density of order $G\mu^6$, where μ is some cut-off scale. Even for μ as low as 1 GeV (for no good reason) this is about 9 orders of magnitude larger than the observational bound (discussed later).

3 Vacuum fluctuations, vacuum energy

Without gravity, we do not care about the absolute energy of the vacuum, because only energy *differences* matter. In particular, differences of vacuum energies are relevant in many instances, whenever a system is studied under varying external conditions. A beautiful example is the *Casimir effect* [21]. In this case the presence of the conducting plates modifies the vacuum energy density in a manner which depends on the separation of the plates. This implies an attractive force between the plates. Precision experiments have recently confirmed the theoretical prediction

[3] A trace of this is in Pauli's Handbuch article [20] on wave mechanics in the section where he discusses the meaning of the zero-point energy of the quantized radiation field.

to high accuracy (for a recent review, see [22]). We shall consider other important examples, but begin with a very simple one which illustrates the main point.

3.1 A simplified model for the van der Waals force

Recall first how the zero-point energy of the harmonic oscillator can be understood on the basis of the canonical commutation relations $[q,p]=i$. These prevent the *simultaneous* vanishing of the two terms in the Hamiltonian

$$H=\frac{1}{2m}p^2+\frac{1}{2}m\omega^2q^2. \tag{9}$$

The lowest energy state results from a compromise between the potential and kinetic energies, which vary oppositely as functions of the width of the wave function. One understands in this way why the ground state has an absolute energy which is not zero (*zero-point-energy* $\omega/2$).

Next, we consider two identical harmonic oscillators separated by a distance R, which are harmonically coupled by the dipole-dipole interaction energy $\frac{e^2}{R^3}q_1q_2$. With a simple canonical transformation we can decouple the two harmonic oscillators and find for the frequencies of the decoupled ones $\omega_i^2=\omega^2\pm\frac{e^2}{m}\frac{1}{R^3}$, and thus for the ground state energy

$$E_0(R)=\frac{1}{2}(\omega_1+\omega_2)\approx\omega-\frac{e^4}{8\omega^3R^6}.$$

The second term on the right depends on R and gives the van der Waals force (which vanishes for $\hbar\to 0$).

3.2 Vacuum fluctuations for the free radiation field

Similar phenomena arise for quantized fields. We consider, as a simple example, the free quantized electromagnetic field $F_{\mu\nu}(x)$. For this we have for the equal times commutators the following nontrivial one (Jordan and Pauli [23]):

$$\Big[E_i(\boldsymbol{x}),B_{jk}(\boldsymbol{x}')\Big]=i\Big(\delta_{ij}\frac{\partial}{\partial x_k}-\delta_{ik}\frac{\partial}{\partial x_j}\Big)\delta^{(3)}(\boldsymbol{x}-\boldsymbol{x}') \tag{10}$$

(all other equal time commutators vanish); here $B_{12}=B_3$, and cyclic. This basic commutation relation prevents the simultaneous vanishing of the electric and magnetic energies. It follows that the ground state of the quantum field (the vacuum) has a non-zero absolute energy, and that the variances of $\boldsymbol{E}$ and $\boldsymbol{B}$ in this state are nonzero. This is, of course, a quantum effect.

In the Schrödinger picture the electric field operator has the expansion

$$\boldsymbol{E}(\boldsymbol{x})=\frac{1}{(2\pi)^{3/2}}\int\frac{d^3k}{\sqrt{2\omega(k)}}\sum_\lambda\Big[i\omega(k)a(\boldsymbol{k},\lambda)\epsilon(\boldsymbol{k},\lambda)\exp(i\boldsymbol{k}\cdot\boldsymbol{x})+h.c.\Big]. \tag{11}$$

(We use Heaviside units and always set $\hbar=c=1$.)

Clearly,

$$< \boldsymbol{E}(\boldsymbol{x}) >_{vac} = 0.$$

The expression $< \boldsymbol{E}^2(\boldsymbol{x}) >_{vac}$ is not meaningful. We smear $\boldsymbol{E}(\boldsymbol{x})$ with a real test function f:

$$\begin{aligned} \boldsymbol{E}_f(\boldsymbol{x}) &= \int \boldsymbol{E}(\boldsymbol{x}+\boldsymbol{x}')f(\boldsymbol{x}')d^3x' \\ &= \frac{1}{(2\pi)^{3/2}} \int \frac{d^3k}{\sqrt{2\omega(k)}} \sum_\lambda \Big[i\omega(k)a(\boldsymbol{k},\lambda)\boldsymbol{\epsilon}(\boldsymbol{k},\lambda)\exp(i\boldsymbol{k}\cdot\boldsymbol{x})\hat{f}(\boldsymbol{k}) + h.c. \Big], \end{aligned}$$

where

$$\hat{f}(\boldsymbol{k}) = \int f(\boldsymbol{x}) \exp(i\boldsymbol{k}\cdot\boldsymbol{x})\, d^3x.$$

It follows immediately that

$$< \boldsymbol{E}_f^2(\boldsymbol{x}) > = 2 \int \frac{d^3k}{(2\pi)^3} \frac{\omega}{2} |\hat{f}(\boldsymbol{k})|^2.$$

For a sharp momentum cutoff $\hat{f}(\boldsymbol{k}) = \Theta(\mathcal{K} - |\boldsymbol{k}|)$, we have

$$< \boldsymbol{E}_f^2(x) >_{vac} = \frac{1}{2\pi^2} \int_0^{\mathcal{K}} \omega^3 d\omega = \frac{\mathcal{K}^4}{8\pi^2}. \tag{12}$$

The vacuum energy density for $|k| \le \mathcal{K}$ is

$$\rho_{vac} = \frac{1}{2} < \boldsymbol{E}^2 + \boldsymbol{B}^2 >_{vac} = < \boldsymbol{E}^2 >_{vac} = \frac{\mathcal{K}^4}{8\pi^2}. \tag{13}$$

Again, without gravity we do not care, but as in the example above, this vacuum energy density becomes interesting when we consider varying external conditions. This leads us to the next example.

3.3 The Casimir effect

This well-known instructive example has already been mentioned. Let us consider the simple configuration of two large parallel perfectly conducting plates, separated by the distance d. The vacuum energy per unit surface of the conductor is, of course, divergent and we have to introduce some intermediate regularization. Then we must subtract the free value (without the plates) for the same volume. Removing the regularization afterwards, we end up with a finite d-dependent result.

Let me give for this simple example the details for two different regularization schemes. If the plates are parallel to the (x_1, x_2)-plane, the vacuum energy per unit surface is (formally):

$$\mathcal{E}_{vac} = \sum_{l=0}^{\infty} \int_{\mathbf{R}^2} \left[k_1^2 + k_2^2 + (\frac{l\pi}{d})^2\right]^{1/2} \frac{d^2k}{(2\pi)^2}. \tag{14}$$

In the first regularization we replace the frequencies ω of the allowed modes (the square roots in eq. (13)) by $\omega \exp(-\frac{\alpha}{\pi}\omega)$, with a parameter α. A polar integration can immediately be done, and we obtain (leaving out the $l = 0$ term, which does not contribute after subtraction of the free case):

$$\begin{aligned} \mathcal{E}_{vac}^{reg} &= \frac{\pi^2}{4} \sum_{l=1}^{\infty} (\frac{l}{d})^3 \int_0^{\infty} \exp(-\alpha\frac{l}{d}\sqrt{1+z})\sqrt{1+z}dz \\ &= \frac{\pi^2}{4} \frac{\partial^3}{\partial\alpha^3} \sum_{l=1}^{\infty} \int_0^{\infty} \exp(-\alpha\frac{l}{d}\sqrt{1+z}) \frac{dz}{1+z}. \end{aligned}$$

In the last expression the sum is just a geometrical series. After carrying out one differentiation the integral can easily been done, with the result

$$\mathcal{E}_{vac}^{reg} = \frac{\pi^2}{2d} \frac{\partial^2}{\partial\alpha^2} \frac{d/\alpha}{e^{\alpha/d} - 1}. \tag{15}$$

Here we use the well-known formula

$$\frac{x}{e^x - 1} = \sum_{n=0}^{\infty} \frac{B_n}{n!} x^n, \tag{16}$$

where the B_n are the Bernoulli numbers. It is then easy to perform the renormalization (subtraction of the free case). Removing afterwards the regularization ($\alpha \to 0$) gives the renormalized result:

$$\mathcal{E}_{vac}^{ren} = -\frac{\pi^2}{d^3} \frac{B_4}{4!} = -\frac{\pi^2}{720} \frac{1}{d^3}. \tag{17}$$

The corresponding force per unit area is

$$\mathcal{F} = -\frac{\pi^2}{240} \frac{1}{d^4} = -\frac{0.013}{(d(\mu m))^4} dyn/cm^2. \tag{18}$$

Next, I describe the ζ-function regularization. This method has found many applications in quantum field theory, and is particularly simple in the present example.

Let me first recall the definition of the ζ-function belonging to a selfadjoint operator A with a purely discrete spectrum, $A = \sum_n \lambda_n P_n$, where the λ_n are the eigenvalues and P_n the projectors on their eigenspaces with dimension g_n. By definition

$$\zeta_A(s) = \sum_n \frac{g_n}{\lambda_n^s}. \tag{19}$$

Assume that A is positive and that the trace of $A^{\frac{1}{2}}$ exists, then

$$TrA^{\frac{1}{2}} = \zeta_A(-1/2). \tag{20}$$

Formally, the sum (13) is — up to a factor 2 — the trace (19) for $A = -\Delta$, where Δ is the Laplace operator for the region between the two plates with the boundary conditions imposed by the ideally conducting plates. (Recall that the term with $l = 0$ is irrelevant.) Since this trace does not exist, we proceed as follows (ζ- function regularization): Use that $\zeta_A(s)$ is well-defined for $\Re s > 2$ and that it can analytically be continued to some region with $\Re s < 2$ including $s = -1/2$ (see below), we can *define* the regularized trace by eq. (19).

The short calculation involves the following steps. For $s > 2$ we have

$$\zeta_{-\Delta}(s) = 2\sum_{l=1}^{\infty}\int_{\mathbf{R}^2} \frac{1}{\left[k_1^2 + k_2^2 + (\frac{l\pi}{d})^2\right]^s}\frac{d^2k}{(2\pi)^2} \tag{21}$$

$$= \frac{1}{2\pi}\frac{\Gamma(s-1)}{\Gamma(s)}\zeta_R(2s-2)(\frac{\pi}{d})^{2(1-s)}, \tag{22}$$

where $\zeta_R(s)$ is the ζ-function of Riemann. For the analytic continuation we make use of the well-known formula

$$\zeta_R(1-s) = \frac{1}{(2\pi)^s}2\Gamma(s)\cos(\frac{\pi s}{2})\zeta_R(s) \tag{23}$$

and find

$$\zeta_{-\Delta}(-1/2) = -\frac{\pi^2}{360}\frac{1}{d^3}. \tag{24}$$

This gives the result (16).

For a mathematician this must look like black magic, but that's the kind of things physicists are doing to extract physically relevant results from mathematically ill-defined formalisms.

One can similarly work out the other components of the energy-momentum tensor, with the result

$$< T^{\mu\nu} >_{vac} = \frac{\pi^2}{720}\frac{1}{d^4}diag(-1,1,1,-3). \tag{25}$$

This can actually be obtained without doing additional calculations, by using obvious symmetries and general properties of the energy-momentum tensor.

By now the literature related to the Casimir effect is enormous. For further information we refer to the recent book [24].

3.4 Radiative corrections to Maxwell's equations

Another very interesting example of a vacuum energy effect was fist discussed by Heisenberg and Euler [25], and later by Weisskopf [26].

When quantizing the electron-positron field one also encounters an infinite vacuum energy (the energy of the Dirac sea):

$$\mathcal{E}_0 = -\sum_{\boldsymbol{p},\sigma} \varepsilon^{(-)}_{\boldsymbol{p},\sigma},$$

where $-\varepsilon^{(-)}_{\boldsymbol{p},\sigma}$ are the negative frequencies of the solutions the Dirac equation. Note that $\mathcal{E}_0$ is *negative*, which already early gave rise to the hope that perhaps fermionic and bosonic contributions might compensate. Later, we learned that this indeed happens in theories with unbroken supersymmetries. The constant $\mathcal{E}_o$ itself again has no physical meaning. However, if an external electromagnetic field is present, the energy levels $\varepsilon^{(-)}_{\boldsymbol{p},\sigma}$ will change. These changes are finite and *physically significant*, in that they alter the equations for the electromagnetic field in vacuum.

The main steps which lead to the correction $\mathcal{L}'$ of Maxwell's Lagrangian $\mathcal{L}_o = -\frac{1}{4}F_{\mu\nu}F^{\mu\nu}$ are the following ones (for details see [27]):

First one shows (Weisskopf) that

$$\mathcal{L}' = -\left[\mathcal{E}_0 - \mathcal{E}_0|_{E=B=0}\right].$$

After a charge renormalization, which ensures that $\mathcal{L}'$ has no quadratic terms, one arrives at a finite correction. For almost homogeneous fields it is a function of the invariants

$$\mathcal{F} = \frac{1}{4}F_{\mu\nu}F^{\mu\nu} = \frac{1}{2}(\boldsymbol{B}^2 - \boldsymbol{E}^2), \tag{26}$$

$$\mathcal{G}^2 = \left(\frac{1}{4}F^*_{\mu\nu}F^{\mu\nu}\right)^2 = (\boldsymbol{E}\cdot\boldsymbol{B})^2. \tag{27}$$

In [27] this function is given in terms of a 1-dimensional integral. For weak fields one finds

$$\mathcal{L}' = \frac{2\alpha^2}{45m^4}\left[(\boldsymbol{E}^2 - \boldsymbol{B}^2)^2 + 7(\boldsymbol{E}\cdot\boldsymbol{B})^2\right] + \cdots. \tag{28}$$

An alternative efficient method to derive this result again makes use of the ζ-function regularization (see, e.g., [28]).

For other fluctuation-induced forces, in particular in condensed matter physics, I refer to the review article [29].

4 Vacuum energy and gravity

When we consider the coupling to gravity, the vacuum energy density acts like a cosmological constant. In order to see this, first consider the vacuum expectation value of the energy-momentum tensor in Minkowski spacetime. Since the vacuum state is Lorentz invariant, this expectation value is an invariant symmetric tensor, hence proportional to the metric tensor. For a curved metric this is still the case, up to higher curvature terms:

$$< T_{\mu\nu} >_{vac} = g_{\mu\nu}\rho_{vac} + \textit{higher curvature terms}. \tag{29}$$

The *effective* cosmological constant, which controls the large scale behavior of the Universe, is given by

$$\Lambda = 8\pi G\rho_{vac} + \Lambda_0, \tag{30}$$

where Λ_0 is a bare cosmological constant in Einstein's field equations.

We know from astronomical observations discussed later in Sect. 5 and 6 that $\rho_\Lambda \equiv \Lambda/8\pi G$ can not be larger than about the critical density:

$$\begin{aligned} \rho_{crit} &= \frac{3H_0^2}{8\pi G} \\ &= 1.88 \times 10^{-29} h_0^2 gcm^{-3} \\ &= 8 \times 10^{-47} h_0^2 GeV^4, \end{aligned} \tag{31}$$

where h_0 is the *reduced Hubble parameter*

$$h_0 = H_0/(100kms^{-1}Mpc^{-1}) \tag{32}$$

and is close to 0.6 [30].

It is a complete mystery as to why the two terms in (29) should almost exactly cancel. This is — more precisely stated — the famous Λ-problem. It is true that we are unable to calculate the vacuum energy density in quantum field theories, like the Standard Model of particle physics. But we can attempt to make what appear to be reasonable order-of-magnitude estimates for the various contributions. This I shall describe in the remainder of this section. The expectations will turn out to be in gigantic conflict with the facts.

Simple estimates of vacuum energy contributions

If we take into account the contributions to the vacuum energy from vacuum fluctuations in the fields of the Standard Model up to the currently explored energy, i.e., about the electroweak scale $M_F = G_F^{-1/2} \approx 300GeV$ (G_F : Fermi coupling constant), we cannot expect an almost complete cancellation, because there is *no symmetry principle* in this energy range that could require this. The only symmetry principle which would imply this is *supersymmetry*, but supersymmetry is broken (if it is realized in nature). Hence we can at best expect a very imperfect cancellation below the electroweak scale, leaving a contribution of the order of M_F^4. (The contributions at higher energies may largely cancel if supersymmetry holds in the real world.)

We would reasonably expect that the vacuum energy density is at least as large as the condensation energy density of the QCD phase transition to the broken phase of chiral symmetry. Already this is far too large: $\sim \Lambda_{QCD}^4/16\pi^2 \sim 10^{-4}\,GeV^4$; this is *more than 40 orders of magnitude larger* than ρ_{crit}. Beside the formation of quark condensates $< \bar{q}q >$ in the QCD vacuum which break chirality, one also expects a gluon condensate $< G_a^{\mu\nu} G_{a\mu\nu} >$ $\sim \Lambda_{QCD}^4$. This produces a significant vacuum energy density as a result of a dilatation anomaly: If Θ_μ^μ denotes the "classical" trace of the energy-momentum tensor, we have [31]

$$T_\mu^\mu = \Theta_\mu^\mu + \frac{\beta(g_3)}{2g_3} G_a^{\mu\nu} G_{a\mu\nu}, \tag{33}$$

where the second term is the QCD piece of the trace anomaly ($\beta(g_3)$ is the β-function of QCD that determines the running of the strong coupling constant). I recall that this arises because a scale transformation is no more a symmetry if quantum corrections are included. Taking the vacuum expectation value of (32), we would again naively expect that $< \Theta_\mu^\mu >$ is of the order M_F^4. Even if this should vanish for some unknown reason, the anomalous piece is cosmologically gigantic. The expectation value $< G_a^{\mu\nu} G_{a\mu\nu} >$ can be estimated with QCD sum rules [32], and gives

$$< T_\mu^\mu >^{anom} \sim (350 MeV)^4, \tag{34}$$

about 45 orders of magnitude larger than ρ_{crit}. This reasoning should show convincingly that the cosmological constant problem is indeed a profound one. (Note that there is some analogy with the (much milder) strong CP problem of QCD. However, in contrast to the Λ-problem, Peccei and Quinn [33] have shown that in this case there is a way to resolve the conundrum.)

Let us also have a look at the Higgs condensate of the electroweak theory. Recall that in the Standard Model we have for the Higgs doublet Φ in the broken phase for $< \Phi^* \Phi > \equiv \frac{1}{2}\phi^2$ the potential

$$V(\phi) = -\frac{1}{2} m^2 \phi^2 + \frac{\lambda}{8}\phi^4. \tag{35}$$

Setting as usual $\phi = v + H$, where v is the value of ϕ where V has its minimum,

$$v = \sqrt{\frac{2m^2}{\lambda}} = 2^{-1/4} G_F^{-1/2} \sim 246 GeV, \tag{36}$$

we find that the Higgs mass is related to λ by $\lambda = M_H^2/v^2$. For $\phi = v$ we obtain the energy density of the Higgs condensate

$$V(\phi = v) = -\frac{m^4}{2\lambda} = -\frac{1}{8\sqrt{2}} M_F^2 M_H^2 = \mathcal{O}(M_F^4). \tag{37}$$

We can, of course, add a constant V_0 to the potential (34) such that it cancels the Higgs vacuum energy in the broken phase — including higher order corrections. This again requires an extreme fine tuning. A remainder of only $\mathcal{O}(m_e^4)$, say, would be catastrophic. This remark is also highly relevant for models of inflation and quintessence.

In attempts beyond the Standard Model the vacuum energy problem so far remains, and often becomes even worse. For instance, in supergravity theories with spontaneously broken supersymmetry there is the following simple relation between the gravitino mass m_g and the vacuum energy density

$$\rho_{vac} = \frac{3}{8\pi G} m_g^2.$$

Comparing this with eq. (30) we find

$$\frac{\rho_{vac}}{\rho_{crit}} \simeq 10^{122} \Big(\frac{m_g}{m_{Pl}} \Big)^2.$$

Even for $m_g \sim 1\ eV$ this ratio becomes 10^{66}. (m_g is related to the parameter F characterizing the strength of the supersymmetry breaking by $m_g = (4\pi G/3)^{1/2} F$, so $m_g \sim 1\ eV$ corresponds to $F^{1/2} \sim 100\ TeV$.)

Also string theory has not yet offered convincing clues why the cosmological constant is so extremely small. The main reason is that a *low energy mechanism* is required, and since supersymmetry is broken, one again expects a magnitude of order M_F^4, which is *at least 50 orders of magnitude too large* (see also [34]). However, non-supersymmetric physics in string theory is at the very beginning and workers in the field hope that further progress might eventually lead to an understanding of the cosmological constant problem.

I hope I have convinced you, that there is something profound that we do not understand at all, certainly not in quantum field theory, but so far also not in string theory. (For other recent reviews, see also [35], [36], and [37]. These contain more extended lists of references.)

This is the moment to turn to the astronomical and astrophysical aspects of our theme. Here, exciting progress can be reported.

5 Luminosity-redshift relation for type Ia supernovas

A few years ago the Hubble diagram for type Ia supernovas gave the first serious evidence for an accelerating Universe. Before presenting and discussing these exciting results we recall some theoretical background.

5.1 Theoretical redshift-luminosity relation

In cosmology several different distance measures are in use. They are all related by simple redshift factors. The one which is relevant in this Section is the *luminosity distance* D_L, defined by

$$D_L = (\mathcal{L}/4\pi\mathcal{F})^{1/2}, \tag{38}$$

where $\mathcal{L}$ is the intrinsic luminosity of the source and $\mathcal{F}$ the observed flux.

We want to express this in terms of the redshift z of the source and some of the cosmological parameters. If the comoving radial coordinate r is chosen such that the Friedmann–Lemaître metric takes the form

$$g = dt^2 - a^2(t)\Big[\frac{dr^2}{1-kr^2} + r^2 d\Omega^2\Big], \quad k = 0, \pm 1, \tag{39}$$

then we have

$$\mathcal{F} dt_0 = \mathcal{L} dt_e \cdot \frac{1}{1+z} \cdot \frac{1}{4\pi (r_e a(t_0))^2}.$$

The second factor on the right is due to the redshift of the photon energy; the indices $0, e$ refer to the present and emission times, respectively. Using also $1+z = a(t_0)/a(t_e)$, we find in a first step:

$$D_L(z) = a_0(1+z)r(z) \quad (a_0 \equiv a(t_0)). \tag{40}$$

We need the function $r(z)$. From

$$dz = -\frac{a_0}{a}\frac{\dot{a}}{a}dt, \qquad dt = -a(t)\frac{dr}{\sqrt{1-kr^2}}$$

for light rays, we see that

$$\frac{dr}{\sqrt{1-kr^2}} = \frac{1}{a_0}\frac{dz}{H(z)} \qquad (H(z) = \frac{\dot{a}}{a}). \tag{41}$$

Now, we make use of the Friedmann equation

$$H^2 + \frac{k}{a^2} = \frac{8\pi G}{3}\rho. \tag{42}$$

Let us decompose the total energy-mass density ρ into nonrelativistic (NR), relativistic (R), Λ, quintessence (Q), and possibly other contributions

$$\rho = \rho_{NR} + \rho_R + \rho_\Lambda + \rho_Q + \cdots . \tag{43}$$

For the relevant cosmic period we can assume that the "energy equation"

$$\frac{d}{da}(\rho a^3) = -3pa^2 \tag{44}$$

also holds for the individual components $X = NR, R, \Lambda, Q, \cdots$. If $w_X \equiv p_X/\rho_X$ is constant,this implies that

$$\rho_X a^{3(1+w_X)} = const. \tag{45}$$

Therefore,

$$\rho = \sum_X (\rho_X a^{3(1+w_X)})_0 \frac{1}{a^{3(1+w_X)}} = \sum_X (\rho_X)_0 (1+z)^{3(1+w_X)}. \tag{46}$$

Hence the Friedmann equation (41) can be written as

$$\frac{H^2(z)}{H_0^2} + \frac{k}{H_0^2 a_0^2}(1+z)^2 = \sum_X \Omega_X (1+z)^{3(1+w_X)}, \tag{47}$$

where Ω_X is the dimensionless density parameter for the species X,

$$\Omega_X = \frac{(\rho_X)_0}{\rho_{crit}}. \tag{48}$$

Using also the curvature parameter $\Omega_K \equiv -k/H_0^2 a_0^2$, we obtain the useful form

$$H^2(z) = H_0^2 E^2(z; \Omega_K, \Omega_X), \tag{49}$$

with

$$E^2(z; \Omega_K, \Omega_X) = \Omega_K (1+z)^2 + \sum_X \Omega_X (1+z)^{3(1+w_X)}. \tag{50}$$

Especially for $z = 0$ this gives

$$\Omega_K + \Omega_0 = 1, \quad \Omega_0 \equiv \sum_X \Omega_X. \tag{51}$$

If we use (48) in (40), we get

$$\int_0^{r(z)} \frac{dr}{\sqrt{1-r^2}} = \frac{1}{H_0 a_0} \int_0^z \frac{dz'}{E(z')} \tag{52}$$

and thus

$$r(z) = \mathcal{S}(\chi(z)), \tag{53}$$

where

$$\chi(z) = |\Omega_K|^{1/2} \int_0^z \frac{dz'}{E(z')} \tag{54}$$

and

$$\mathcal{S} = \begin{cases} \sin\chi & : \quad k = 1 \\ \chi & : \quad k = 0 \\ \sinh\chi & : \quad k = 1. \end{cases} \tag{55}$$

Inserting this in (39) gives finally the relation we were looking for

$$D_L(z) = \frac{1}{H_0} \mathcal{D}_L(z; \Omega_K, \Omega_X), \tag{56}$$

with

$$\mathcal{D}_L(z;\Omega_K,\Omega_X) = (1+z)\frac{1}{|\Omega_K|^{1/2}}\mathcal{S}(|\Omega_K|^{1/2}\int_0^z \frac{dz'}{E(z')}). \tag{57}$$

Note that for a flat universe, $\Omega_K = 0$ or equivalently $\Omega_0 = 1$, the "Hubble-constant-free" luminosity distance is

$$\mathcal{D}_L(z) = (1+z)\int_0^z \frac{dz'}{E(z')}. \tag{58}$$

Astronomers use as logarithmic measures of $\mathcal{L}$ and $\mathcal{F}$ the *absolute and apparent magnitudes* [4], denoted by M and m, respectively. The conventions are chosen such that the *distance modulus* $m - M$ is related to D_L as follows

$$m - M = 5\log\Big(\frac{D_L}{1Mpc}\Big) + 25. \tag{59}$$

Inserting the representation (55), we obtain the following relation between the apparent magnitude m and the redshift z:

$$m = \mathcal{M} + 5\log\mathcal{D}_L(z;\Omega_K,\Omega_X), \tag{60}$$

where, for our purpose, $\mathcal{M} = M - 5\log H_0 - 25$ is an uninteresting fit parameter. The comparison of this theoretical *magnitude redshift relation* with data will lead to interesting restrictions for the cosmological Ω-parameters. In practice often only Ω_M and Ω_Λ are kept as independent parameters, where from now on the subscript M denotes (as in most papers) nonrelativistic matter.

The following remark about *degeneracy curves* in the Ω-plane is important in this context. For a fixed z in the presently explored interval, the contours defined by the equations $\mathcal{D}_L(z;\Omega_M,\Omega_\Lambda) = const$ have little curvature, and thus we can associate an approximate slope to them. For $z = 0.4$ the slope is about 1 and increases to 1.5-2 by $z = 0.8$ over the interesting range of Ω_M and Ω_Λ. Hence even quite accurate data can at best select a strip in the Ω-plane, with a slope in the range just discussed. This is the reason behind the shape of the likelihood regions shown later (Fig. 2).

In this context it is also interesting to determine the dependence of the *deceleration parameter*

$$q_0 = -\Big(\frac{a\ddot{a}}{\dot{a}^2}\Big)_0 \tag{61}$$

on Ω_M and Ω_Λ. At an any cosmic time we obtain from (3) and (45)

$$-\frac{\ddot{a}a}{\dot{a}^2} = \frac{1}{2}\frac{1}{E^2(z)}\sum_X \Omega_X(1+z)^{3(1+w_X)}(1+3w_X). \tag{62}$$

[4] Beside the (bolometric) magnitudes m, M, astronomers also use magnitudes $m_B, m_V, \ldots$ referring to certain wavelength bands B (blue), V (visual), and so on.

For $z = 0$ this gives

$$q_0 = \frac{1}{2} \sum_X \Omega_X (1 + 3w_X) = \frac{1}{2}(\Omega_M - 2\Omega_\Lambda + \cdots). \tag{63}$$

The line $q_0 = 0$ ($\Omega_\Lambda = \Omega_M/2$) separates decelerating from accelerating universes at the present time. For given values of Ω_M, Ω_Λ, etc, (61) vanishes for z determined by

$$\Omega_M(1+z)^3 - 2\Omega_\Lambda + \cdots = 0. \tag{64}$$

This equation gives the redshift at which the deceleration period ends (coasting redshift).

5.2 Type Ia supernovas as standard candles

It has long been recognized that supernovas of type Ia are excellent standard candles and are visible to cosmic distances [38] (the record is at present at a redshift of about 1.7). At relatively closed distances they can be used to measure the Hubble constant, by calibrating the absolute magnitude of nearby supernovas with various distance determinations (e.g., Cepheids). There is still some dispute over these calibration resulting in differences of about 10% for H_0. (For a review see, e.g., [30].)

In 1979 Tammann [39] and Colgate [40] independently suggested that at higher redshifts this subclass of supernovas can be used to determine also the deceleration parameter. In recent years this program became feasible thanks to the development of new technologies which made it possible to obtain digital images of faint objects over sizable angular scales, and by making use of big telescopes such as Hubble and Keck.

There are two major teams investigating high-redshift SNe Ia, namely the 'Supernova Cosmology Project' (SCP) and the 'High-Z Supernova search Team' (HZT). Each team has found a large number of SNe, and both groups have published almost identical results. (For up-to-date information, see the home pages [41] and [42].)

Before discussing these, a few remarks about the nature and properties of type Ia SNe should be made. Observationally, they are characterized by the absence of hydrogen in their spectra, and the presence of some strong silicon lines near maximum. The immediate progenitors are most probably carbon-oxygen white dwarfs in close binary systems, but it must be said that these have not yet been clearly identified. [5]

In the standard scenario a white dwarf accretes matter from a nondegenerate companion until it approaches the critical Chandrasekhar mass and ignites carbon burning deep in its interior of highly degenerate matter. This is followed by an outward-propagating nuclear flame leading to a total disruption of the white

[5] This is perhaps not so astonishing, because the progenitors are presumably faint compact dwarf stars.

dwarf. Within a few seconds the star is converted largely into nickel and iron. The dispersed nickel radioactively decays to cobalt and then to iron in a few hundred days. A lot of effort has been invested to simulate these complicated processes. Clearly, the physics of thermonuclear runaway burning in degenerate matter is complex. In particular, since the thermonuclear combustion is highly turbulent, multidimensional simulations are required. This is an important subject of current research. (One gets a good impression of the present status from several articles in [43]. See also the recent review [44].) The theoretical uncertainties are such that, for instance, predictions for possible evolutionary changes are not reliable.

It is conceivable that in some cases a type Ia supernova is the result of a merging of two carbon-oxygen-rich white dwarfs with a combined mass surpassing the Chandrasekhar limit. Theoretical modelling indicates, however, that such a merging would lead to a collapse, rather than a SN Ia explosion. But this issue is still debated.

In view of the complex physics involved, it is not astonishing that type Ia supernovas are not perfect standard candles. Their peak absolute magnitudes have a dispersion of 0.3–0.5 mag, depending on the sample. Astronomers have, however learned in recent years to reduce this dispersion by making use of empirical correlations between the absolute peak luminosity and light curve shapes. Examination of nearby SNe showed that the peak brightness is correlated with the time scale of their brightening and fading: slow decliners tend to be brighter than rapid ones. There are also some correlations with spectral properties. Using these correlations it became possible to reduce the remaining intrinsic dispersion to $\simeq 0.17 mag$. (For the various methods in use, and how they compare, see [45], and references therein.) Other corrections, such as Galactic extinction, have been applied, resulting for each supernova in a corrected (rest-frame) magnitude. The redshift dependence of this quantity is compared with the theoretical expectation given by eqs. (59) and (56).

5.3 Results

In Fig.1 the Hubble diagram for the high-redshift supernovas, published by the SCP and HZT teams [46], [47], [48] is shown. All data have been normalized by the same (Δm_{15}) method [49]. In both panels the magnitude differences relative to an empty universe are plotted. The upper panel shows the data for both teams separately. These can roughly be summarized by the statement that distant supernovas are in the average *about 0.20 magnitudes fainter than in an empty Friedmann universe.* In the lower panel the data are redshift binned, and the result for the very distant SN 1999ff at $z \simeq 1.7$ is also shown.

The main result of the analysis is presented in Fig. 2. Keeping only Ω_M and Ω_Λ in eq. (56) (whence $\Omega_K = 1 - \Omega_M - \Omega_\Lambda$) in the fit to the data of 79 SNe Ia, and adopting the same luminosity width correction method (Δm_{15}) for all of them, it shows the resulting confidence regions corresponding to 68.3%, 95.4%, and 99.7% probability in the $(\Omega_M, \Omega_\Lambda)$-plane. Taken at face value, this result excludes $\Omega_\Lambda = 0$

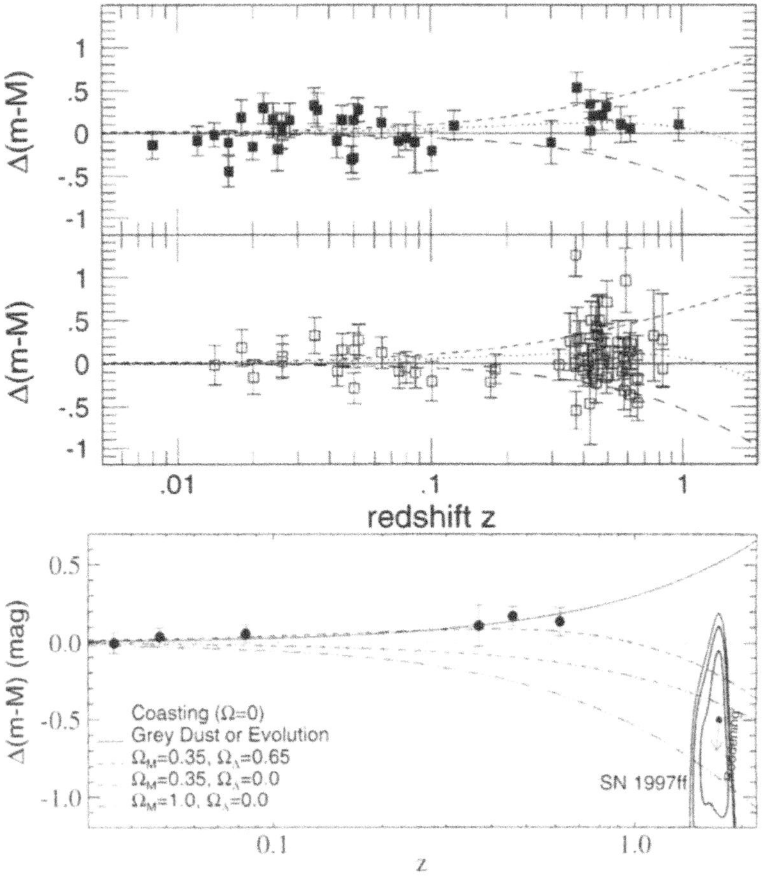

Figure 1: Hubble diagram of Type Ia Supernovas minus an empty (Ω_0) Universe compared to cosmological models. All data in the upper panel have been normalized with the same (Δm_{15}) method (Leibundgut [49]). The filled squares are the data from HZT [47], and those of SCP [46] are shown as open squares. The parameters (Ω_M, Ω_Λ) of the cosmological models are: (1,0) (long dashes), (0,1) (dashed line), (0.3,0.7) (dotted line). In the lower panel the points are redshift-binned data from both teams [51]. A typical curve for grey dust evolution is also shown. In spite of the large uncertainties of SN 1999ff at $z \simeq 1.7$, simple grey dust evolution seems to be excluded.

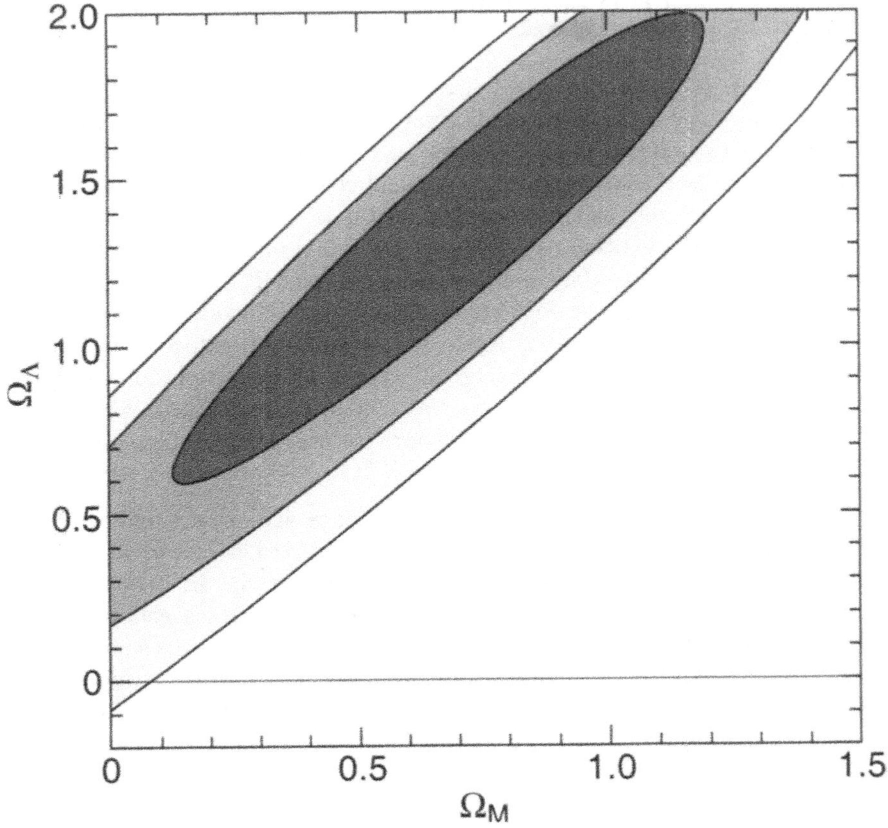

Figure 2: Likelihood regions in the $\Omega_M - \Omega_\Lambda$ plane for the data in Fig. 1. Contours give the 68.3%, 95.4%, and 99.7% statistical confidence regions (adapted from [49]).

for values of Ω_M which are consistent with other observations (e.g., of clusters of galaxies). This is certainly the case if a flat universe is assumed. The probability regions are inclined along $\Omega_\Lambda \approx 1.3\Omega_M + (0.3 \pm 0.2)$. It will turn out that this information is largely complementary to the restrictions we shall obtain in Sect. 6 from the CMB anisotropies.

5.4 Systematic uncertainties

Possible systematic uncertainties due to astrophysical effects have been discussed extensively in the literature. The most serious ones are (i) *dimming* by intergalactic dust, and (ii) *evolution* of SNe Ia over cosmic time, due to changes in progenitor mass, metallicity, and C/O ratio. I discuss these concerns only briefly (see also [49], [50]).

Concerning extinction, detailed studies show that high-redshift SN Ia suffer little reddening; their B-V colors at maximum brightness are normal. However, it can a priori not be excluded that we see distant SNe through a grey dust with grain sizes large enough as to not imprint the reddening signature of typical interstellar extinction. One argument against this hypothesis is that this would also imply a larger dispersion than is observed. The discovery [51] of SN 1997ff with the very high redshift $z \approx 1.7$ led to the conclusion that its redshift and distance estimates are inconsistent with grey dust. Perhaps this statement is too strong, because a pair of galaxies in the foreground of SN 1997ff at $z = 0.56$ may induce a magnification due to gravitational lensing of $\sim 0.3mag$ [52]. With more examples of this type the issue could be settled. eq. (63) shows that at redshifts $z \geq (2\Omega_\Lambda/\Omega_M)^{1/3} - 1 \simeq 1.2$ the Universe is *decelerating*, and this provides an almost unambiguous signature for Λ, or some effective equivalent.

The same SN has provided also some evidence against a simple luminosity evolution that could mimic an accelerating Universe. Other empirical constraints are obtained by comparing subsamples of low-redshift SN Ia believed to arise from old and young progenitors. It turns out that there is no difference within the measuring errors, *after* the correction based on the light-curve shape has been applied. Moreover, spectra of high-redshift SNe appear remarkably similar to those at low redshift. This is very reassuring. On the other hand, there seems to be a trend that more distant supernovas are bluer. It would, of course, be helpful if evolution could be predicted theoretically, but in view of what has been said earlier, this is not (yet) possible.

In conclusion, none of the investigated systematic errors appear to reconcile the data with $\Omega_\Lambda = 0$ and $q_0 \geq 0$. But further work is necessary before we can declare this as a really established fact.

To improve the observational situation a satellite mission called SNAP ("Supernovas Acceleration Probe") has been proposed [53]. According to the plans this satellite would observe about 2000 SNe within a year and much more detailed studies could then be performed. For the time being some scepticism with regard to the results that have been obtained is not out of place.

Finally, I mention a more theoretical complication. In the analysis of the data the luminosity distance for an ideal Friedmann universe was always used. But the data were taken in the real inhomogeneous Universe. This may not be good enough, especially for high-redshift standard candles. The simplest way to take this into account is to introduce a filling parameter which, roughly speaking, represents matter that exists in galaxies but not in the intergalactic medium. For a constant filling parameter one can determine the luminosity distance by solving the Dyer–Roeder equation. But now one has an additional parameter in fitting the data. For a flat universe this was recently investigated in [54].

6 Microwave background anisotropies

By observing the cosmic microwave background (CMB) we can directly infer how the Universe looked at the time of recombination. Besides its spectrum, which is Planckian to an incredible degree [55], we also can study the temperature fluctuations over the "cosmic photosphere" at a redshift $z \approx 1100$. Through these we get access to crucial cosmological information (primordial density spectrum, cosmological parameters, etc). A major reason for why this is possible relies on the fortunate circumstance that the fluctuations are tiny ($\sim 10^{-5}$) at the time of recombination. This allows us to treat the deviations from homogeneity and isotropy for an extended period of time perturbatively, i.e., by linearizing the Einstein and matter equations about solutions of the idealized Friedmann–Lemaître models. Since the physics is effectively *linear*, we can accurately work out the *evolution* of the perturbations during the early phases of the Universe, given a set of cosmological parameters. Confronting this with observations, tells us a lot about the initial conditions, and thus about the physics of the very early Universe. Through this window to the earliest phases of cosmic evolution we can, for instance, test general ideas and specific models of inflation.

6.1 On the physics of CMB

Long before recombination (at temperatures $T > 6000K$, say) photons, electrons and baryons were so strongly coupled that these components may be treated together as a single fluid. In addition to this there is also a dark matter component. For all practical purposes the two interact only gravitationally. The investigation of such a two-component fluid for small deviations from an idealized Friedmann behavior is a well-studied application of cosmological perturbation theory. (For the basic equations and a detailed analytical study, see [56] and [57].)

At a later stage, when decoupling is approached, this approximate treatment breaks down because the mean free path of the photons becomes longer (and finally 'infinite' after recombination). While the electrons and baryons can still be treated as a single fluid, the photons and their coupling to the electrons have to be described by the general relativistic Boltzmann equation. The latter is, of course, again linearized about the idealized Friedmann solution. Together with the linearized fluid equations (for baryons and cold dark matter, say), and the linearized Einstein equations one arrives at a complete system of equations for the various perturbation amplitudes of the metric and matter variables. There exist widely used codes [58], [59] that provide the CMB anisotropies — for given initial conditions — to a precision of about 1%.

A lot of qualitative and semi-quantitative insight into the relevant physics can be gained by looking at various approximations of the 'exact' dynamical system. Below I shall discuss some of the main points. (For well-written papers on this aspect I recommend [60], [61].)

For readers who want to skip this somewhat technical discussion and proceed directly to the observational results (Sect.6.2), the following qualitative remarks may be useful. A characteristic scale, which is reflected in the observed CMB anisotropies, is the sound horizon at last scattering, i.e., the distance over which a pressure wave can propagate until η_{dec}. This can be computed within the unperturbed model and subtends about one degree on the sky for typical cosmological parameters. For scales larger than this sound horizon the fluctuations have been laid down in the very early Universe. These have been detected by the COBE satellite. The (brightness) temperature perturbation $\Theta = \Delta T/T$ (defined precisely in eq. (88) below) is dominated by the combination of the intrinsic temperature fluctuations and gravitational redshift or blueshift effects. For example, photons that have to climb out of potential wells for high-density regions are redshifted. In Sect.6.1.5 it is shown that these effects combine for adiabatic initial conditions to $\frac{1}{3}\Psi$, where Ψ is the gravitational Bardeen potential (see eq. (73)). The latter, in turn, is directly related to the density perturbations. For scale-free initial perturbations the corresponding angular power spectrum of the temperature fluctuations turns out to be nearly flat (Sachs–Wolfe plateau in Fig. 3). The C_l plotted in Fig. 3 are defined in (109) as the expansion coefficients of the angular correlation function in terms of Legendre polynomials.

On the other hand, inside the sound horizon (for $\eta \leq \eta_{dec}$), acoustic, Doppler, gravitational redshift, and photon diffusion effects combine to the spectrum of small angle anisotropies shown in Fig. 3. These result from gravitationally driven acoustic oscillations of the photon-baryon fluid, which are damped by photon diffusion (Sect. 6.1.4).

6.1.1 Cosmological perturbation theory

Unavoidably, the detailed implementation of what has just been outlined is somewhat complicated, because we are dealing with quite a large number of dynamical variables. This is not the place to develop cosmological perturbation theory in any detail [6], but I have to introduce some of it.

Mode decomposition

Because we are dealing with slightly perturbed Friedmann spacetimes we may regard the various perturbation amplitudes as time dependent functions on a three-dimensional Riemannian space (Σ, γ) of constant curvature K. Since such a space is highly symmetrical we are invited to perform two types of decompositions.

In a first step we split the perturbations into *scalar, vector*, and *tensor* contributions. This is based on the following decompositions of vector and symmetric

[6] There is by now an extended literature on cosmological perturbation theory. Beside the recent book [62], the review articles [63], [64], and [65] are recommended. Especially [63] is still useful for the general (gauge invariant) formalism for multi-component systems. Unpublished lecture notes by the author [66] are planned to become available.

tensor fields on Σ : A vector field ξ is a unique sum of a gradient and a vector field ξ_* with vanishing divergence,

$$\xi = \xi_* + \nabla f, \quad \nabla \cdot \xi_* = 0. \tag{65}$$

(If Σ is noncompact we have to impose some fall-off conditions.) The first piece ξ_* is the 'vector' part, and ∇f is the 'scalar' part of ξ. This is a special case of the Hodge decomposition for differential forms. For a symmetric tensor field S_{ij} we have correspondingly :

$$S_{ij} = S_{ij}^{(scalar)} + S_{ij}^{(vector)} + S_{ij}^{(tensor)}, \tag{66}$$

with

$$S_{ij}^{(scalar)} = \gamma_{ij} S^k{}_k + (\nabla_i \nabla_j - \frac{1}{3}\gamma_{ij}\Delta) f, \tag{67}$$

$$S_{ij}^{(vector)} = \nabla_i \xi_j + \nabla_j \xi_i, \tag{68}$$

with $\nabla_k \xi^k = 0$, and where $S_{ij}^{(tensor)}$ is a symmetric tensor field with vanishing trace and zero divergence.

The main point is that these decompositions respect the covariant derivative ∇ on (Σ, γ). For example, if we apply the Laplacian on (64) we readily obtain

$$\Delta \xi = \Delta \xi_* + \nabla(\Delta f + 2Kf),$$

and here the first term has vanishing divergence. For this reason the different components in the perturbation equations *do not mix.*

In a second step we can perform a *harmonic decomposition*, in expanding all amplitudes in terms of generalized spherical harmonics on (Σ, γ). For $K = 0$ this is just Fourier analysis. Again the various modes do not mix, and very importantly, the perturbation equations become for each mode *ordinary* differential equations. (From the Boltzmann equation we get an infinite hierarchy; see below.)

Gauge transformations, gauge invariant amplitudes

In general relativity the diffeomorphism group of spacetime is an *invariance group.* This means that the physics is not changed if we replace the metric g and all the matter variables simultaneously by their diffeomorphically transformed objects. For small amplitude departures from some unperturbed situation, $g = g^{(0)} + \delta g$, etc., this implies that we have the *gauge freedom*

$$\delta g \longrightarrow \delta g + L_\xi g^{(0)}, \quad etc., \tag{69}$$

where L_ξ is the Lie derivative with respect to any vector field ξ. Sets of metric and matter perturbations which differ by Lie derivatives of their unperturbed values are physically equivalent. Such gauge transformations induce changes in the various perturbation amplitudes. It is clearly desirable to write all independent perturbation equations in a *manifestly gauge invariant* manner. Then one gets rid of uninteresting gauge modes, and misinterpretations of the formalism are avoided.

Let me show how this works for the metric. The most general *scalar* perturbation δg of the Friedmann metric

$$g^{(0)} = dt^2 - a^2(t)\gamma = a^2(t)\Big[d\eta^2 - \gamma\Big] \tag{70}$$

can be parameterized as follows

$$\delta g = 2a^2(\eta)\Big[A d\eta^2 + B_{,i}dx^i d\eta - (D\gamma_{ij} + E_{|ij})dx^i dx^j\Big]. \tag{71}$$

The functions $A(\eta, x^i), B, D, E$ are the scalar perturbation amplitudes; $E_{|ij}$ denotes the second covariant derivative $\nabla_i \nabla_j E$ on (Σ, γ). It is easy to work out how A, B, D, E change under a gauge transformation (68) for a vector field ξ of 'scalar' type: $\xi = \xi^0 \partial_0 + \xi^i \partial_i$, with $\xi^i = \gamma^{ij}\xi_{|j}$. From the result one can see that the following *Bardeen potentials* [63]

$$\Psi = A - \frac{1}{a}\Big[a(B + E')\Big]', \tag{72}$$

$$\Phi = D - \mathcal{H}(B + E') \tag{73}$$

are *gauge invariant.* Here, a prime denotes the derivative with respect to the conformal time η, and $\mathcal{H} = a'/a$. The potentials Ψ and Φ are the only independent gauge invariant metric perturbations of scalar type. One can always chose the gauge such that only the A and D terms in (70) are present. In this so-called *longitudinal* or *conformal Newtonian* gauge we have $\Psi = A, \Phi = D$, hence the metric becomes

$$g = a^2(\eta)\Big[(1 + 2\Psi)d\eta^2 - (1 + 2\Phi)\gamma_{ij}dx^i dx^j\Big]. \tag{74}$$

Boltzmann hierarchy

Boltzmann's description of kinetic theory in terms of a one particle distribution function finds a natural setting in general relativity. The metric induces a diffeomorphism between the tangent bundle TM and the cotangent bundle T^*M over the spacetime manifold M. With this the standard symplectic form on T^*M can be pulled back to TM. In natural bundle coordinates the diffeomorphism is: $(x^\mu, p^\alpha) \mapsto (x^\mu, p_\alpha = g_{\alpha\beta}p^\beta)$, hence the symplectic form on TM is given by

$$\omega_g = dx^\mu \wedge d(g_{\mu\nu}p^\nu). \tag{75}$$

The *geodesic spray* is the Hamiltonian vector field X_g on (TM, ω_g) belonging to the "Hamiltonian function" $L = \frac{1}{2}g_{\mu\nu}p^\mu p^\nu$. Thus, in standard notation,

$$i(X_g)\,\omega_g = dL. \tag{76}$$

In bundle coordinates

$$X_g = p^\mu \frac{\partial}{\partial x^\mu} - \Gamma^\mu{}_{\alpha\beta} p^\alpha p^\beta \frac{\partial}{\partial p^\mu}. \tag{77}$$

The integral curves of this vector field satisfy the canonical equations

$$\frac{dx^\mu}{d\lambda} = p^\mu, \tag{78}$$

$$\frac{dp^\mu}{d\lambda} = -\Gamma^\mu{}_{\alpha\beta} p^\alpha p^\beta. \tag{79}$$

The *geodesic flow* is the flow of X_g. The Liouville volume form Ω_g is proportional to the fourfold wedge product of ω_g, and has the bundle coordinate expression

$$\Omega_g = (-g) dx^{0123} \wedge dp^{0123}, \tag{80}$$

where $dx^{0123} \equiv dx^0 \wedge dx^1 \wedge dx^2 \wedge dx^3$, etc. .

The one-particle phase space for particles of mass m is the submanifold $\Phi_m = \{v \in TM \mid g(v,v) = m^2\}$ of TM. This is invariant under the geodesic flow. The restriction of X_g to Φ_m will also be denoted by X_g. Ω_g induces a volume form Ω_m on Φ_m, which is remains invariant under X_g, thus $L_{X_g}\Omega_m = 0$. A simple calculation shows that $\Omega_m = \eta \wedge \Pi_m$, where η is the standard volume form of (M,g), $\eta = \sqrt{-g}dx^{0123}$, and $\Pi_m = \sqrt{-g}dp^{123}/p_0, p_0$ being determined by $g_{\mu\nu}p^\mu p^\nu = m^2$.

Let f be a distribution function on Φ_m. The particle number current density is

$$n^\mu(x) = \int_{P_m(x)} f p^\mu \Pi_m, \tag{81}$$

where $P_m(x)$ is the fiber over x in Φ_m (all momenta with $g(p,p) = m^2$). Similarly, the energy-momentum tensor is

$$T^{\mu\nu} = \int f p^\mu p^\nu \Pi_m. \tag{82}$$

One can show that

$$\nabla_\mu n^\mu = \int_{P_m} (L_{X_g} f) \Pi_m, \tag{83}$$

and

$$\nabla_\nu T^{\mu\nu} = \int_{P_m} p^\mu (L_{X_g} f) \Pi_m. \tag{84}$$

The *Boltzmann equation* has the form

$$L_{X_g} f = C[f], \tag{85}$$

where $C[f]$ is the collision term. If this is (for instance) inserted into (83), we get an expression for the divergence of $T^{\mu\nu}$ in terms of a collision integral. For collisionless particles (neutrinos) this vanishes, of course.

Turning to perturbation theory, we set again $f = f^{(0)} + \delta f$, where $f^{(0)}$ is the unperturbed distribution function of the Friedmann model. For the perturbation δf we choose as independent variables η, x^i, q, γ^j, where q is the magnitude and the γ^j the directional cosines of the momentum vector relative to an orthonormal triad field $\hat{e}_i (i = 1, 2, 3)$ of the unperturbed spatial metric γ on Σ.

From now on we consider always the massless case (photons). By investigating the gauge transformation behavior of δf [67] one can define a gauge invariant perturbation $\mathcal{F}$ which reduces in the longitudinal gauge to δf (there are other choices possible [67]), and derive with some effort the following linearized Boltzmann equation for photons:

$$(\partial_\eta + \gamma^i \hat{e}_i)\mathcal{F} - \hat{\Gamma}^i{}_{jk}\gamma^j\gamma^k \frac{\partial \mathcal{F}}{\partial \gamma^i} - q\Big[\Phi' + \gamma^i \hat{e}_i \Psi\Big]\frac{\partial f^{(0)}}{\partial q} =$$
$$a x_e n_e \sigma_T \Big[< \mathcal{F} > - \mathcal{F} - q\frac{\partial f^{(0)}}{\partial q}\gamma^i \hat{e}_i V_b + \frac{3}{4} Q_{ij}\gamma^i\gamma^j \Big]. \qquad (86)$$

On the left, the $\hat{\Gamma}^i{}_{jk}$ denote the Christoffel symbols of (Σ, γ) relative to the triad $\hat{e}_i$. On the right, $x_e n_e$ is the unperturbed free electron density (x_e = ionization fraction), σ_T the Thomson cross section, and V_b the gauge invariant scalar velocity perturbation of the baryons. Furthermore, we have introduced the spherical averages

$$< \mathcal{F} > = \frac{1}{4\pi}\int_{S^2} \mathcal{F} d\Omega_\gamma, \qquad (87)$$
$$Q_{ij} = \frac{1}{4\pi}\int_{S^2} [\gamma_i\gamma_j - \frac{1}{3}\delta_{ij}]\mathcal{F} d\Omega_\gamma. \qquad (88)$$

In our applications to the CMB we work with the gauge invariant *brightness temperature* perturbation

$$\Theta(\eta, x^i, \gamma^j) = \int \mathcal{F} q^3 dq \Big/ 4\int f^{(0)} q^3 dq. \qquad (89)$$

(The factor 4 is chosen because of the Stephan Boltzmann law, according to which $\delta\rho/\rho = 4\delta T/T$.) It is simple to translate (85) to the following equation for Θ

$$(\Theta + \Psi)' + \gamma^i \hat{e}_i(\Theta + \Psi) - \hat{\Gamma}^i{}_{jk}\gamma^j\gamma^k \frac{\partial}{\partial \gamma^i}(\Theta + \Psi) =$$
$$(\Psi' - \Phi') + \dot{\tau}(\theta_0 - \Theta + \gamma^i \hat{e}_i V_b + \frac{1}{16}\gamma^i\gamma^j \Pi_{ij}), \qquad (90)$$

with $\dot{\tau} = x_e n_e \sigma_T a/a_0$, $\theta_0 = < \Theta >$ (spherical average),

$$\frac{1}{12}\Pi_{ij} = \frac{1}{4\pi}\int [\gamma_i\gamma_j - \frac{1}{3}\delta_{ij}]\Theta \, d\Omega_\gamma. \qquad (91)$$

Let me from now on specialize to the spatially flat case ($K = 0$). In a mode decomposition (Fourier analysis of the x^i-dependence), and introducing the *brightness moments* $\theta_l(\eta)$ by

$$\Theta(\eta, k^i, \gamma^j) = \sum_{l=0}^{\infty}(-i)^l\theta_l(\eta, k)P_l(\mu), \quad \mu = \hat{\boldsymbol{k}} \cdot \gamma, \tag{92}$$

we obtain

$$\Theta' + ik\mu(\Theta + \Psi) = -\Phi' + \dot{\tau}[\theta_0 - \Theta - i\mu V_b - \frac{1}{10}\theta_2 P_2(\mu)]. \tag{93}$$

It is now straightforward to derive from the last two equations the following hierarchy of ordinary differential equations for the brightness moments[7] $\theta_l(\eta)$:

$$\theta_0' = -\frac{1}{3}k\theta_1 - \Phi', \tag{94}$$

$$\theta_1' = k\Big(\theta_0 + \Psi - \frac{2}{5}\theta_2\Big) - \dot{\tau}(\theta_1 - V_b), \tag{95}$$

$$\theta_2' = k\Big(\frac{2}{3}\theta_1 - \frac{3}{7}\theta_3\Big) - \dot{\tau}\frac{9}{10}\theta_2, \tag{96}$$

$$\theta_l' = k\Big(\frac{l}{2l-1}\theta_{l-1} - \frac{l+1}{2l+3}\theta_{l+1}\Big), \quad l > 2. \tag{97}$$

The complete system of perturbation equations

Without further ado I collect below the complete system of perturbation equations. For this some additional notation has to be fixed.

Unperturbed *background* quantities: ρ_α, p_α denote the densities and pressures for the species $\alpha = b$ (baryon and electrons), γ (photons), c (cold dark matter); the total density is the sum $\rho = \sum_\alpha \rho_\alpha$, and the same holds for the total pressure p. We also use $w_\alpha = p_\alpha/\rho_\alpha, w = p/\rho$. The sound speed of the baryon-electron fluid is denoted by c_b, and R is the ratio $3\rho_b/4\rho_\gamma$.

Here is the list of gauge invariant *scalar perturbation* amplitudes (for further explanations see [64]):

- δ_α, δ : density perturbations ($\delta\rho_\alpha/\rho_\alpha, \delta\rho/\rho$ in the longitudinal gauge); clearly: $\rho\,\delta = \sum \rho_\alpha\delta_\alpha$.
- V_α, V : velocity perturbations; $\rho(1 + w)V = \sum_\alpha \rho_\alpha(1 + w_\alpha)V_\alpha$.
- θ_l, N_l : brightness moments for photons and neutrinos.
- Π_α, Π : anisotropic pressures; $\Pi = \Pi_\gamma + \Pi_\nu$. For the lowest moments the following relations hold:

$$\delta_\gamma = 4\theta_0, \;\; V_\gamma = \theta_1, \;\; \Pi_\gamma = \frac{12}{5}\theta_2, \tag{98}$$

[7] In the literature the normalization of the θ_l is sometimes chosen differently: $\theta_l \to (2l+1)\theta_l$.

and similarly for the neutrinos.

- Ψ, Φ: Bardeen potentials for the metric perturbation.

As independent amplitudes we can choose: $\delta_b, \delta_c, V_b, V_c, \Phi, \Psi, \theta_l, N_l$. The basic evolution equations consist of three groups.

- Fluid equations:

$$
\begin{aligned}
\delta_c' &= -kV_c - 3\Phi', && (99)\\
V_c' &= -aHV_c + k\Psi; && (100)\\
\delta_b' &= -kV_b - 3\Phi', && (101)\\
V_b' &= -aHV_b + kc_b^2\delta_b + k\Psi + \dot{\tau}(\theta_1 - V_b)/R. && (102)
\end{aligned}
$$

- Boltzmann hierarchies for photons (eqs. (93)–(96)) and the collisionless neutrinos.

- Einstein equations : We only need the following algebraic ones for each mode:

$$
\begin{aligned}
k^2\Phi &= 4\pi Ga^2\rho\Big[\delta + 3\frac{aH}{k}(1+w)V\Big], && (103)\\
k^2(\Phi+\Psi) &= -8\pi Ga^2 p\,\Pi. && (104)
\end{aligned}
$$

In arriving at these equations some approximations have been made which are harmless [8], except for one: We have ignored polarization effects in Thomson scattering. For quantitative calculations these have to be included. Moreover, polarization effects are highly interesting, as I shall explain later.

6.1.2 Angular correlations of temperature fluctuations

The system of evolution equations has to be supplemented by initial conditions. We can not hope to be able to predict these, but at best their statistical properties (as, for instance, in inflationary models). Theoretically, we should thus regard the brightness temperature perturbation $\Theta(\eta, x^i, \gamma^j)$ as a random field. Of special interest is its angular correlation function at the present time η_0. Observers measure only one realization of this, which brings unavoidable *cosmic variances*.

For further elaboration we insert (91) into the Fourier expansion of Θ, obtaining

$$\Theta(\eta, \boldsymbol{x}, \gamma) = (2\pi)^{-\frac{3}{2}} \int d^3k \sum_l \theta_l(\eta, k) G_l(\boldsymbol{x}, \gamma; \boldsymbol{k}), \tag{105}$$

where

$$G_l(\boldsymbol{x}, \gamma; \boldsymbol{k}) = (-i)^l P_l(\hat{\boldsymbol{k}} \cdot \gamma)\exp(i\boldsymbol{k} \cdot \boldsymbol{x}). \tag{106}$$

[8] In the notation of [64] we have set $q_\alpha = \Gamma_\alpha = 0$, and are thus ignoring certain intrinsic entropy perturbations within individual components.

Hence we have

$$\Theta(\eta, \boldsymbol{x}, \boldsymbol{\gamma}) = \sum_{lm} a^*_{lm} Y_{lm}(\boldsymbol{\gamma}), \tag{107}$$

with

$$a_{lm} = (2\pi)^{-\frac{3}{2}} \int d^3k\, \theta_l(\eta, k)\, i^l \frac{4\pi}{2l+1} Y_{lm}(\hat{\boldsymbol{k}}) \exp(-i\boldsymbol{k}\cdot\boldsymbol{x}). \tag{108}$$

We expect on the basis of rotation invariance that the two-point correlation of the random variables a_{lm} has the form

$$< a_{lm} a^*_{l'm'} >= C_l \delta_{ll'} \delta_{mm'}. \tag{109}$$

From (106) and (108) we see that the angular correlation function of Θ in **x**-space is

$$< \Theta(\boldsymbol{\gamma})\Theta(\boldsymbol{\gamma}') >= \sum_l \frac{2l+1}{4\pi} C_l P_l(\boldsymbol{\gamma}\cdot\boldsymbol{\gamma}'). \tag{110}$$

If different modes in **k**-space are uncorrelated, we obtain from (107)

$$\frac{2l+1}{4\pi} C_l = \frac{1}{2\pi^2} \int_0^\infty \frac{dk}{k} \frac{k^3 |\theta_l(k)|^2}{2l+1}. \tag{111}$$

Cosmic variance

The C_l are the expectation values of the stochastic variable

$$Z = \frac{1}{2l+1} \sum_m a_{lm} a^*_{lm}.$$

If the a_{lm} are Gaussian random variables, as in simple inflationary models, then the variance of Z, and thus of C_l, is easily found to be given by

$$\sigma(C_l) = \sqrt{\frac{2}{2l+1}}. \tag{112}$$

This is a serious limitation for low multipoles that cannot be overcome. For large l the measured C_l should be accurately described by (110), taken at the present time.

6.1.3 Brightness moments in sudden decoupling

The linearized Boltzmann equation in the form (92) as an inhomogeneous linear differential for the η-dependence has the 'solution'

$$(\Theta + \Psi)(\eta_0, \mu; k) = \int_0^{\eta_0} d\eta \Big[\dot{\tau}(\theta_0 + \Psi - i\mu V_b - \frac{1}{10}\theta_2 P_2) + \Psi' - \Phi'\Big] e^{-\tau(\eta,\eta_0)} e^{ik\mu(\eta-\eta_0)}, \tag{113}$$

where

$$\tau(\eta, \eta_0) = \int_\eta^{\eta_0} \dot{\tau} d\eta \tag{114}$$

is the *optical depth*. The combination $\dot{\tau}e^{-\tau}$ is the (conformal) *time visibility function*. It has a simple interpretation: Let $p(\eta, \eta_0)$ be the probability that a photon did not scatter between η and today (η_0). Clearly, $p(\eta - d\eta, \eta_0) = p(\eta, \eta_0)(1 - \dot{\tau} d\eta)$. Thus $p(\eta, \eta_0) = e^{-\tau(\eta, \eta_0)}$, and the visibility function times $d\eta$ is the probability that a photon last scattered between η and $\eta + d\eta$. The visibility function is therefore *strongly peaked* near decoupling. This is very useful, both for analytical and numerical purposes.

In order to obtain an integral representation for the multipole moments θ_l, we insert in (112) for the μ-dependent factors standard expansions in terms of Legendre polynomials. For $l \geq 2$ we find the following useful formula:

$$\frac{\theta_l(\eta_0)}{2l+1} = \int_0^{\eta_0} d\eta e^{-\tau(\eta)} \Big[(\dot{\tau}\theta_0 + \dot{\tau}\Psi + \Psi' - \Phi') j_l(k(\eta_0 - \eta)) + \dot{\tau} V_b j_l' + \dot{\tau}\frac{1}{20}\theta_2(3j_l'' + j_l) \Big]. \tag{115}$$

In a reasonably good approximation we can replace the visibility function by a δ-function, and obtain (with $\Delta\eta \equiv \eta_0 - \eta_{dec}$, $V_b(\eta_{dec}) \simeq \theta_1(\eta_{dec})$:

$$\frac{\theta_l(\eta_0, k)}{2l+1} \simeq [\theta_0 + \Psi](\eta_{dec}, k) j_l(k\Delta\eta) + \theta_1(\eta_{dec}, k) j_l'(k\Delta\eta) + ISW + Quad. \tag{116}$$

Here, the quadrupole contribution (last term) is not important. ISW denotes the *integrated Sachs–Wolfe effect*:

$$ISW = \int_{\eta_{dec}}^{\eta_0} d\eta (\Psi' - \Phi') j_l(k\Delta\eta), \tag{117}$$

which only depends on the time variations of the Bardeen potentials between recombination and the present time.

The interpretation of the first two terms in (115) is quite obvious: The first describes the fluctuations of the *effective* temperature $\theta_0 + \Psi$ on the cosmic photosphere, as we would see them for free streaming between there and us, — if the gravitational potentials would not change in time. (Ψ includes blue- and redshift effects.) The dipole term has to be interpreted, of course, as a Doppler effect due to the velocity of the baryon-photon fluid.

In this approximate treatment we only have to know the effective temperature $\theta_0 + \Psi$ and the velocity moment θ_1 at decoupling. The main point is that eq. (115) provides a good understanding of the physics of the CMB anisotropies. Note that the individual terms are all gauge invariant. In gauge dependent methods interpretations would be ambiguous.

6.1.4 Acoustic oscillations

In this subsection we derive from the Boltzmann hierarchy (93)–(96) an approximate equation for the effective temperature fluctuation $\Delta T \equiv \theta_0 + \Psi$, which will teach us a lot.

As long as the mean free path of photons is much shorter than the wavelength of the fluctuation, the optical depth through a wavelength $\sim \dot{\tau}/k$ is large. Thus the evolution equations may be expanded in $k/\dot{\tau}$.

In lowest order we obtain $\theta_1 = V_b$, $\theta_l = 0$ for $l \geq 2$, thus $\delta_b' = 3\theta_0'$. Going to the first order, we can replace on the right of the following form of eq. (94),

$$\theta_1 - V_b = \frac{R}{\dot{\tau}}[V_b' + \frac{a'}{a}V_b - k\Psi], \tag{118}$$

V_b by θ_1:

$$\theta_1 - V_b = \frac{R}{\dot{\tau}}[\theta_1' + \frac{a'}{a}\theta_1 - k\Psi]. \tag{119}$$

We insert this in (94), and set in first order $\theta_2 = 0$. Using also $a'/a = R'/R$ we get

$$\theta_1' = \frac{1}{1+R}k\theta_0 + k\Psi - \frac{R'}{1+R}\theta_1. \tag{120}$$

Together with (93) we find the driven oscillator equation

$$\theta_0'' + \frac{R'}{1+R}\frac{a'}{a}\theta_0' + c_s^2 k^2 \theta_0 = F(\eta), \tag{121}$$

where

$$c_s^2 = \frac{1}{3(1+R)}, \quad F(\eta) = -\frac{k^2}{3}\Psi - \frac{R}{1+R}\frac{a'}{a}\Phi' - \Phi''. \tag{122}$$

The damping term is due to expansion. In second order one finds an additional damping term proportional to θ_0':

$$\frac{1}{3}\frac{k^2}{\dot{\tau}}\left[(\frac{R}{1+R})^2 + \frac{8}{9}\frac{1}{1+R}\right]\theta_0'. \tag{123}$$

This describes the damping due to photon diffusion (Silk damping).

We discuss here only the first order equation, which we rewrite in the more suggestive form ($m_{eff} \equiv 1 + R$)

$$(m_{eff}\theta_0')' + \frac{k^2}{3}(\theta_0 + m_{eff}\Psi) = -(m_{eff}\Phi')'. \tag{124}$$

This equation may be interpreted as follows: The change in momentum of the photon-baryon fluid is determined by a competition between pressure restoring and gravitational driving forces.

Let us, in a first step, ignore the time dependence of m_{eff} (i.e., of the baryon-photon ratio R), then we get a forced harmonic oscillator equation

$$m_{eff}\theta_0'' + \frac{k^2}{3}\theta_0 = -\frac{k^2}{3}m_{eff}\Psi - (m_{eff}\Phi')'. \tag{125}$$

The effective mass $m_{eff} = 1 + R$ accounts for the inertia of baryons. Baryons also contribute gravitational mass to the system, as is evident from the right hand side of the last equation. Their contribution to the pressure restoring force is, however, negligible.

We now ignore in (124) also the time dependence of the gravitational potentials Φ, Ψ. With (121) this then reduces to

$$\theta_0'' + k^2 c_s^2 \theta_0 = -\frac{1}{3}k^2\Psi. \tag{126}$$

This simple harmonic oscillator under constant acceleration provided by gravitational infall can immediately be solved:

$$\theta_0(\eta) = [\theta_0(0) + (1+R)\Psi]\cos(kr_s) + \frac{1}{kc_s}\dot{\theta}_0(0)\sin(kr_s) - (1+R)\Psi, \tag{127}$$

where $r_s(\eta)$ is the comoving sound horizon $\int c_s d\eta$.

One can show that for *adiabatic* initial conditions there is only a cosine term. In this case we obtain for ΔT:

$$\Delta T(\eta, k) = [\Delta T(0,k) + R\Psi]\cos(kr_s(\eta)) - R\Psi. \tag{128}$$

Discussion

In the radiation dominated phase ($R = 0$) this reduces to $\Delta T(\eta) \propto \cos kr_s(\eta)$, which shows that the oscillation of θ_0 is displaced by gravity. The zero point corresponds to the state at which gravity and pressure are balanced. The displacement $-\Psi > 0$ yields hotter photons in the potential well since gravitational infall not only increases the number density of the photons, but also their energy through gravitational blue shift. However, well after last scattering the photons also suffer a redshift when climbing out of the potential well, which precisely cancels the blue shift. Thus the effective temperature perturbation we see in the CMB anisotropies is — as remarked in connection with eq. (115) — indeed $\Delta T = \theta_0 + \Psi$.

It is clear from (127) that a characteristic wave-number is $k = \pi/r_s(\eta_{dec}) \simeq \pi/c_s\eta_{dec}$. A spectrum of k-modes will produce a sequence of peaks with wave numbers

$$k_m = m\pi/r_s(\eta_{dec}), \quad m = 1, 2, ... \tag{129}$$

Odd peaks correspond to the compression phase (temperature crests), whereas even peaks correspond to the rarefaction phase (temperature troughs) inside the potential wells. Note also that the characteristic length scale $r_s(\eta_{dec})$, which is reflected in the peak structure, is determined by the underlying unperturbed Friedmann model. This comoving sound horizon at decoupling depends on cosmological parameters, but not on Ω_Λ. Its role will further be discussed in Sect. 6.2.

Inclusion of baryons not only changes the sound speed, but gravitational infall leads to greater compression of the fluid in a potential well, and thus to a further displacement of the oscillation zero point (last term in (127). This is not compensated by the redshift after last scattering, since the latter is not affected by the baryon content. As a result all peaks from compression are enhanced over those from rarefaction. Hence, the relative heights of the first and second peak is a sensitive measure of the baryon content. We shall see that the inferred baryon abundance from the present observations is in complete agreement with the results from big bang nucleosynthesis.

What is the influence of the slow evolution of the effective mass $m_{eff} = 1+R$? Well, from the adiabatic theorem we know that for a slowly varying m_{eff} the ratio energy/frequency is an adiabatic invariant. If A denotes the amplitude of the oscillation, the energy is $\frac{1}{2}m_{eff}\omega^2 A^2$. According to (121) the frequency $\omega = kc_s$ is proportional to $m_{eff}^{-1/2}$. Hence $A \propto \omega^{-1/2} \propto m_{eff}^{1/4} \propto (1+R)^{-1/4}$.

6.1.5 Angular power spectrum for large scales

The *angular power spectrum* is defined as $l(l+1)C_l$ versus l. For large scales, i.e., small l, observed with COBE, the first term in eq. (115) dominates. Let us have a closer look at this so-called Sachs-Wolfe contribution.

For large scales (small k) we can neglect in the first equation (93) of the Boltzmann hierarchy the term proportional to k: $\theta_0' \approx -\Phi' \approx \Psi'$, neglecting also Π (i.e., θ_2) on large scales. Thus

$$\theta_0(\eta) \approx \theta_0(0) + \Psi(\eta) - \Psi(0). \tag{130}$$

To proceed, we need a relation between $\theta_0(0)$ and $\Psi(0)$. This can be obtained by looking at superhorizon scales in the tight coupling limit. (Alternatively, one can investigate the Boltzmann hierarchy in the radiation dominated era.) For adiabatic initial perturbations one easily finds $\theta_0(0) = -\frac{1}{2}\Psi(0)$, while for the isocurvature case one gets $\theta_0(0) = \Psi(0) = 0$. Using this in (129), and also that $\Psi(\eta) = \frac{9}{10}\Psi(0)$ in the matter dominated era, we find for the effective temperature fluctuations at decoupling

$$(\theta_0 + \Psi)(\eta_{dec}) = \frac{1}{3}\Psi(\eta_{dec}) \tag{131}$$

for the adiabatic case. For initial isocurvature fluctuations the result is *six times larger*. Eq. (130) is due to Sachs and Wolfe. It allows us to express the angular CMB power spectrum on large scales in terms of the power spectrum of density fluctuations at decoupling. If the latter has evolved from a scale free primordial

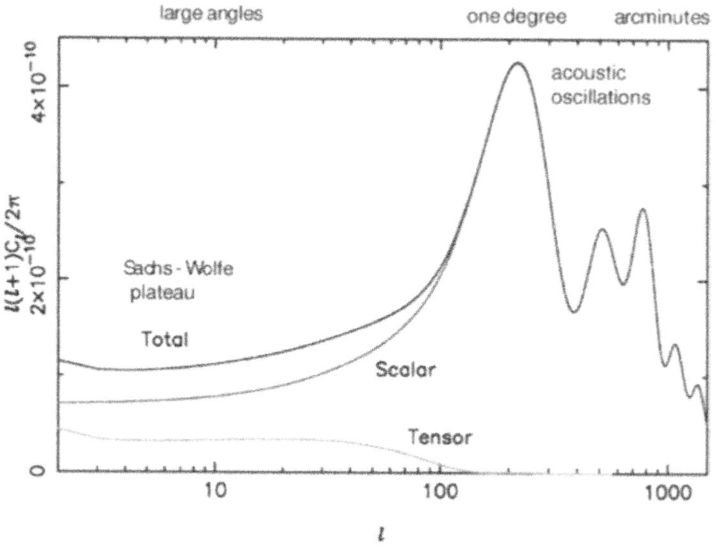

Figure 3: Theoretical angular power spectrum for adiabatic initial perturbations and typical cosmological parameters. The scalar and tensor contributions to the anisotropies are also shown.

spectrum, it turns out that $l(l+1)C_l$ is *constant* for small $l's$. It should be emphasized that on these large scales the power spectrum remains close to the primordial one.

Having discussed the main qualitative aspects, we show in Fig. 3 a typical theoretical CMB power spectrum for scale free adiabatic initial conditions.

6.2 Observational results

CMB anisotropies had been looked for ever since Penzias and Wilson's discovery of the CMB, but had eluded all detection until the *Cosmic Background Explorer* (COBE) satellite discovered them on large angular scales in 1992 [68]. It is not at all astonishing that it took so long in view of the fact that the temperature perturbations are only one part in 10^{-5} (after subtraction of the obvious dipole anisotropy). There are great experimental difficulties to isolate the cosmologically interesting signal from foreground contamination. The most important of these are: (i) galactic dust emission; (ii) galactic thermal and synchrotron emission; (iii) discrete sources; (iv) atmospheric emission, in particular at frequencies higher than $\sim$10 GHz.

After 1992 a large number of ground and balloon-borne experiments were set up to measure the anisotropies on smaller scales. Until quite recently the measuring

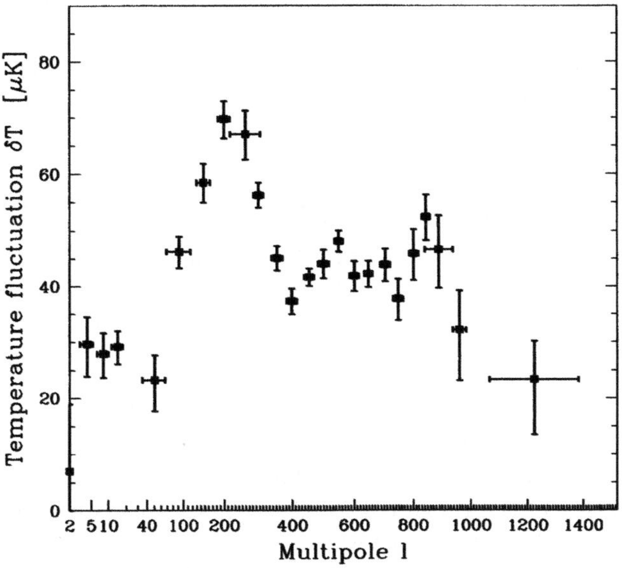

Figure 4: Band-averaged CMB power spectrum, together with the $\pm 1\sigma$ errors (Fig. 2 of Ref. [70]).

errors were large and the data had a considerable scatter, but since early 2001 the situation looks much better. Thanks to the experiments BOOMERanG [69], MAXIMA [70] and DASI [71] we now have clear evidence for multiple peaks in the angular power spectrum at positions and relative heights that were expected on the basis of the simplest inflationary models and big bang nucleosynthesis.

Wang et al. [72] have compressed all available data into a single band-averaged set of estimates of the CMB power spectrum. Their result, together with the $\pm 1\sigma$ errors, is reproduced in Fig. 4. These data provide tight constraints for the cosmological parameters. However, the CMB anisotropies alone do not fix them all because there are unavoidable degeneracies, especially when tensor modes (gravity waves) are included. This degeneracy is illustrated in Fig. 9 of Ref. [70] by three best fits that are obtained by fixing $\Omega_b h_0^2$ in a reasonable range.

Such degeneracies can only be lifted if other cosmological information is used. Beside the supernova results, discussed in Sect. 5, use has been made, for instance, of the available information for the galaxy power spectrum. In [73] the CMB data have been combined with the power spectrum of the $2dF$ (2-degree-Field) Galaxy

Redshift Survey (2dFGRS). The authors summarize their results of the combined likelihood analysis in Table 1 of their paper. Here, I quote only part of it. The Table below shows the $\pm 2\sigma$ parameter ranges for some of the cosmological parameters, for two types of fits. In the first only the CMB data are used (but tensor modes are included), while in the second these data are combined with the 2dFGRS power spectrum (assuming adiabatic, Gaussian initial conditions described by power laws).

Table 1

Parameter	CMB alone	CMB and 2dFGRS
$\Omega_b h_0^2$	0.016-0.045	0.018-0.034
$\Omega_c h_0^2$	0.03-0.18	0.07-0.13
Ω_K	-0.68-0.06	-0.05-0.04
Ω_Λ	<0.88	0.65-0.85

Note that Ω_K is not strongly constraint by CMB alone. However, if h_0 is assumed a priori to be within a reasonable range, then Ω_K has to be close to zero (flat universe). It is very satisfying that the combination of the CMB and 2dFGRS data constrain Ω_Λ in the range $0.65 \leq \Omega_\Lambda \leq 0.85$. This is independent of — but consistent with — the supernova results.

Another beautiful result has to be stressed. For the baryon parameter $\Omega_b h_0^2$ there is now full agreement between the CMB results and the BBN prediction. Earlier speculations in connection with possible contradictions now have evaporated. The significance of this consistency cannot be overemphasized.

All this looks very impressive. It is, however, not forbidden to still worry about possible complications located in the initial conditions, for which we have no established theory. For example, an isocurvature admixture cannot be excluded and the primordial power spectrum may have unexpected features.

Temperature measurements will not allow us to isolate the contribution of gravitational waves. This can only be achieved with future sensitive polarization experiments. Polarization information will provide crucial clues about the physics of the very early Universe. It can, for instance, be used to discriminate between models of inflation. With the *Planck* satellite, currently scheduled for launch in February 2007, it will be possible to detect gravitational waves even if they contribute only 10 percent to the anisotropy signal.

7 Quintessence

For the time being, we have to live with the mystery of the incredible smallness of a gravitationally effective vacuum energy density. For most physicists it is too much to believe that the vacuum energy constitutes the missing two thirds of the average energy density of the *present* Universe. This would really be bizarre. The goal of quintessence models is to avoid such an extreme fine-tuning. In many ways people thereby repeat what has been done in inflationary cosmology. The main motivation there was, as is well-known, to avoid excessive fine tunings of standard big bang cosmology (horizon and flatness problems).

In concrete models the exotic missing energy with negative pressure is again described by a scalar field, whose potential is chosen such that the energy density of the homogeneous scalar field adjusts itself to be comparable to the matter density today for quite generic initial conditions, and is dominated by the potential energy. This ensures that the pressure becomes sufficiently negative. It is not simple to implement this general idea such that the model is phenomenologically viable. For instance, the success of BBN should not be spoiled. CMB and large scale structure impose other constraints. One also would like to understand why cosmological acceleration started at about $z \sim 1$, and not much earlier or in the far future. There have been attempts to connect this with some characteristic events in the post-recombination Universe. On a fundamental level, the origin of a quintessence field that must be extremely weakly coupled to ordinary matter, remains in the dark.

There is already an extended literature on the subject. Refs. [74]–[80] give a small selection of important early papers and some recent reviews. I conclude by emphasizing again that on the basis of the vacuum energy problem we would expect a huge additive constant for the quintessence potential that would destroy the hole picture. Thus, assuming for instance that the potential approaches zero as the scalar field goes to infinity, has (so far) no basis. Fortunately, future more precise observations will allow us to decide whether the presently dominating exotic energy density satisfies $p/\rho = -1$ or whether this ratio is somewhere between -1 and $-1/3$. Recent studies (see [81], [82], and references therein) which make use of existing CMB data, SN Ia observations and other information do not yet support quintessence with $w_Q > -1$.

However, even if convincing evidence for this should be established, we will not be able to predict the distant future of the Universe. Eventually, the quintessence energy density may perhaps become negative. This illustrates that we may never be able to predict the asymptotic behavior of the most grandiose of all dynamical systems. Other conclusions are left to the reader.

References

[1] S. Weinberg, *Rev. Mod. Phys.* **61**, 1 (1989).

[2] A. Einstein, *Sitzungsber. Preuss. Akad. Wiss.* phys.-math. Klasse VI, 142 (1917). See also: Ref. [2], Vol. 6, p. 540, Doc. 43.

[3] A. Einstein, *The Collected Papers of Albert Einstein.* Princeton University Press; Vol. 8, Part A, p. 386, Doc. 294.

[4] A. Pais, *'Subtle is the Lord...': The Science and the Life of Albert Einstein.* Oxford University Press (1982). See especially Sect. 15e.

[5] W. de Sitter, *Proc. Acad. Sci.*, **19**, 1217 (1917); and **20**, 229 (1917).

[6] A.S. Eddington, *The Mathematical Theory of Relativity.* Chelsea Publishing Company (1924). Third (unaltered) Edition (1975). See especially Sect. 70.

[7] J.D. North, *The Measure of the Universe: A History of Modern Cosmology.* Oxford: Clarendon Press (1965). Reprinted: Dover (1990).

[8] H. Weyl, *Raum, Zeit, Materie.* Fifth Edition. Springer-Verlag, Berlin (1923). See especially Appendix III.

[9] A. Friedmann, *Z. Phys.* **10**, 377 (1922); **21**, 326 (1924).

[10] G. Lemaître, *Ann. Soc. Sci. Brux. A* **47**, 49 (1927).

[11] A. Einstein, *S.B. Preuss. Akad. Wiss.* (1931), 235.

[12] A. Einstein, Appendix to the 2nd edn. of *The Meaning of Relativity*, (1945); reprinted in all later editions.

[13] W. Pauli, *Theory of Relativity.* Pergamon Press (1958); Supplementary Note **19**.

[14] O. Heckmann, *Theorien der Kosmologie*, berichtigter Nachdruck, Springer-Verlag (1968).

[15] N. Straumann, *General Relativity and Relativistic Astrophysics*, Springer-Verlag (1984).

[16] V. Petrosian, E.E. Salpeter, and P. Szekeres, *Astrophys. J.* **147**, 1222 (1967).

[17] W. Pauli, *Pauli Lectures on Physics*; Ed. C.P. Enz. MIT Press (1973); Vol. 4, especially Sect. 20.

[18] C.P. Enz, and A. Thellung, *Helv. Phys. Acta* **33**, 839 (1960).

[19] Y.B. Zel'dovich, *JETP letters* **6**, 316 (1967); *Soviet Physics Uspekhi* **11**, 381 (1968).

[20] W. Pauli, *Die allgemeinen Prinzipien der Wellenmechanik.* Handbuch der Physik, Vol. XXIV (1933). New edition by N. Straumann, Springer-Verlag (1990); see Appendix III, p. 202.

[21] H.B.G. Casimir, *Kon. Ned. Akad. Wetensch.* **B51**, 793 (1948).

[22] S.K. Lamoreaux, quant-ph/9907076.

[23] P. Jordan and W. Pauli, *Z. Phys.* **47**, 151 (1928).

[24] M. Bordag, U. Mohideen, and V.M. Mostepanenko *New Developments in the Casimir Effect*, quant-ph/0106045.

[25] W. Heisenberg and H. Euler, *Z. Phys.* **38**, 714 (1936).

[26] V.S. Weisskopf, Kongelige Danske Videnskabernes Selskab, Mathematisk-fysiske Meddelelser XIV, No.6 (1936).

[27] L.D. Landau and E.M. Lifshitz, *Quantum Electrodynamics*, Vol. 4, second edition, Pergamon Press (1982); Sect. 129.

[28] W. Dittrich and M. Reuter, *Lecture Notes in Physics*, Vol.220 (1985).

[29] M. Kardar and R. Golestanian, *Rev. Mod. Phys.* **71**, 1233 (1999).

[30] G.A. Tammann, A. Sandage, and A. Saha, astro-ph/0010422.

[31] C.G. Callan, S. Coleman, and R. Jackiw, *Ann.Phys.* **59**, 42 (1970).

[32] T. Schäfer and E.V. Shuryak, *Rev. Mod. Phys.* **70**, 323 (1998).

[33] R.D. Peccei and H. Quinn, *Phys. Rev. Lett*, **38**, 1440 (1977); *Phys. Rev.* **D16**, 1791 (1977).

[34] E. Witten, hep-ph/0002297.

[35] S.M. Caroll, *Living Reviews in Relativity*, astro-ph/0004075.

[36] N. Straumann, in *Dark Matter in Astro- and Particle Physics*, Edited by H.V. Klapdor–Kleingrothaus, Springer (2001), p. 110.

[37] S.E. Rugh and H. Zinkernagel, hep-th/0012253.

[38] W. Baade, *Astrophys. J.* **88**, 285 (1938).

[39] G.A. Tammann. In*Astronomical Uses of the Space Telescope*, eds. F. Macchetto, F. Pacini and M. Tarenghi, p. 329. Garching: ESO.

[40] S. Colgate, *Astrophys. J.* **232**, 404 (1979).

[41] SCP-Homepage: http://www-supernova.LBL.gov

[42] HZT-Homepage:
http://cfa-www.harvard.edu/cfa/oir/Research/supernova
/HighZ.html

[43] Cosmic Explosions, eds. S. Holt and W. Zhang, AIP, in press.

[44] W. Hillebrandt and J.C. Niemeyer, *Ann. Rev. Astron. Astrophys.* **38**, 191–230 (2000).

[45] B. Leibundgut, *Astron. Astrophys.* **10**, 179 (2000).

[46] S. Perlmutter, et al., *Astrophys. J.* **517**, 565 (1999).

[47] B. Schmidt, et al., *Astrophys. J.* **507**, 46 (1998).

[48] A.G. Riess, et al., *Astron. J.* **116**, 1009 (1998).

[49] B. Leibundgut, *Ann. Rev. Astron. Astrophys.* **39**, 67 (2001).

[50] A. Fillipenko and A. Riess, astro-ph/9905049; astro-ph/0008057.

[51] A.G. Riess, et al., astro-ph/0104455.

[52] G.F. Lewis and R.A. Ibata, astro-ph/0104254.

[53] Snap-Homepage: http://snap.lbl.gov

[54] R. Kantowski and R.C. Thomas, *Astrophys. J.* **561**, 491 (2001); astro-ph/0011176.

[55] D.J. Fixsen, et al., *Astrophys. J.* **473**, 576 (1996).

[56] H. Kodama and M. Sasaki, *Internat. J. Mod. Phys.* **A1**, 265 (1986).

[57] H. Kodama and M. Sasaki, *Internat. J. Mod. Phys.* **A2**, 491 (1987).

[58] U. Seljak, and M. Zaldarriaga, *Astrophys.. J.* **469**, 437 (1996). (See also *http://www.sns.ias.edu/matiasz/CMBFAST/cmbfast.html*)

[59] A. Lewis, A. Challinor and A. Lasenby, *Astrophys. J.* **538**, 473 (2000).

[60] W. Hu and N. Sugiyama, *Astrophys. J.* **444**, 489 (1995).

[61] W. Hu and N. Sugiyama, *Phys. Rev.* **D51**, 2599 (1995).

[62] A.R. Liddle and D.H. Lyth, *Cosmological Inflation and Large-Scale Structure.* Cambridge University Press (2000).

[63] J. Bardeen, *Phys. Rev.* **D22**, 1882 (1980).

[64] H. Kodama and M. Sasaki, *Progr. Theor. Phys. Suppl.* **78**, 1 (1984).

[65] R. Durrer, *Fund. of Cosmic Physics* **15**, 209 (1994).

[66] N. Straumann, *Cosmological Perturbation Theory.* Unpublished lecture notes, to be made available.

[67] R. Durrer and N. Straumann, *Helvetica Phys. Acta* **61**, 1027 (1988).

[68] G.F. Smoot, et al., *Astrophys. J.* **396**, L1 (1992).

[69] C.B. Netterfield, et al., astro-ph/0104460; S. Masi, et al., astro-ph/0201137.

[70] A.T. Lee, et al., astro-ph/0104459.

[71] N.W. Halverson, et al., astro-ph/0104489.

[72] X. Wang M. Tegmark and M. Zaldarriaga, astro-ph/0105091.

[73] G. Efstathiou, et al., astro-ph/0109152.

[74] C. Wetterich, *Nucl. Phys.* **B302**, 668 (1988).

[75] B. Ratra and P.J.E. Peebles, *Astrophys. J. Lett.* **325**, L17 (1988); *Phys. Rev.* **D37**, 3406 (1988).

[76] R.R. Caldwell, R. Dave and P.J. Steinhardt, *Phys. Rev. Lett.* **80**, 1582 (1998).

[77] P.J. Steinhardt, L. Wang and I. Zlatev, *Phys. Rev. Lett.* **82**, 896 (1999); *Phys. Rev.* **D59**, 123504 (1999).

[78] C. Armendariz–Picon, V. Mukhanov and P.J. Steinhardt, astro-ph/0004134 and astro-ph/0006373.

[79] P. Binétruy, hep-ph/0005037.

[80] C. Wetterich, University of Heidelberg preprint, (2001).

[81] R. Bean, S.H. Hansen, and A. Melchiorri, astro-ph/0201127.

[82] C. Baccigalupi et al., astro-ph/0109097.

Norbert Straumann
Institute for Theoretical Physics
University of Zurich
CH–8057 Zurich, Switzerland

Poincaré Seminar 2002, 53 – 69

Introduction à l'effet Casimir

Bertrand Duplantier

LA SOMME

Devant la chaux d'un mur que rien
ne nous défend d'imaginer comme infini
un homme s'est assis et songe
à tracer d'une touche rigoureuse
sur le mur blanc le monde entier :
portes, balances, jacinthes et tartares,
anges, bibliothèques, labyrinthes,
ancres, Uxmal, l'infini, le zéro.
J.-L. Borges.

Constantes physiques

$$\text{Constante de Planck : } \hbar = h/2\pi \simeq 1{,}055 \times 10^{-34} \text{ J s ;}$$
$$\text{vitesse de la lumière : } c \simeq 3 \times 10^{8} \text{ m s}^{-1} \text{ ;}$$
$$\text{constante de Boltzmann : } k_B \simeq 1{,}381 \times 10^{-23} \text{ J K}^{-1}.$$

D'après Planck (1900) [1], l'énergie d'un mode stationnaire du champ électromagnétique, de vecteur d'onde $\mathbf{k}$, est quantifiée, avec naturellement $\hbar\omega_{\mathbf{k}} = h\nu$ par photon présent. En raison de la structure quadratique de l'énergie du champ électromagnétique, il existe une transformation canonique qui associe un hamiltonien d'oscillateur harmonique (à une dimension) à chaque mode. Au niveau classique, cette vision était déjà présente chez Rayleigh, et Planck a précisément découvert (ou même inventé) la quantification des niveaux d'énergie de ces oscillateurs. En effet, en mécanique quantique, nous savons que les niveaux d'un oscillateur harmonique sont quantifiés de manière régulière et donnés par

$$E_m = \hbar\omega\,(m + \frac{1}{2}),\ m \in \mathbf{N}\ . \tag{1}$$

Il existe une énergie fondamentale non nulle $\frac{1}{2}\hbar\omega$, reflet au principe d'incertitude d'Heisenberg. La théorie quantique du champ s'appuie précisément sur cette description par oscillateurs, identifiant le niveau m de l'oscillateur au nombre de photons dans le mode de pulsation ω. A priori, l'énergie du fondamental est bien présente dans la construction théorique. On parle de l'*énergie du vide quantique*, ici électromagnétique. On pourrait alors croire que l'énergie du vide, énergie de référence, est inobservable. Or, il n'en est rien, et le physicien hollandais H. B. G. Casimir a montré en 1948 dans un article célèbre [2] qu'une force *macroscopique*, a priori mesurable, pouvait être engendrée par les fluctuations du vide électromagnétique. Dans la géométrie originellement considérée par Casimir, cette force s'exerce entre deux plaques conductrices *non chargées*. Lorsque la température est non nulle, s'y ajoutent les effets du rayonnement du **corps noir** en équilibre thermique avec les surfaces conductrices. Lors d'expériences récentes en 1997-98 ([3, 4]), la force résultante entre une sphère et un plan, tous deux métallisées, a pu être mesurée avec précision à l'aide d'un microscope à force atomique.

Nous donnons ici un aperçu du calcul de Casimir, ainsi que sa généralisation aux effets de température. Remarquons tout d'abord qu'un argument dimensionnel permet d'anticiper la forme du résultat. Entre deux plaques très grandes, on attend ainsi l'existence d'une pression (force par unité de surface) quantique électromagnétique, dépendante donc a priori de $\hbar$, c et de la distance L entre plaques. Pour construire une pression, mesurée en unités de pression $P \propto \mathrm{N\ m^{-2}} = \mathrm{J\ m^{-3}}$, avec $\hbar \propto \mathrm{J\ s}$, $c \propto \mathrm{m\ s^{-1}}$, et $L \propto \mathrm{m}$, il faut clairement multiplier $\hbar$ par c, ce qui fait disparaître le temps, et enfin diviser par la puissance appropriée de L, c'est-à-dire L^{-4}. On trouve donc la seule forme possible a priori : $P_{\mathrm{Casimir}} \propto \frac{\hbar c}{L^4}$. Par anticipation, on peut dire que la surprise sera que le coefficient soit non nul !

Le calcul de Casimir

Considérons deux plaques planes identiques, toutes deux parallèles au plan yOz et de grande aire $\mathcal{A} = L_y \times L_z$, séparées d'une distance L dans la direction orthogonale Ox.

Les plaques sont supposées parfaitement conductrices, et les modes stationnaires du champ électro- magnétique entre ces plaques sont décrits par un vecteur d'onde $(k_x,\mathbf{k})$, où $\mathbf{k} = (k_y,k_z)$ est parallèle aux plaques ; dans la direction perpendiculaire aux plaques, les conditions aux limites électromagnétiques de conducteur parfait fixent la suite discrète de valeurs : $k_x = \frac{\pi n}{L}$, où n est un entier positif ou nul, tandis que dans les directions parallèles aux plaques, des conditions aux limites périodiques (fictives*) donnent pour les deux composantes de $\mathbf{k}$ des valeurs de la forme $(k_y,k_z) = \left(\frac{2\pi n_y}{L_y},\frac{2\pi n_z}{L_z}\right)$, où n_y,n_z sont deux entiers relatifs : $n_y,n_z \in \mathbf{Z}$.

---

* On sait que le résultat asymptotique pour de grandes plaques ne dépend pas de ces conditions.

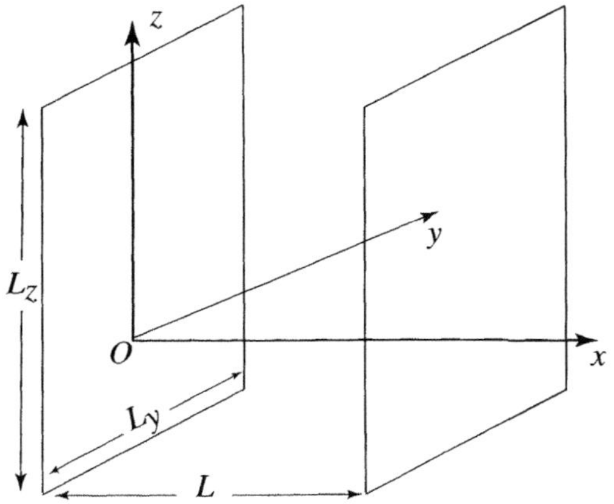

Figure 1: Configuration des plaques conductrices.

Ce mode de vecteur d'onde (k_x,k_y,k_z), noté $(n,\mathbf{k})$, oscille avec une pulsation ω (correspondant à une fréquence $\nu = \omega/2\pi$) :

$$\omega = \omega_n(\mathbf{k}) = c\sqrt{k_x^2 + k_y^2 + k_z^2} = c\sqrt{\frac{\pi^2 n^2}{L^2} + \mathbf{k}^2}. \tag{2}$$

Chaque mode est doublement dégénéré, du fait des deux polarisations possibles de l'onde électro- magnétique (à l'exception du mode $n = 0$ qui n'est pas dégénéré).

À température nulle, l'énergie électromagnétique de la cavité est la somme $\mathcal{E}_0$ des énergies de point zéro de chaque mode :

$$\mathcal{E}_0 = \sum_{\text{modes }(n,\mathbf{k})} \varepsilon_0[\omega_n(\mathbf{k})] \; , \; \varepsilon_0(\omega) = \frac{1}{2}\hbar\omega. \tag{3}$$

Cette suite de modes est non bornée et la somme précédente *diverge*. La physique nous guide pour lui donner un sens : dans l'établissement des modes stationnaires dans une cavité, les conditions aux limites interviennent, et l'on a considéré que l'enceinte était un conducteur parfait. Or aucun conducteur n'est parfait jusqu'aux fréquences infinies : pour de très hautes fréquences, le matériau conducteur de la plaque devient diélectrique et transparent au rayonnement. Dans ce cas, les conditions aux limites ne s'appliquent plus, et les modes de très hautes fréquences ne vont pas contribuer à la force résultante. Ceci conduit à modifier la somme (3) en coupant les hautes fréquences, c'est-à-dire à introduire l'expression *régularisée* :

$$\mathcal{E}_0 = \sum_{\text{modes }(n,\mathbf{k})} \varepsilon_0[\omega_n(\mathbf{k})] \; \chi\left(\frac{\omega_n(\mathbf{k})}{\omega_c}\right), \tag{4}$$

où $\chi(\omega/\omega_c)$ est une fonction de coupure, telle que $\chi(0) = 1$, et régulière au voisinage de l'origine. Elle s'annule, ainsi que toutes ses dérivées, pour ω/ω_c tendant vers l'infini, suffisamment vite pour que la somme régularisée soit convergente. La pulsation de coupure ω_c apparaît dans χ pour des raisons dimensionnelles ; elle dépend des caractéristiques microscopiques du matériau. (La limite du conducteur *parfait* va correspondre à $\omega_c \to +\infty$, pour laquelle $\chi(\omega/\omega_c) \to 1$ pour tout ω fini).

Dans la limite où l'on considère des plaques de grandes dimensions L_y, L_z, on peut remplacer la somme sur les vecteurs d'onde *parallèles* aux plaques par une intégrale. La distance L entre plaques, quant à elle, reste finie. Dans cette limite, on a†

$$\sum_{\text{modes }(n,\mathbf{k})} \cdots = 2\frac{\mathcal{A}}{(2\pi)^2} \sum_{n=0}^{\infty}{}' \int_{\mathbf{R}^2} \mathrm{d}^2\mathbf{k} \cdots , \tag{5}$$

où le signe *prime* sur la somme signifie que le mode $n = 0$ est affecté d'un poids $1/2$. Posons $\varepsilon(\omega) = \varepsilon_0(\omega)\ \chi(\omega/\omega_c)$. L'énergie (4) s'écrit alors à l'aide de (5)

$$\mathcal{E}_0(L) = 2\frac{\mathcal{A}}{(2\pi)^2} \sum_{n=0}^{\infty}{}' \int_{\mathbf{R}^2} \mathrm{d}^2\mathbf{k}\ \varepsilon\left[\omega_n(\mathbf{k})\right].$$

Comme l'on a par définition :

$$\omega^2 = \omega_n^2(\mathbf{k}) = c^2\left(\frac{\pi^2 n^2}{L^2} + \mathbf{k}^2\right),$$

on a, à n fixé, la différentielle $\omega \mathrm{d}\omega = c^2 k \mathrm{d}k$, où $k = |\mathbf{k}|$. On a alors par intégration simple sur les vecteurs d'onde parallèles $\mathbf{k}$:

$$\int_{\mathbf{R}^2} \mathrm{d}^2\mathbf{k}\, \varepsilon\left[\omega_n(\mathbf{k})\right] = \int_0^{+\infty} 2\pi k\, \mathrm{d}k\ \varepsilon\left[\omega_n(k)\right] = \int_{\omega_n(\mathbf{0})}^{+\infty} 2\pi c^{-2}\omega\, \mathrm{d}\omega\ \varepsilon(\omega), \tag{6}$$

et l'on trouve donc :

$$\mathcal{E}_0 = \mathcal{A}\frac{1}{\pi c^2} \sum_{n=0}^{\infty}{}' \int_{\omega_n}^{\infty} \mathrm{d}\omega\ \omega\varepsilon_0(\omega)\chi\left(\frac{\omega}{\omega_c}\right), \quad \omega_n \equiv \omega_n(\mathbf{0}) = \pi c n/L. \tag{7}$$

---

† Dans le cas de la géométrie de condensateur plan, $L_y, L_z \to +\infty$, la somme sur les entiers n_y, n_z, divisée par le produit $L_y L_z$, est une somme de Riemann, et tend vers l'intégrale sur les mêmes variables :

$$\frac{2\pi}{L_y}\frac{2\pi}{L_z} \sum_{n_y, n_z \in \mathbf{Z}^2} \cdots \to \frac{2\pi}{L_y}\frac{2\pi}{L_z} \int_{\mathbf{R}^2} \mathrm{d}n_y \mathrm{d}n_z \cdots = \int_{\mathbf{R}^2} \mathrm{d}k_y \mathrm{d}k_z \cdots ,$$

soit encore

$$\sum_{n_y, n_z \in \mathbf{Z}^2} \cdots \to \frac{\mathcal{A}}{(2\pi)^2} \int_{\mathbf{R}^2} \mathrm{d}^2\mathbf{k} \cdots ,$$

qui est la mesure habituelle bidimensionnelle sur les vecteurs d'ondes dans une grande boîte. Par ailleurs, dans la direction Ox où la longueur L reste finie, la somme $2 \sum_{n=0}^{\infty}{}' \cdots$ reste discrète et décrit les indices n de la partie transverse des modes électromagnétiques, y compris la dégénérescence 2 globale de polarisation, sauf pour le mode $n = 0$ qui est unique.

La force associée, X_0, se trouve par dérivation :

$$X_0 = -\frac{\partial \mathcal{E}_0}{\partial L} = -\mathcal{A}\frac{1}{\pi c^2}\sum_{n=0}^{\infty}{}' \frac{1}{L}\omega_n^2\, \varepsilon_0(\omega_n)\,\chi\left(\frac{\omega_n}{\omega_c}\right), \tag{8}$$

soit encore :

$$X_0 = -\mathcal{A}\frac{\pi^2\hbar c}{2L^4}\sum_{n=0}^{\infty}{}' g(n),\ g(n) = n^3\chi\left(\frac{\omega_n}{\omega_c}\right).$$

L'équivalent X_0^∞ de X_0 dans la limite L grand est donné, comme dans toute limite continue, par l'intégrale sur n au lieu de la somme "prime" sur n, soit

$$X_0^\infty = -\mathcal{A}\frac{\pi^2\hbar c}{2L^4}\int_0^\infty \mathrm{d}n\ g(n). \tag{9}$$

Pour obtenir la force totale de température nulle, venant de la seule énergie de point zéro $\varepsilon_0(\omega) = \frac{1}{2}\hbar\omega$, et s'exerçant sur la plaque considérée, on doit aussi prendre en compte la force (en sens contraire) exercée par le vide électromagnétique (infini) à l'*extérieur* du condensateur. Cette force n'est autre que l'opposée de (9), d'où la *force résultante* $\tilde{X}_0 = X_0 - X_0^\infty$. Cette résultante est donc la différence entre une série et son intégrale associée :

$$\tilde{X}_0 = -\mathcal{A}\frac{\pi^2\hbar c}{2L^4}\left[\sum_{n=0}^{\infty}{}' g(n) - \int_0^\infty \mathrm{d}n g(n)\right],\ g(n) = n^3\chi\left(\frac{\omega_n}{\omega_c}\right).$$

Pour évaluer une telle différence existe alors la formule d'*Euler-Maclaurin* :

$$\sum_{n=0}^{\infty}{}' g(n) - \int_0^{+\infty} dn\ g(n) = -\frac{1}{12}g'(0) + \frac{1}{6!}g'''(0) + O\left(g^{[5]}(0)\right), \tag{10}$$

qui contient toutes les dérivées d'ordre impair de g, prises à l'origine, et qui est valable pour une fonction g s'annulant à l'infini, ainsi que toutes ses dérivées. En calculant les dérivées successives, on trouve ici

$$g'(0) = 0,\ g'''(0) = 6\chi(0) = 6,\ g^{[p]}(0) = O\left({\omega_c}^{-(p-3)}\right),\ p \geq 3.$$

On en déduit donc la valeur finie‡

$$\sum_{n=0}^{\infty}{}' g(n) - \int_0^\infty dng(n) = \frac{1}{5!} + O\left({\omega_c}^{-2}\right).$$

‡ On peut remarquer que c'est la présence du facteur n^3, rapidement variable avec n, qui a donné la valeur finie $g'''(0) = 3!$ par dérivation. En l'*absence* d'un tel facteur, la formule d'Euler-Maclaurin aurait commencé par $g'(0) = O\left({\omega_c}^{-1}\right)$, et la différence entre somme et intégrale aurait disparu dans la limite du conducteur parfait.

La force résultante à température nulle possède donc une limite universelle pour des conducteurs parfaits, c'est-à-dire lorsque $\omega_c \to +\infty$. La pression limite, trouvée par H. B. G. Casimir en 1948, est :

$$\frac{1}{\mathcal{A}}\tilde{X}_0 = -\frac{\pi^2}{240}\frac{\hbar c}{L^4}. \tag{11}$$

La force de Casimir est *attractive*, et l'on retrouve bien la forme analytique prévue pour la force par unité d'aire en fonction de $\hbar$, c et de la longueur L. Seul le coefficient numérique était à déterminer : $-\pi^2/240$, et le fait remarquable est qu'il est non nul et *universel*, c'est-à-dire indépendant de la nature des conducteurs parfaits. À la force de Casimir est associée une *énergie de point zéro* qui s'obtient par simple intégration

$$\tilde{X}_0 = -\frac{\partial\tilde{\mathcal{E}}_0}{\partial L}, \ \frac{1}{\mathcal{A}}\tilde{\mathcal{E}}_0 = -\frac{\pi^2}{720}\frac{\hbar c}{L^3}. \tag{12}$$

Énergie libre électromagnétique d'un condensateur plan

Nous considérons dans la suite les effets de température[§], afin d'évaluer ceux-ci en regard de l'effet de point zéro. À une température donnée T, des photons vont être présents dans la cavité, qui vont suivre la distribution statistique du "rayonnement du corps noir". D'après la théorie de la quantification du champ électromagnétique, chaque mode propre *classique* $(n,\mathbf{k})$ est décrit par un hamiltonien d'oscillateur harmonique *quantique*, à une dimension, de pulsation $\omega = \omega_n(\mathbf{k})$, donnée par (2). Les niveaux d'énergie de ce hamiltonien sont alors $\varepsilon_m = \hbar\omega(m + 1/2)$, où $m \geq 0$ est le nombre de photons, chacun d'énergie $\hbar\omega = h\nu$, présents dans le mode considéré.

En se restreignant d'abord à un seul mode de pulsation ω, l'énergie libre de cet oscillateur quantique à l'équilibre à la température T se met alors sous la forme

$$f(\omega) = \varepsilon_0(\omega) + f_{\mathrm{T}}(\omega), \tag{13}$$

avec

$$\varepsilon_0(\omega) = \frac{1}{2}\hbar\omega, \ f_{\mathrm{T}}(\omega) = \beta^{-1}\varphi\left(\beta\hbar\omega\right), \tag{14}$$

où $f_{\mathrm{T}}(\omega)$ est la partie d'énergie libre *thermique* du mode, $\beta = 1/k_BT$, où k_B est la *constante de Boltzmann*, et où $\varphi(x)$ est une fonction simple. En effet, la fonction de partition d'un oscillateur harmonique est bien connue :

$$Z = \sum_{m=0}^{\infty} e^{-\beta E_m} = \sum_{m=0}^{\infty} e^{-\beta\hbar\omega(m+\frac{1}{2})} = e^{-\frac{1}{2}\beta\hbar\omega}\frac{1}{1 - e^{-\beta\hbar\omega}},$$

§ Les premiers calculs sont dus à Fierz [5] et Mehra [6].

ce qui donne l'énergie libre :

$$f(\omega) = -\frac{1}{\beta}\ln Z = \frac{1}{2}\hbar\omega + \frac{1}{\beta}\ln\left(1 - e^{-\beta\hbar\omega}\right),$$

soit pour la fonction φ :

$$\varphi(x) = \ln\left(1 - e^{-x}\right), \ [\varphi(x) \leq 0].$$

Notons que par construction la limite de température nulle de $f_{\mathrm{T}}(\omega)$ est nulle.

L'énergie libre électromagnétique purement *thermique* de la cavité située entre les plaques est alors définie, en sommant la partie thermique de l'énergie libre de chaque mode, comme :

$$\mathcal{F}_{\mathrm{T}} = \sum_{\text{modes } (n,\mathbf{k})} f_{\mathrm{T}}[\omega_n(\mathbf{k})]. \tag{15}$$

À la différence de la somme sur les modes (3) associée au vide, la somme (15) associée au rayonnement thermique est *convergente*. On peut l'écrire, en utilisant (5),

$$\mathcal{F}_{\mathrm{T}}(L) = \sum_{\text{modes } (n,\mathbf{k})} \beta^{-1}\varphi\left[\beta\hbar\omega_n(\mathbf{k})\right] = 2\frac{\mathcal{A}}{(2\pi)^2}\sum_{n=0}^{\infty}{}' \int_{\mathbf{R}^2} \mathrm{d}^2\mathbf{k}\ \beta^{-1}\varphi\left[\beta\hbar\omega_n(\mathbf{k})\right]. \tag{16}$$

À n fixé, on a d'après l'expression (2) la différentielle $\omega \mathrm{d}\omega = c^2 k \mathrm{d}k$, où $k = |\mathbf{k}|$. Dans l'énergie libre thermique (16) on a alors par simple intégration sur les vecteurs d'onde parallèles $\mathbf{k}$:

$$\int_{\mathbf{R}^2} \mathrm{d}^2\mathbf{k}\, \varphi[\beta\hbar\omega_n(\mathbf{k})] = \int_0^{+\infty} 2\pi k\, \mathrm{d}k\ \varphi[\beta\hbar\omega_n(k)] = \int_{\omega_n(\mathbf{0})}^{+\infty} 2\pi c^{-2}\omega\, \mathrm{d}\omega\ \varphi(\beta\hbar\omega). \tag{17}$$

On introduit la variable $x = \beta\hbar\omega$, qui est sans dimension, ainsi que les bornes $u_n = \beta\hbar\omega_n(\mathbf{0})$, et l'on obtient l'énergie libre comme une série simple :

$$\mathcal{F}_{\mathrm{T}}(L) = 2\frac{\mathcal{A}}{2\pi\beta}\frac{1}{(\beta\hbar c)^2}\sum_{n=0}^{\infty}{}'\psi(u_n), \tag{18}$$

où

$$\psi(u) = \int_u^{+\infty} \mathrm{d}x\ x\varphi(x). \tag{19}$$

Les bornes inférieures successives u_n s'expriment en fonction du paramètre fondamental sans dimension α :

$$\alpha = \beta\pi\hbar c/L,\ u_n = n\ \alpha. \tag{20}$$

En comparant les énergies $\hbar\omega$ de photons appartenant à deux modes consécutifs, on peut préciser, à l'aide du paramètre α, le domaine de température ou d'espacement des plaques pour lequel le caractère discret des modes disparaît. Pour simplifier, on prend des vecteurs d'onde parallèles nuls pour ces deux modes : $\mathbf{k} = \mathbf{0}$. Pour de tels modes, l'écart entre les énergies de deux photons associés est donc $\hbar\Delta\omega = \hbar[\omega_{n+1}(\mathbf{0}) - \omega_n(\mathbf{0})] = \hbar\pi c/L = \alpha k_B T$. Le caractère discret des modes disparaitra si cette différence d'énergie est petite devant l'énergie thermique $k_B T$, soit $\alpha \leq 1$. À température ordinaire, $T \simeq 300\,\mathrm{K}$, cela donne $L \geq 24\,\mu\mathrm{m}$, et à plus courte distance le caractère discret sera détectable.

On peut maintenant calculer, dans la limite L grand, l'équivalent de l'énergie libre $\mathcal{F}_{\mathrm{T}}(L)$ (18), que l'on note alors $\mathcal{F}_{\mathrm{T}}^{\infty}(L)$. D'après la discussion précédente, dans la limite de grandes distances (ou de haute température) $\alpha = \beta\pi\hbar c/L \ll 1$, on est en présence d'un continuum de modes. Dans cette limite, les u_n varient continûment. La somme sur n dans $\mathcal{F}_{\mathrm{T}}$, si on la multiplie par $\alpha \propto 1/L$, devient une somme de Riemann, et tend vers l'intégrale convergente avec comme mesure $\alpha dn = du$.¶

On a donc:

$$\alpha\, \mathcal{F}_{\mathrm{T}}^{\infty} \sim \lim_{\alpha \to 0} \alpha\, \mathcal{F}_{\mathrm{T}} = 2\frac{\mathcal{A}}{2\pi\beta}\frac{1}{(\beta\hbar c)^2}\int_0^{+\infty} du\ \psi(u).$$

Une intégration par parties donne immédiatement : $\int_0^{+\infty} du\ \psi(u) = -\int_0^{+\infty} du\ u\psi'(u)$ puisque la fonction ψ s'annule à l'infini. D'où

$$\int_0^{+\infty} du\ \psi(u) = \int_0^{+\infty} dx\ x^2\varphi(x) = -I_2 = -2\zeta(4) = -\frac{\pi^4}{45},$$

où l'on a utilisé l'intégrale

$$I_p = -\int_0^{+\infty} dx\ x^p \ln(1 - e^{-x}) = p!\ \zeta(p+2),\ p \in \mathbf{N}, \tag{21}$$

et la fonction ζ de Riemann

$$\zeta(p) = \sum_{n=1}^{\infty} n^{-p},\ \zeta(4) = \frac{\pi^4}{90}.$$

Après division par α, l'énergie libre ainsi obtenue,

$$\mathcal{F}_{\mathrm{T}}^{\infty} = 2\frac{\mathcal{A}}{2\pi\beta}\frac{1}{(\beta\hbar c)^2}\left(-\frac{2}{\alpha}\zeta(4)\right) = -\frac{\mathcal{A}\, L}{\beta}\frac{1}{(\beta\hbar c)^3}\frac{\pi^2}{45}, \tag{22}$$

¶ Il faut ici remarquer que la présence du "prime" dans la somme sur n et du facteur $\frac{1}{2}$ pour le mode $n = 0$ ne joue pas de rôle dans la limite L grand. La différence entre les deux sommes est en effet finie, alors que l'on évalue ici ces sommes à l'ordre $O(L)$, supposé grand.

est précisément l'*énergie libre du corps noir* dans un volume $\Omega = \mathcal{A} \times L$. La limite $L \to \infty$ a restauré la symétrie entre les trois directions de l'espace, et l'on retrouve la limite des grands volumes.

Pour le calcul des forces, il va être utile d'introduire l'énergie libre thermique, *relative* ou encore *soustraite*, définie comme :

$$\tilde{\mathcal{F}}_{\mathrm{T}} = \mathcal{F}_{\mathrm{T}} - \mathcal{F}_{\mathrm{T}}^{\infty}. \tag{23}$$

Son expression explicite se trouve par (18) et (22)

$$\tilde{\mathcal{F}}_{\mathrm{T}} = 2\frac{\mathcal{A}}{2\pi\beta}\frac{1}{(\beta\hbar c)^2}\left[\sum_{n=0}^{\infty}{}'\psi(u_n) - \frac{1}{\alpha}\int_0^{+\infty} \mathrm{d}u\ \psi(u)\right] \tag{24}$$

$$= 2\frac{\mathcal{A}}{2\pi\beta}\frac{1}{(\beta\hbar c)^2}\left[\sum_{n=0}^{\infty}{}'\psi(\alpha n) + \frac{2}{\alpha}\zeta(4)\right]. \tag{25}$$

Forces dues au rayonnement thermique

Le rayonnement compris dans cette enceinte, à l'équilibre thermique, exerce une pression sur les plaques. Comme en thermodynamique, la pression se calcule comme la dérivée de l'énergie libre par rapport au volume, soit $P_{\mathrm{T}} = -\dfrac{\partial \mathcal{F}_{\mathrm{T}}}{\partial \Omega}$. Ici, pour un déplacement de la plaque de droite, par exemple, on peut écrire $\mathrm{d}\Omega = \mathcal{A}\ \mathrm{d}x = \mathcal{A}\ \mathrm{d}L$. D'où $P_{\mathrm{T}} = -\dfrac{1}{\mathcal{A}}\dfrac{\partial \mathcal{F}_{\mathrm{T}}}{\partial L}$, ce qui donne la formule pour la force

$$X_{\mathrm{T}}(L) = -\frac{\partial \mathcal{F}_{\mathrm{T}}(L)}{\partial L}. \tag{26}$$

Elle se calcule à partir de l'expression (18) de l'énergie libre, à l'aide des équations élémentaires

$$-\frac{\partial \psi(u_n)}{\partial L} = -\frac{\partial u_n}{\partial L}\psi'(u_n) = \frac{\partial u_n}{\partial L}\ u_n\varphi(u_n),$$

où $\partial u_n/\partial L = -u_n/L$. Soit :

$$X_{\mathrm{T}} = -2\frac{\mathcal{A}}{2\pi\beta L}\frac{1}{(\beta\hbar c)^2}\sum_{n=0}^{\infty}{}'u_n^2\varphi(u_n). \tag{27}$$

Cette force, perpendiculaire à la plaque, est positive ($\varphi(x) \leq 0$), donc répulsive. C'est la force de pression du corps noir en géométrie finie, où le caractère discret des modes apparaît clairement par la présence même de la série.

Dans la limite $L \to \infty$, on doit retrouver la force de pression du corps noir de volume infini, soit

$$X_{\mathrm{T}}^{\infty} = -\frac{\partial \mathcal{F}_{\mathrm{T}}^{\infty}}{\partial L} = \frac{\mathcal{A}}{\beta}\frac{1}{(\beta\hbar c)^3}\frac{\pi^2}{45}. \tag{28}$$

C'est bien en effet la limite obtenue pour (27) en y remplaçant la somme sur n par l'intégrale correspondante.

En fait, une plaque métallique maintenue à la température T est en équilibre thermique avec le rayonnement existant de chaque côté. Par conséquent, cette plaque va être soumise également à l'action des photons extérieurs à la cavité. La force due au rayonnement électromagnétique situé à l'extérieur (infini) du condensateur est alors simplement la force de pression du corps noir de volume infini, que nous venons de calculer, qui est négative pour la plaque de droite, et donc simplement égale à $-X_{\mathrm{T}}^{\infty}$.

En définitive, la force *résultante* thermique, notée $\tilde{X}_{\mathrm{T}}(L)$, exercée par l'*ensemble du rayonnement thermique* sur la plaque considérée, ici de droite, est simplement la force :

$$\tilde{X}_{\mathrm{T}} = X_{\mathrm{T}} - X_{\mathrm{T}}^{\infty} = -\frac{\partial\left(\mathcal{F}_{\mathrm{T}} - \mathcal{F}_{\mathrm{T}}^{\infty}\right)}{\partial L} = -\frac{\partial \tilde{\mathcal{F}}_{\mathrm{T}}}{\partial L}. \tag{29}$$

Développement de courte distance ou de basse température

L'expression en série (27) de la force X_{T}, due au rayonnement thermique situé à l'intérieur des plaques, fournit un développement naturel de *basse température* ou, de manière équivalente de *courte distance*, où $\alpha = \beta\pi\hbar c/L \gg 1$. En effet, on remarque tout d'abord que le terme $n = 0$ ne contribue pas à la force, car $\lim_{x\to 0} x^2 \ln x = 0$. Pour $n \geq 1$ $u_n = n\alpha = n\beta\pi\hbar c/L \gg 1$, et $\varphi(x) \sim -e^{-x}$ pour x grand, d'où $u_n^2\varphi(u_n)_{\alpha\gg 1} \simeq -n^2\alpha^2 e^{-n\alpha}$. Le mode $n = 1$ contribue ainsi une force répulsive dominante, exponentiellement petite. On a ainsi:

$$X_{\mathrm{T}} = -2\frac{\mathcal{A}}{2\pi\beta L}\frac{1}{(\beta\hbar c)^2}\left[\alpha^2\varphi(\alpha) + 4\alpha^2\varphi(2\alpha) + \cdots\right],$$

soit, après réduction :

$$X_{\mathrm{T}} = \mathcal{A}\frac{1}{\beta}\frac{\pi}{L^3}\left[e^{-\alpha} + O\left(e^{-2\alpha}\right)\right], \ \alpha \gg 1.$$

Dans la force *résultante thermique* (29) $\tilde{X}_{\mathrm{T}}$, la force dominante est la force de poussée due au rayonnement extérieur du corps noir, tandis que le mode interne $n = 1$ contribue une force répulsive, exponentiellement petite. On a plus précisément :

$$\tilde{X}_{\mathrm{T}} = -2\frac{\mathcal{A}}{2\pi\beta}\frac{1}{(\beta\hbar c)^2}\frac{1}{L}\left\{2\zeta(4)\alpha^{-1} - \alpha^2\left[e^{-\alpha} + O\left(e^{-2\alpha}\right)\right]\right\}, \ \alpha = \frac{\pi\beta\hbar c}{L}.$$

Soit explicitement :

$$\frac{\tilde{X}_{\mathrm{T}}}{\mathcal{A}} = -\frac{\pi^2}{45}\frac{1}{\beta}\frac{1}{(\beta\hbar c)^3} + \frac{1}{\beta}\frac{\pi}{L^3}\left[e^{-\alpha} + O\left(e^{-2\alpha}\right)\right]. \tag{30}$$

Énergie libre totale

Introduisons finalement l'*énergie libre totale* $\tilde{\mathcal{F}}$ associée à la résultante totale des forces, et provenant de l'énergie du vide et de l'énergie libre thermique: ‖

$$\tilde{\mathcal{F}} = \tilde{\mathcal{E}}_0 + \tilde{\mathcal{F}}_{\mathrm{T}} = \tilde{\mathcal{E}}_0 + \mathcal{F}_{\mathrm{T}} - \mathcal{F}_{\mathrm{T}}^{\infty}. \tag{31}$$

Il est alors utile de réécrire l'énergie libre thermique (24) sous la forme plus explicite en la variable α:

$$\tilde{\mathcal{F}}_{\mathrm{T}}(L,\alpha) = \mathcal{A}\frac{\pi^2\hbar c}{L^3}\mathcal{G}(\alpha) \tag{32}$$

$$\mathcal{G}(\alpha) = \frac{1}{\alpha^3}\left[\sum_{n=0}^{\infty}{}'\psi(\alpha n) + \frac{2}{\alpha}\zeta(4)\right]. \tag{33}$$

L'énergie libre totale est donc simplement dans ces notations (voir (12))

$$\tilde{\mathcal{F}} = \tilde{\mathcal{E}}_0 + \tilde{\mathcal{F}}_{\mathrm{T}} = \mathcal{A}\frac{\pi^2\hbar c}{L^3}\left[-\frac{1}{720} + \mathcal{G}(\alpha)\right]. \tag{34}$$

Le *développement de basse température* de $\mathcal{G}$ est pour $\alpha \gg 1$:

$$\mathcal{G}(\alpha) = \frac{1}{\alpha^3}\left[\frac{1}{2}\psi(0) + \psi(\alpha) + \cdots + \frac{2}{\alpha}\zeta(4)\right]. \tag{35}$$

Nous avons, d'après les définitions (19) de ψ et (21)

$$\psi(0) = -I_1 = -\zeta(3) \tag{36}$$

$$\psi(\alpha) = -(\alpha+1)\left[e^{-\alpha} + O\left(e^{-2\alpha}\right)\right],\ \alpha \gg 1, \tag{37}$$

d'où explicitement:

$$\tilde{\mathcal{F}} = \mathcal{A}\frac{\pi^2\hbar c}{L^3}\left\{-\frac{1}{720} + \frac{1}{\alpha^3}\left[-\frac{1}{2}\zeta(3) + \frac{2}{\alpha}\zeta(4) - (\alpha+1)\left[e^{-\alpha} + O\left(e^{-2\alpha}\right)\right]\right]\right\}. \tag{38}$$

avec $\alpha = \beta\pi\hbar c/L \gg 1$.

Le *développement de haute température* peut être obtenu à partir de l'expression (32) et de la formule de Poisson. On peut aussi utiliser une formule remarquable de *dualité basse et haute températures*, satisfaite par la fonction $\mathcal{G}$[7]

$$\alpha^2\mathcal{G}(\alpha) = \alpha'^2\mathcal{G}(\alpha') \tag{39}$$

$$\alpha\alpha' = (2\pi)^2, \tag{40}$$

‖ Cette énergie libre totale $\tilde{\mathcal{F}}$ est donc soustraite, en ce sens qu'elle tient compte de l'effet de l'extérieur des parois, et elle est identique à celle qui apparaît dans la contribution suivante de R. Balian, où elle est notée $F(T)$. On a donc aussi la correspondance: $\tilde{\mathcal{F}}_{\mathrm{T}} = F(T) - F(0)$.

ce qui donne pour l'énergie libre thermique (32) : **

$$\tilde{\mathcal{F}}_{\mathrm{T}}(L,\alpha) = \left(\frac{2\pi}{\alpha}\right)^4 \tilde{\mathcal{F}}_{\mathrm{T}}\left(L, \frac{(2\pi)^2}{\alpha}\right). \tag{41}$$

On déduit alors de (38) la limite haute température de l'énergie libre totale :

$$\tilde{\mathcal{F}} = -\mathcal{A}\frac{\zeta(3)}{8\pi\beta L^2} + O\left(\beta^{-2} e^{-\frac{4\pi^2}{\alpha}}\right). \tag{42}$$

avec $\alpha = \beta\pi\hbar c/L \ll 1$.

Comparaison des effets thermiques et de point zéro

Nous avons vu que la force résultante de Casimir à température nulle possède une limite universelle pour des conducteurs parfaits :

$$\frac{1}{\mathcal{A}}\tilde{X}_0 = -\frac{\pi^2}{240}\frac{\hbar c}{L^4}.$$

La force résultante totale, $\tilde{X}$, est définie à partir de l'énergie libre totale (31) comme :

$$\tilde{X} = -\frac{\partial \tilde{\mathcal{F}}}{\partial L}. \tag{43}$$

En rassemblant les résultats précédents (30), on trouve que la force résultante totale s'exerçant par unité d'aire sur la plaque, est à basse température ou à courte distance :

$$\frac{1}{\mathcal{A}}\tilde{X} = \frac{1}{\mathcal{A}}\left(\tilde{X}_0 + \tilde{X}_{\mathrm{T}}\right) = -\frac{\pi^2}{240}\frac{\hbar c}{L^4} - \frac{\pi^2}{45}\frac{1}{\beta}\frac{1}{(\beta\hbar c)^3} + \frac{1}{\beta}\frac{\pi}{L^3}e^{-\alpha} + \cdots. \tag{44}$$

La force résultante est dominée par la force de Casimir et par celle du corps noir, toutes deux attractives, la première correction due aux modes internes discrets étant exponentiellement petite.

Pour $L = 1\mu$m, on a $\alpha \simeq 24$. On est donc loin déjà dans le domaine discret pour les modes intérieurs aux plaques : le premier ne contribue pas à la force thermique, si ce n'est avec un facteur e^{-24} ! Le rapport $\gamma = \frac{\tilde{X}_{\mathrm{T}}}{\tilde{X}_0}$ est donc, d'après (44), égal à :

$$\gamma \simeq -\frac{X_{\mathrm{T}}^{\infty}}{\tilde{X}_0} = \frac{240}{45}\frac{L^4}{(\beta\hbar c)^4} = \frac{1}{3}\left(\frac{2\pi}{\alpha}\right)^4. \tag{45}$$

Pour $L = 500\,\mathrm{nm}$, $\alpha = 48$, on trouve $\gamma = 0{,}98 \times 10^{-4}$. On voit donc que, même à température ordinaire, l'effet des fluctuations du vide domine largement la force de corps noir. Tout se passe comme si l'on était à température nulle, et des expériences sensibles vont être capables de détecter l'effet quantique du vide électromagnétique.

** Cette dualité n'est valable que pour la configuration des plaques parallèles ; voir la contribution de R. Balian dans le même séminaire.

Force entre plan et sphère, et approximation de Derjaguin

• Force entre plan et sphère : la situation expérimentale réelle est décrite sur la figure. Une sphère métallisée de rayon R est placée en regard d'un plan conducteur, à une distance d'approche $OO' = L$ de celui-ci.

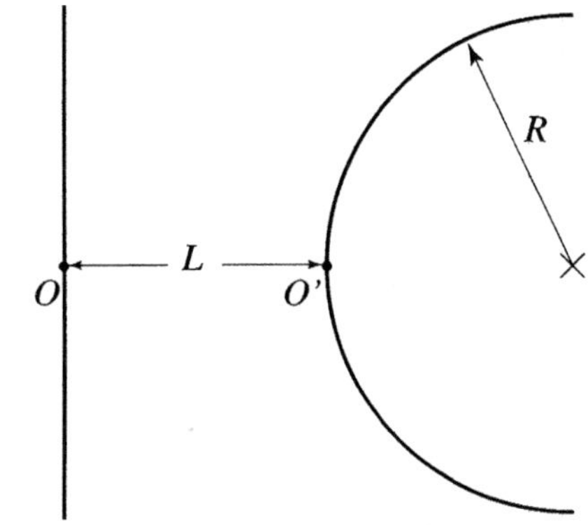

Figure 2: Configuration expérimentale des surfaces conductrices.

Dans les expériences menées en 1998 [4], une sphère de polystyrène est en effet montée sur le bras d'un microscope à force atomique, et approchée d'une surface plane polie. Sur chacune de ces surfaces est déposée une couche d'aluminium de quelques centaines de nanomètres d'épaisseur. Elles sont ensuite couvertes, pour éviter toute corrosion, d'une très fine couche d'alliage, qui est transparente au rayonnement impliqué. Le rayon total de la sphère ainsi métallisée est de l'ordre de $R = 98{,}0 \pm 0{,}25$ μm. Les distances d'approche varient dans le domaine 120 nm $\leq L \leq$ 500 nm (1nm $= 10^{-9}$m). Les mesures sont faites à la *température ambiante.*

Une méthode d'approximation, due à Derjaguin (1934) [8], permet en fait, dans la limite $L \ll R$, de calculer l'interaction sphère-plan en terme de l'interaction entre deux plans. Elle consiste à remplacer, au voisinage du point d'approche minimal O' de la sphère, les tranches élémentaires successives de sphère par leur projection orthogonale dans la direction de la surface conductrice plane. La force résultante sur la sphère s'écrit alors :

$$X^{\text{sph}}(L) = 2\pi R \int_L^{+\infty} \frac{1}{\mathcal{A}} \tilde{X}(x)\, \mathrm{d}x, \tag{46}$$

où $\frac{1}{\mathcal{A}}\tilde{X}(x)$ est la force résultante totale par unité d'aire (43), s'exerçant entre deux *plans* séparés d'une distance x.

Dans le domaine de mesures : $120\,\text{nm} \leq L \leq 500\,\text{nm}$, la valeur minimale de α est $\alpha_{\min} = \alpha(L = 500\,\text{nm}) = 48$. D'après la discussion suivant l'équation (45) on a donc $\gamma = \tilde{X}_{\mathrm{T}}/\tilde{X}_0 \leq 10^{-4}$. En première approximation, on remplace donc, dans l'expression intégrale pour la sphère, la force totale entre plans par la force de Casimir de température nulle.[††] La force de Casimir sur la sphère s'écrit donc dans cette limite :

$$X^{\rm sph}(L) \simeq X_0^{\rm sph}(L) = 2\pi R\, \frac{1}{\mathcal{A}} \int_L^{+\infty} \tilde{X}_0(x)\, {\rm d}x = 2\pi R \frac{1}{\mathcal{A}} \tilde{\mathcal{E}}_0(L)\ , \tag{47}$$

où $\tilde{\mathcal{E}}_0(L)$ est l'énergie dont dérive la force de Casimir (11) entre deux plans :

$$\tilde{X}_0 = -\frac{\partial \tilde{\mathcal{E}}_0}{\partial L},\ \ \frac{1}{\mathcal{A}}\tilde{\mathcal{E}}_0 = -\frac{\pi^2}{720}\frac{\hbar c}{L^3}. \tag{48}$$

La force de Casimir sur la sphère s'écrit donc finalement :

$$X_0^{\rm sph}(L) = -\frac{\pi^3}{360} R \frac{\hbar c}{L^3}. \tag{49}$$

Avec un rayon de sphère $R = 98{,}0 \pm 0{,}25\,\mu$m, et pour $L = 200\,$nm, on trouve par exemple : $X^{\rm sph} \simeq -33{,}4 \pm 0{,}09 \times 10^{-12}\,$N. Cette force est donc de l'ordre de la dizaine de pico-newtons, une force parfaitement mesurable et, par ailleurs, comparable aux forces mises en jeu dans les systèmes biologiques, par exemple lors de micromanipulations de molécules d'ADN.

Corrections de température dans la formule de Derjaguin

À partir de l'expression (46) et de la définition (43) de $\tilde{X}$ nous trouvons immédiatement :

$$\begin{aligned} X^{\rm sph}(L) &= 2\pi R \frac{1}{\mathcal{A}} \left[\tilde{\mathcal{F}}(L) - \tilde{\mathcal{F}}(+\infty)\right] \qquad (50)\\ &= 2\pi R \frac{1}{\mathcal{A}} \tilde{\mathcal{F}}(L), \qquad (51)\end{aligned}$$

où nous utilisons le fait que l'énergie libre complète $\tilde{\mathcal{F}}$ s'annule à grande distance, comme le montre l'équivalent (42). Nous pouvons alors reprendre pour le domaine

[††] Remarque : Il faut faire ici un peu attention, car dans l'intégrale (46) le domaine d'intégration va en principe jusqu'à l'infini. Dans ce domaine, qui est équivalent à celui de haute température, la force thermique $\tilde{X}_{\rm T}$ (qui est en fait la force de van der Waals entre plans) finit par l'emporter sur $\tilde{X}_0$, et l'on ne peut plus invoquer l'argument précédent. En fait, l'approximation de Derjaguin consistant à remplacer, au voisinage du point d'approche minimal O' de la sphère, les tranches élémentaires successives de sphère par leur projection orthogonale dans la direction de l'autre surface conductrice plane, on conçoit physiquement que seul le domaine de distances $x \ll R$ soit important dans cette approximation, et que l'approximation par la force de température nulle ou de courte distance soit valable. On peut par ailleurs remarquer que pour x très grand, la force thermique résultante $\tilde{X}_{\rm T}(x) = X_{\rm T}(x) - X_{\rm T}^{\infty}(x)$ tend vers sa limite de grande distance, qui est nulle par construction, et que l'intégrale de Derjaguin converge rapidement. L'analyse des corrections thermiques, donnée dans le paragraphe suivant, confirme ce point.

expérimental l'équivalent de courte distance (38) de $\tilde{\mathcal{F}}(L)$, qui nous donne l'expression de la force complète sur la sphère :

$$\begin{aligned} X^{\rm sph}(L) &= -\tfrac{\pi^3}{360} R \tfrac{\hbar c}{L^3} \left\{1 - \tfrac{720}{\alpha^3}\left[-\tfrac{1}{2}\zeta(3) + \tfrac{2}{\alpha}\zeta(4)\right.\right. \\ &\quad \left.\left. -(\alpha+1)\left[e^{-\alpha} + O\left(e^{-2\alpha}\right)\right]\right]\right\}. \end{aligned} \tag{52}$$

En ne gardant que les termes non exponentiellement petits, nous pouvons finalement écrire

$$X^{\rm sph}(L) = -\frac{\pi^3}{360} R \frac{\hbar c}{L^3} \left\{1 + 720\left[\frac{1}{2\alpha^3}\zeta(3) - \frac{2}{\alpha^4}\zeta(4) + O\left(\alpha^{-2}e^{-\alpha}\right)\right]\right\}. \tag{53}$$

La correction thermique dans la formule (53) est en fait dominée par le terme en $\zeta(3)$, ce qui donne dans le domaine expérimental considéré ($L \leq 500\,\mathrm{nm}, \alpha_{\rm min} = 48$) une correction thermique relative de l'ordre de 4×10^{-3}, et de même signe que la force de Casimir (49). La formule (53) est utilisée par les expérimentateurs [4].

La formule théorique simple (49) de point zéro est donc comparée sur la figure 3 aux résultats expérimentaux. La courbe théorique (49) est indiquée en tirets. On voit qu'un bon accord existe. Cependant, l'écart avec la courbe expérimentale croît quand L diminue. Il est de l'ordre de la dizaine de %. Les corrections de température ne peuvent expliquer cet écart avec les résultats expérimentaux. De même signe que la force de Casimir, elles accroissent, quoique qu'insensiblement, l'écart théorie-expérience. L'accord théorie-expérience est restauré lorsque des corrections de conductivité finie et de rugosité de surfaces sont apportées [4]. (Voir la description complète de ces expériences et l'analyse de ces corrections dans la contribution de S. Reynaud dans le même séminaire.) Remarquons enfin qu'il faudrait également effectuer l'analyse théorique des *corrections géométriques* à la formule de Derjaguin (46), et pour ce faire utiliser les méthodes générales introduites en [9], qui sont décrites dans la contribution suivante de R. Balian à ce séminaire.

En conclusion, on peut dire que l'*on observe donc, dans ces expériences, des forces macroscopiques électromagnétiques, engendrées par les fluctuations quantiques du vide, proportionnelles à $\hbar c$, et ceci en l'absence de toute charge et de tout photon dans la cavité !*‡‡ Planck, lorsqu'il inventa sa fameuse formule de quantification pour le corps noir, et introduisit la constante h, certes sans pouvoir encore imaginer le complément d'énergie du vide à la théorie du corps noir, avait vraiment déchiré un coin du voile.

Aujourd'hui, la littérature sur l'effet Casimir et ses divers développements physiques et mathématiques est devenue énorme, allant de l'énergie du vide pour les différents champs quantiques jusqu'aux effets de taille finie dans les systèmes critiques. Faisons donc le choix ici de ne citer que les ouvrages récemment parus, qui pourront guider le lecteur ou la lectrice intéressé(e) vers d'autres articles de recherche [10, 11, 12, 13, 14, 15].

‡‡ En fait, il y a bien sûr une infinité de photons mous longitudinaux, avec $n = 0$, qui ne contribuent pas à la force ; en revanche, les photons à $n \geq 1$ sont essentiellement absents de la cavité, ce qui permet d'observer l'effet de l'énergie du vide.

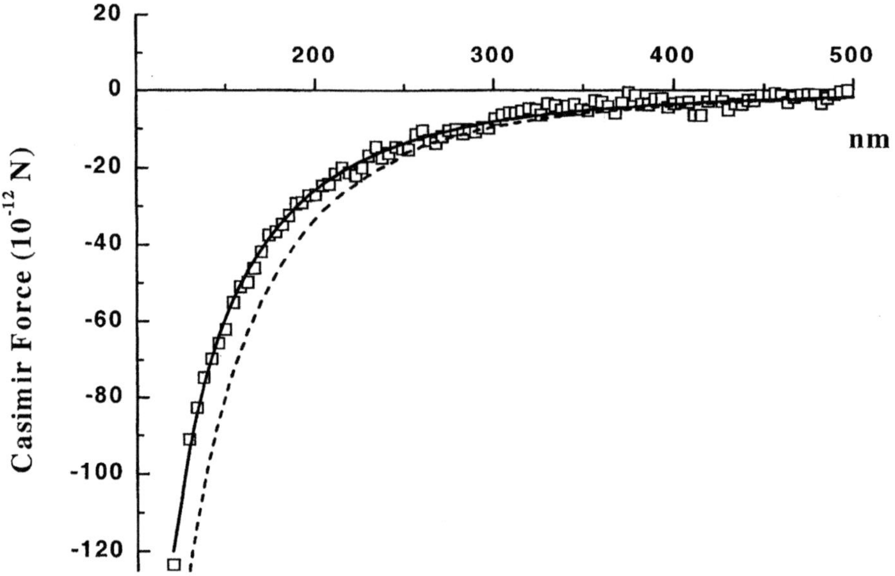

Figure 3: Comparaison de la force calculée (49) (en trait pointillé) avec les résultats expérimentaux ; la courbe en trait plein tient compte des corrections de conductivité finie et de rugosité de surface (expériences de U. Mohideen et A. Roy, Physical Review Letters, **81**, 4549 (1998)).

Références

[1] M. Planck, *Ann. d. Phys.* **4**, 553 (1901).

[2] H.B.G. Casimir, *Proc. Kon. Nederl. Akad. Wetensch.* B**51**, 793 (1948).

[3] S.K. Lamoreaux, *Phys. Rev. Lett.* **78**, 5 (1997) ; *ibid* **81**, 5475 (1998).

[4] U. Mohideen et A. Roy, *Phys. Rev. Lett.* **81**, 4549 (1998).

[5] M. Fierz, *Helv. Phys. Acta* **33**, 855 (1960).

[6] J. Mehra, *Physica* **37**, 145 (1967).

[7] L. S. Brown et G. J. Maclay, *Phys. Rev.* **184**, 1272 (1969).

[8] B Derjaguin, *Kolloid Z.* **69**, 155 (1934).

[9] R. Balian et B. Duplantier, *Ann. Phys.* **104**, 300 (1977) ; *Ann. Phys.* **112**, 165 (1978).

[10] F. S. Levin et D. A. Micha *eds.*, *Long-Range Casimir Forces*, Plenum Press, New York (1993).

[11] P. W. Milonni, *The Quantum Vacuum*, Academic Press, San Diego (1994).

[12] V. M. Mostepanenko et N. N. Trunov, *The Casimir Effect and its Applications*, Clarendon Press, Oxford (1997).

[13] E. Elizalde *et al.*, *Zeta Regularization Techniques with Applications*, World Scientific, Singapour (1994).

[14] M. Krech, *The Casimir Effect in Critical Systems*, World Scientific, Singapour (1994).

[15] J. G. Brankov, D. M. Danchev, N. S. Tonchev, *Theory of Critical Phenomena in Finite-Size Systems*, World Scientific, Singapour (2000).

Bertrand Duplantier
Service de Physique Théorique*
Orme des Merisiers
CEA, Saclay
F-91191 Gif-sur-Yvette Cedex

* Unité de recherche associée au CNRS

Poincaré Seminar 2002, 71 – 92

Casimir Effect and Geometry

Roger Balian

Abstract. When vacuum is partitioned by material boundaries with arbitrary shape, one can define the zero-point energy and the free energy of the electromagnetic waves that are located in it, independently of the nature of the boundaries, in the limit when they become perfect conductors and provided their curvature is finite. An explicit expression for the zero-point energy and the free energy is given in terms of an integral kernel acting on the boundaries; it can be expanded into a convergent series interpreted as a succession of an even number of scatterings of a wave. The quantum and the thermal fluctuations of vacuum then appear as a purely geometric property. The Casimir effect thus defined exists only owing to the electromagnetic nature of the field. It does not exist for thin foils with sharp folds, but Casimir forces between solid wedges are finite. Various applications are worked out: low temperature, high temperature where wrinkling constraints appear, stability of a plane foil, transfer of energy from one side of a curved boundary to the other, forces between distant conductors, special shapes (parallel plates, sphere, cylinder, honeycomb).

1 Introduction

In its original form, the Casimir effect describes an attraction between two plane, parallel, perfectly conducting plates, which is explained by the occurence of a virtual electromagnetic field in the vacuum separating the plates and in the vacuum lying outside them. Indeed, even in the ground state of the system, when no real photon is present, the so-called zero-point motion of each mode m of the field (with frequency $\omega_m/2\pi$) yields a contribution $\frac{1}{2}\hbar\omega_m$ to the energy, which depends on the distance l between the plates. The variation with l of the overall zero-point energy manifests itself as the Casimir force. It is remarkable that, in spite of the divergence which appears when we sum the zero-point energies of the various modes over all of them, we find a finite value for this force.

It is somewhat puzzling to regard the Casimir effect as a property of vacuum containing a virtual electromagnetic field. It may look more natural to attribute it to the matter of the plates. Actually it is possible through Maxwell's equations to express the electromagnetic field, and hence its energy, in terms of the charge and current densities of the particles which can move within the conducting plates and which are the sources for the field. The zero-point motion of these particles then yields a non-vanishing interaction energy although the expectation values of the charge and current densities vanish. The force between the conducting plates at zero temperature can then be interpreted as a result of the interaction between the virtual zero-point currents that must exist in the ground state of matter owing to Heisenberg's inequality. However, this viewpoint is plagued by the fact that the interaction between charged particles is not instantaneous, but retarded (see the contribution of A. Aspect and J. Dalibard in this volume). On the other hand, the evaluation of the force as a result of this interaction would rely on the specific structure of matter. In contrast, Casimir's viewpoint, which focuses on the field rather than on the matter of the plates, shows that the effect is nearly independent of the properties of matter (provided the plates are good conductors); moreover, as function of the electromagnetic field $\mathbf{E}, \mathbf{B}$ the energy of this field at a given time is simply expressed as

$$\mathcal{E}_{\mathrm{f}} = \frac{1}{2} \int d^3r [\epsilon_0 \mathbf{E}^2(r) + \mu_0^{-1} \mathbf{B}^2(r)] \tag{1}$$

in terms of the field at the same time, whereas $\mathbf{E}(r)$ and $\mathbf{B}(r)$ depend on the charges and currents at earlier times.

In the present contribution, we shall exhibit general aspects of the Casimir effect, for walls with arbitrary shape and at arbitrary temperatures. We wish to answer a few theoretical questions. Can the energy associated with the quantized electromagnetic field in the empty regions bounded by conducting walls be defined independently of the properties of these walls? How does it vary with the temperature of the photon gas that constitutes the field? Does the existence of the Casimir effect depend on the specific features of electromagnetism?

We shall rely on two detailed articles [1,2] which deal with the above questions. This will allow us to leave aside the technicalities and to focus on the various ideas. The main result is embedded in section 6. For a bibliography on the Casimir effect we refer the reader to the other articles published in the present volume.

2 Statement of the problem

We wish to study the energy $\mathcal{E}_{\mathrm{f}}$ of the electromagnetic field in empty regions of space limited by boundaries with arbitrary shape, under circumstances when the expectation value $< \mathbf{E} >, < \mathbf{B} >$ of the quantum field vanishes at any point. The energy (1) may be non-zero for two reasons.

On the one hand, the uncertainty relations prevent the *quantum fluctuations* of $\mathbf{E}$ and $\mathbf{B}$ from vanishing, since these operators do not commute; hence the expression (1) of the energy has a positive minimum, the zero-point energy of the field. (We shall see that this value is not only positive, but even infinite, but that its variations are finite and physically meaningful.)

On the other hand, the vacuum where the field is considered is bounded by walls with which the field can be in thermodynamic equilibrium, at some temperature T. The field therefore presents random *thermal fluctuations* around its vanishing expectation value; they contribute to the energy (1) and to the entropy. We shall study this problem by evaluating the free energy of the field, wherefrom all thermodynamic equilibrium properties follow; in particular its variations with the shape provide the constraints on the walls at fixed temperature. This free energy includes the zero-point energy, to which it reduces at $T = 0$. We shall thus treat simultaneously the *Casimir effect* proper, associated with the *zero-point* energy, that is, to virtual photons, and the *radiation pressure* effects for a *black body* with arbitrary shape, which are associated with real photons in equilibrium with the walls that act as a thermal bath.

Since general relativity is of no relevance in the present problem, energy is defined within an additive constant. We are interested only in its variations and can thus get rid of divergences in the theory by substracting some constant that will tend to infinity.

We wish to define the Casimir effect as a property belonging only to the *field in vacuum*, independently of the matter of the walls. In general the zero-point energy of a field depends on the matter to which this field is coupled. We shall use the term "Casimir effect", contrary to several authors, only when this energy can be defined separately. For electromagnetic fields within real materials, the interaction between the field and the charges does not allow us in general to separate out the energy (1) of the field alone; moreover the presence of a material affects the field even outside it. However, a complete decoupling is achieved in the limit of *perfectly conducting* boundaries, which can be approached experimentally by use of *superconductors*. Both the electric and the magnetic fields $\mathbf{E}$ and $\mathbf{B}$ vanish inside them. Outside them, they simply impose the boundary conditions

$$\mathbf{E}_{\mathrm{t}} = 0 \quad , \quad \mathbf{B}_{\mathrm{n}} = 0 \tag{2}$$

on the tangential (t) and normal (n) components of the field. The presence of material bodies then has only a mere *geometric effect*. The part $-\int \mathbf{j}.\mathbf{A}$ of the energy, which involves the currents and the vector potential and which determines the matter-radiation coupling in the equations of motion, can be assigned to the matter and left aside while $\mathcal{E}_{\mathrm{f}}$ is assigned to the vacuum. Anyhow, for perfect conductors, in a gauge where $\mathbf{E} = -\partial\mathbf{A}/\partial t$, this coupling energy vanishes on average since $\mathbf{A}$ is perpendicular to the surface current $\mathbf{j}$.

In this idealized model, the field and the matter of the boundaries do not exchange any energy, even in time-dependent situations, although the coupling between the potentials and the charged particles relates $\mathbf{E}$ and $\mathbf{B}$ to the charge and current densities through Maxwell's equations of motion. Indeed, the rate of decrease of the energy (1) in some empty region of space is the outgoing flux of the Poynting vector $\mu_0^{-1}\mathbf{E} \times \mathbf{B}$ across the boundary of this region. The conditions (2) imply that the Poynting vector is tangent to a perfectly conducting wall, and hence that no energy can flow across such a wall. The establishment of thermal equilibrium in a vacuum is ensured only by the fact that real conductors are never perfect; this allows energy transfers between field and matter. Like for a perfect gas, the present model is too crude to describe the establishment of equilibrium, but it is adequate for equilibrium properties.

Let us first consider a single connected region v limited by perfectly conducting boundaries. The electromagnetic field in v can be analyzed in terms of the eigenmodes m, obtained by solving Maxwell's equations

$$\begin{cases} \text{curl } \mathbf{E}_m = \mathrm{i}\omega_m \mathbf{B}_m \ , \ c^2 \text{ curl } \mathbf{B}_m = -\mathrm{i}\omega \mathbf{E}_m \ , \\ \text{div } \mathbf{E}_m = 0 \ , \ \text{div } \mathbf{B}_m = 0, \end{cases} \tag{3}$$

with the boundary conditions (2). We shall keep aside the electrostatic and magnetostatic solutions with zero frequency, which do not contribute to the Casimir effect. Each mode m behaves as a harmonic oscillator with frequency $\nu_m = \omega_m/2\pi = cq_m/2\pi$ where q_m has the dimension of an inverse wavelength. Its associated energy (1) may take the quantized values $\varepsilon(q_m, n) = \hbar c q_m(n + \frac{1}{2})$. At finite temperature it yields a contribution $f(q_m)$ to the free energy, where

$$f(q_m) = -T\ln \sum_{n=0}^{\infty} \mathrm{e}^{-\hbar c q_m (n+\frac{1}{2})/T} = T\ln[2 \ \mathrm{sh}(\hbar c q_m/2T)] \ . \tag{4}$$

At zero temperature the corresponding contribution to the energy of vacuum is $\frac{1}{2}\hbar c q_m$, the limit of (4) as $T \to 0$. For the high frequency modes such that $h\nu \gg T$, the free energy $f(q)$ is dominated by this zero-point energy. For the low frequency modes such that $h\nu \ll T$,

$$f(q) \sim -T\ln(T/\hbar c q) \tag{5}$$

is dominated by the classical $-T\ln T$ behavior.

The spectrum of eigenfrequencies, or equivalently of eigenwavenumbers $q_m^{(v)}$ in the considered region v is characterized by the *density of modes*

$$\rho^{(v)}(q) = \sum_m \delta(q - q_m^{(v)}) \ , \tag{6}$$

and the *free energy* of this region is formally equal to

$$F^{(v)} = \int_0^{\infty} dq \ \rho^{(v)}(q) f(q) \ . \tag{7}$$

A first difficulty arises if the domain v is infinite, since the spectrum is then *continuous*. We therefore imagine that the full system is enclosed in a *large box* Σ, with volume V, which will eventually tend to infinity. By assuming this outermost boundary to be perfectly conducting, we not only discretize the spectrum m in the open regions v, but also confine the field and ensure that no energy is radiated outwards.

A second difficulty is associated with the fact that the spectra are *not bounded*. Indeed, for large q, the distribution (6) has the asymptotic expansion

$$\rho^{(v)}(q) \approx \frac{v}{\pi^2}q^2 - \frac{2}{3\pi^2}\int \frac{d^2\alpha}{R} + \frac{1}{12\pi^2}\int ds \frac{(\pi-\theta)(\pi-5\theta)}{\theta} + \mathcal{O}(\frac{1}{q^2}) + \mathcal{O}(q^{5/2} \times \mathrm{osc}) \ . \tag{8}$$

The dominant term is proportional to the *volume* v of the considered region; it is the only one which contributes to the black-body radiation in the thermodynamic limit. The second one is a *curvature* term; it is the integral over the boundaries of v of the average curvature $R^{-1} = \frac{1}{2}(R_1^{-1} + R_2^{-1})$, where R_1 and R_2 are the two main curvature radii at the point α, oriented towards the interior of v. It is supplemented, in case the boundary has not a finite curvature everywhere and includes wedges with a dihedral angle θ at the point s of the edge, by the next, *wedge* term, integrated along the wedge; we have for instance $\theta = \pi/2$ if v is the interior of a cube, $\theta = 3\pi/2$ if it is the exterior. Even if it is smoothed, the distribution (8) finally includes *oscillatory* terms with an amplitude which increases with q. Since $f(q) \sim \frac{1}{2}\hbar cq$ for $q \to \infty$, all the terms exhibited in (8) lead to *divergences* in the free energy (7).

3 Regularization of the free energy

We encounter here the simplest example of the divergences that plague quantum field theory. We deal with them by using the standard technique. We first *regularize* the divergent formulae by means of *cut-offs*. We then deduce, from the resulting finite expressions, quantities that are physically *observable* at least theoretically. We finally *renormalize* these quantities by letting therein the cut-off parameters go to infinity. The theory is renormalizable if we get a finite limit for the physical quantities.

We have already regularized the "*infrared*" divergence associated with the infinite size of vacuum by introducing the *box* Σ. We shall deal with the "*ultraviolet*" divergence associated with the high frequencies $cq \to \infty$ in the integral (7) by introducing a *cut-off factor* $\chi(q)$ close to 1 for $q \ll Q$ and decreasing sufficiently fast for $q \gg Q$ so as to restore convergence of the integral. Our final goal is to construct a renormalized free energy F, finite in the limit as $\Sigma \to \infty$, $Q \to \infty$. If this is feasible, it will mean that the ideal model of the electromagnetic field outside a set of perfectly conducting boundaries is renormalizable. In other words the Casimir effect exists as a property of the field proper, conditioned by the sole geometry of the boundaries S.

Let us first see how one can get rid of the most severe divergence, associated with the first term of (8) which after regularization with $\chi(q)$ yields $F^{(v)} \propto v\hbar cQ^4$. Consider, for instance, the Casimir force between two spheres with volumes v_1 and v_2. There is here a single empty domain v_0, which lies outside the spheres and inside Σ. Its volume is $v_0 = V - v_1 - v_2$. The only universal and natural way to cancel the corresponding divergence consists in replacing the solid conductors by *hollow thin conducting shells*, and in taking as a reference the free energy of the empty enclosure Σ, which is itself divergent. We now have three empty domains, v_0, v_1, v_2, the volumes of which sum up to V. The most divergent term thus disappears if we substract the free energy of the empty space within Σ from the total free energy in the presence of the two spherical shells, as

$$F^{(v_0)} + F^{(v_1)} + F^{(v_2)} - F^{(\Sigma)} . \tag{9}$$

More generally and more precisely, we denote as S the set of two-dimensional surfaces which bound the considered conductors. They partition the whole space (within the enclosure Σ) into a set of connected regions v, some of which coincide with the actual vacuum (as v_0 above), the other ones with the interiors of the conductors (as v_1 and v_2 above). We then define the *regularized free energy* associated with the *whole space partitioned by* S as

$$F_{\text{reg}} = \int_0^\infty dq \left[\sum_v \rho^{(v)}(q) - \rho^{(\Sigma)}(q) \right] f(q)\chi(q) \equiv \int_0^\infty dq \; \delta\rho(q) f(q) \chi(q) \; . \tag{10}$$

This expression is finite owing to the cut-offs Σ and $\chi(q)$. If, as indicated above, it has a *finite limit* F as $\Sigma \to \infty$ and $\chi(q) \to 1$, independently of the shapes of Σ and $\chi(q)$, the Casimir effect will appear as a universal property characterized by the free energy F for the boundaries S, which will depend only on the *geometry of* S and on the temperature. This will provide us with a generating function for all mechanical and thermal properties in thermodynamic equilibrium, in two idealized circumstances.

On the one hand, the expression (10) can be directly interpreted as the change in the free energy of the vacuum when a system S of closed, extremely *thin conducting foils* is introduced. The variations of F under deformations of such foils determine the *constraints* induced on them by virtual (for $T = 0$) or real photons (for $T \neq 0$). For instance, for a single sphere separating two regions v_1 and v_2, the dependence of (10) on the radius determines the pressure exerted on the skin of this hollow sphere by the internal and the external field. We shall also encounter below constraints which tend to corrugate such thin sheets, by studying how F changes under periodic deformations.

On the other hand, the expression (10) is also suited for the study of forces between *bulky indeformable conductors*, as we now show. Let us return to the above example of two spheres. After regularization the force between them is associated with the variation of $F^{(v_0)}$ when they are shifted apart. The regularized quantities $F^{(v_1)}$, $F^{(v_2)}$ which enter (9) are not physically relevant to the present problem where we deal with bulky rather than empty spheres, but they do not depend on the distance between these two spheres. Actually, a perfectly conducting skin of a sphere behaves as a perfect screen and the electromagnetic fields, inside and outside, are independent. Thus the force evaluated from (9) is the same as that evaluated from $F^{(v_0)}$, and it is preferable to use (9) because the divergences are expected to be eliminated by this combination. More generally, whenever solid conducting bodies can be *displaced but not deformed*, we can derive the forces between them from the Casimir free energy (10) for which the interior of each body is replaced by vacuum. This trick will allow us to renormalize F. (However, for thermal properties, one should leave aside the contributions of real photons within the conductors.)

Note that the cut-off factor $\chi(q)$ which regularizes the integral (10) for large q has a physical meaning. At high frequency, real conductors are never perfect. Electromagnetic waves can penetrate them, and go freely across them if they are thin. The objects S become *transparent* and the modes within Σ tend to be the same, whether S is present or absent. Thus, for imperfect conductors, the factor $\delta\rho(q)$ in (10) would decrease for large q. In our model we simulate *imperfect conduction or transparency* at high frequency by evaluating $\delta\rho(q)$ for perfectly conducting sheets S and multiplying by $\chi(q)$.

We now proceed and study the behaviour of (10) when the boundary Σ is pushed away to infinity and when the conducting sheets S tend to become perfect with $\chi(q) \to 1$.

4 Fields in the presence of perfect conductors

Our strategy will rely on the following ideas.

(i) We replace the solution of eqs. (2), (3), which define the modes in each region v, by the determination of the *Green functions* associated with these partial differential equations and boundary conditions. Such a Green function contains in a synthetic way the whole information on the modes. It is a function of a complex variable k, *analytic* in the upper half-plane.

(ii) We express the distribution of modes $\rho^{(v)}(q)$ in terms of the *boundary value* for $k \to q + \mathrm{i}0$ of the Green functions.

(iii) This will allow us to regard (10) as an integral in the complex plane k along the half-line $k = q + \mathrm{i}0$, $q > 0$, and to *deform this contour* towards the pure imaginary axis $k = \mathrm{i}y$, $y > 0$ where the Green functions are more regular than along the real axis (they have an infinity of poles at $k = \pm q_m$).

(iv) We determine the Green functions by means of *Neumann's method*, which expresses them as solutions of two-dimensional integral equations over the boundaries S and Σ. The regularized free energy will thereby be expressed through the kernel of these integral equations in terms of the geometry of S and Σ.

(v) Along the new integration contour $k = \mathrm{i}y$, we can solve these integral equations by *iteration*. The resulting series are convergent, and can be interpreted physically as describing *multiple scattering* of an electromagnetic wave on the walls, involving *successively induced currents*.

(vi) Convergence of the multiple scattering expansion allows us finally to control the limit $\Sigma \to \infty$, $\chi \to 1$ and to find an explicit expression for the limit F of (10).

We shall content ourselves here with a sketch of this programme. Detailed proofs can be found in [1,2]. Given the symmetry between the fields $\mathbf{E}$ and $\mathbf{B}$ in eqs.(3), it is convenient to introduce *two Green functions*, a *magnetic* one $\mathbf{\Gamma}(r,r')$ and an electric one $\mathbf{\Xi}(r,r')$, which are tensors with two indices at r and r'. The first one represents the magnetic field created at the point r, in the presence of the conducting boundaries S and Σ, by a magnetic dipole lying at r' and oscillating as e^{-ikct} at the complex frequency $ck/2\pi$. The current density associated with this source is (within the factor μ_0)

$$\mathbf{j}_0(r,r') = \mathrm{curl}_r\ [\delta^3(r-r')\mathbf{1}] \tag{11}$$

where $\mathbf{1}$ is the unit tensor. In each region $v, \Gamma(r,r';k)$ is the solution of the partial differential equation

$$(\nabla^2 + k^2)\mathbf{\Gamma} = -\mathrm{curl}\,\mathbf{j}_0 \quad , \quad \mathrm{div}\,\mathbf{\Gamma} = 0, \tag{12}$$

obtained by eliminating $\mathbf{E}$ from the eqs.(3), with the boundary conditions

$$\mathbf{\Gamma}_{\mathrm{n}} = 0 \quad , \quad (\mathrm{curl}\,\mathbf{\Gamma})_{\mathrm{t}} = 0. \tag{13}$$

The Green function $\mathbf{\Gamma}(r,r')$ is the magnetic field generated at the point r by the source $\mathbf{j}_0$ at r' and the currents $\mathbf{j}(\alpha,r')$ that it induces at the points α of the conducting surfaces S,Σ. We shall denote as n_α the normal vector at α, oriented towards the region v when r and r' lie. Both $\mathbf{j}_0$ and $\mathbf{j}$ are tensors with two indices; the first one refers to the direction of the current, the second one to the orientation of the dipole at r'. Using the formalism of retarded potentials, we can express the magnetic field created by each elementary current $\mathbf{j}$ as $M\mathbf{j}$, where M is the kernel

$$M(r,r') = \mathrm{curl}_r\ [G_0(|r-r'|)\mathbf{1}]\ . \tag{14}$$

A product like $M\mathbf{j}$ stands for integration over space and summation over a tensor index. The scalar Green function

$$G_0(r) = \frac{\mathrm{e}^{ikr}}{4\pi r} \tag{15}$$

is the solution of the equation $(\nabla^2 + k^2)G_0(r) = -\delta^3(r)$ that vanishes at infinity for Im$k > 0$. The magnetic Green function $\mathbf{\Gamma}$ is thus expressed as

$$\mathbf{\Gamma}(r,r') = \int d^3r'' M(r,r'')\,\mathbf{j}_0(r'',r') + \int_{S,\Sigma} d^2\alpha\ M(r,\alpha)\,\mathbf{j}(\alpha,r')\ ; \tag{16}$$

its first term $\mathbf{\Gamma}_0$ is the field produced by the dipole (11) in the infinite space. The as yet unkown currents $\mathbf{j}$ satisfy the integral equation on the boundary

$$\mathbf{j} = \mathbf{j}_1 + K\mathbf{j} \quad , \quad \mathbf{j}_1(\alpha,r') = \int d^3r K(\alpha,r)\,\mathbf{j}_0(r,r'), \tag{17}$$

where the kernel K between two points α and β of the boundary is the tensor

$$K(\alpha,\beta) = 2n_\alpha \wedge \mathrm{curl}_\alpha[G_0(|\alpha-\beta|)\mathbf{1}]. \tag{18}$$

In eq.(17) the product $K\mathbf{j}$ stands for summation on the tensor index and integration over α on the boundary. The proof of (17), (18) relies on an extension of Neumann's method, based upon the discontinuity of the surface integral in (16) when r crosses the boundary.

Altogether the solution of (17) determines $\mathbf{j}$, and $\mathbf{\Gamma}$ follows from (16), taking into account the various definitions (11), (14), (15), (18). The solution of the partial differential equation (12) with the boundary conditions (13) thus amounts to the solution of the *integral equation* (17) *on the boundary.*

The Green function $\mathbf{\Gamma}$ is a *generating function for the modes* m defined by (2), (3) in the connected region v where the source (11) lies. Indeed, in terms of the complex variable k, its poles are the real points $k = \pm q_m = \pm\omega_m/c$ and the corresponding residues are given by

$$\mathbf{\Gamma}(r,r';k) = \sum_m \frac{q_m^2}{q_m^2-k^2}\mathbf{B}_m(r)\otimes\mathbf{B}_m(r') \ , \tag{19}$$

where the magnetic field $\mathbf{B}_m(r)$ for the mode m is the real solution of (2), (3) normalized according to $\int_v d^3r\mathbf{B}^2(r) = 1$. The *spectral density* (6) is thus related to $\mathbf{\Gamma}$ through

$$\rho^{(v)}(q) = \frac{2}{\pi q}\int_v d^3r \ \mathrm{tr} \ \mathrm{Im}\ \mathbf{\Gamma}(r,r;q+i0) \ , \tag{20}$$

where the trace tr refers to the tensor indices.

We can likewise introduce an *electric Green function* Ξ by interchanging magnetic and electric fields, which amounts to interchanging the boundary conditions (13). It is the electric field created at the point r by a source with current density $\mathrm{curl}\mathbf{j}_0/ick$ in the presence of the boundaries S,Σ. Taking (11) into account, we see that, except at $r=r'$, this source produces the same electric field as an electric dipole with current density $-ik[\delta^3(r-r')\mathbf{1}]/c$.

Provided r and r' lie in the same region v, the electric Green function $\Xi(r,r';k)$ can be represented by an expression analogous to (16), (17), within the mere *change in sign of* $\mathbf{j}_1$. (However, whereas $\mathbf{\Gamma}$ is expressed by means of (16) in terms of a true electric current density, the similar expression for Ξ involves a fictitious current, without physical meaning. Moreover, if we take r and r' on different sides of a boundary S, the expression (16) for $\mathbf{\Gamma}$ vanishes as it should, whereas the similar representation for Ξ provides an unphysical non-zero value. Such a behaviour is currently found in the books of mathematics that deal with Neumann's method; in fact, paradoxically, single-layer potentials are used there to represent Green functions when they vanish on the boundary, double-layer potentials when their normal derivative vanishes, whereas it is natural in electrostatics to make the converse choice. It is the application of Neumann's method in two different ways which allowed us to find the above simple relation between the representations of $\mathbf{\Gamma}$ and $\mathbf{\Xi}$.)

Since $\mathbf{\Xi}$ has the same spectral representation (19) as $\mathbf{\Gamma}$ within the replacement of $\mathbf{B}_m$ by $\mathbf{E}_m$, we can equivalently express the distribution of eigenmodes (20) as

$$\rho^{(v)}(q) = \frac{1}{\pi q} \int_v d^3r \ \mathrm{tr}\ \mathrm{Im}[\mathbf{\Gamma} + \mathbf{\Xi}(r, r; q + \mathrm{i}0)] \ , \tag{21}$$

and simplications will appear owing to this combination.

In fact, the *iteration* of the integral equation (17) provides

$$\mathbf{j} = \mathbf{j}_1 + K\mathbf{j}_1 + K^2\mathbf{j}_1 + \cdots \ , \tag{22}$$

a series which exhibits the surface current $\mathbf{j}$ as the sum of *successively induced currents*: $\mathbf{j}_1$ is according to (17) a current directly induced by the dipolar source on the conducting boundaries; it induces in turn through the propagator K a secondary current $K\mathbf{j}_1$, and so on. The expression (18) of K and the behaviour of the free Green function (15) show that this propagator decreases exponentially at large distances (while oscillating) for $\mathrm{Im}k > 0$. Moreover, $K(\alpha, \beta)$ vanishes when the point β lies in the plane tangent at α to the boundary. Thus a current circulating on a plane boundary does not induce through K any secondary current on the same plane. At short distances K vanishes for a smooth boundary and is proportional to its curvature; this ensures the convergence of the integrals in $K\mathbf{j}_1$, $K^2\mathbf{j}_1$, etc. By relying on these properties, we can show that the expansion (22) is *convergent* at least in the region $\mathrm{Im}k > |\mathrm{Re}k|$. (For $k = 0$ describing static fields, the convergence depends on the *topology* of the boundaries.) The general theory of the Casimir effect will make use of this convergence.

The series for $\mathbf{\Gamma}$ which results from (16) and (22) reads

$$\mathbf{\Gamma} = M \frac{1}{1 - K} \ \mathbf{j}_0 = M \ \mathbf{j}_0 + MK \ \mathbf{j}_0 + MK^2 \ \mathbf{j}_0 + MK^3 \ \mathbf{j}_0 \ + \ \cdots \ , \tag{23}$$

where M defined by (14) describes the propagation of a wave issued from a unit element of current, and where K defined by (18) describes a similar propagation followed by the creation of an induced current. We can thus interpret (23) as a *multiple scattering expansion*: the wave issued from the source at r' may reach directly r (first term); it may propagate from r' to a point α of the boundary where it is scattered to reach r (second term); it may scatter successively twice on the boundary before reaching r (third term); and so on. The grazing scatterings vanish, so that the series (23) reduces to its first two terms for a single plane boundary, the second one describing the reflected wave.

The electric and magnetic Green functions are related to each other through

$$\mathbf{\Xi}(r, r'; k) = k^{-2} \ \mathrm{curl}_r \ \mathrm{curl}_{r'} \ [\mathbf{\Gamma}(r, r'; k) - \mathbf{\Gamma}(r, r'; 0)], \tag{24}$$

and conversely. However the alternate series

$$\mathbf{\Xi} = M \frac{1}{1 + K} \ \mathbf{j}_0 = M\mathbf{j}_0 - MK \ \mathbf{j}_0 + MK^2\mathbf{j}_0 - \cdots \ , \tag{25}$$

which results from the integral equation over S, Σ for Ξ does not correspond term by term to (23), (24).

This expansion has the same interpretation in terms of successive scatterings as (23) in the physical situation when r and r' lie in the same connected region v. (However, if r and r' lie in two different regions v separated by a boundary S, the series (25) converges towards some non-zero value, and thus does not represent the actual electric field produced at r by the source at r'. On the contrary the various terms of the expansion (23) interfere destructively in such a configuration.)

Adding (23) and (25) as in (21) cancels all the odd terms of the expansion and yields a geometric series in K^2. Only survive in the evaluation of the distribution of modes the terms describing an *even number of scatterings*. On the other hand, in the summation over the regions v, we shall have to transfer the two points r and r' (with $r = r'$) from one side to the other of the surface S. The kernel K changes its sign in this operation, since its definition (18) involves the normal vector n_α oriented in the direction of the domain v where the Green function is evaluated. After addition of $\mathbf{\Gamma}$ and Ξ in (21), the eigenmodes on both sides of S are evaluated with the *same integrand*. This remark will allow us to perform explicitly the summation over v and the integration over r in (26).

5 Limiting process

We are now in position to express the regularized free energy (10) in terms of the kernel K on the boundaries S, Σ. In order to switch the integration over q towards the complex plane k, we introduce the *generating function of the modes*:

$$\delta\Phi(k) = \frac{1}{2}\int d^3r \ \mathrm{tr} \lim_{r'\to r}\left[\sum_v (\mathbf{\Gamma}^{(v)} + \Xi^{(v)}) - \mathbf{\Gamma}^{(\Sigma)} - \Xi^{(\Sigma)}(r, r'; k)\right] . \tag{26}$$

As in (9) or (10), we have summed over the various regions v bounded by S and enclosed in Σ, and substracted the contribution of the empty box; the integral is therefore carried over the whole interior of Σ. The Green functions are singular for $r' \to r$, but their singular part, which arises only from the first term $\mathbf{\Gamma}_0(r - r') = \Xi_0(r - r') = M\mathbf{j}_0$ of (23) and (25), is cancelled in (26) by the substraction of $\mathbf{\Gamma}^{(\Sigma)} + \Xi^{(\Sigma)}$. Thus no divergence appears in (26).

Like the $\mathbf{\Gamma}$ and Ξ's, the function $\delta\Phi(k)$ has no other singularity than poles at the points $k = \pm q_m$ of the real axis, with residues $\mp\frac{1}{2}q_m$. Hence, according to (21), we have

$$\delta\rho(q) = \frac{2}{\pi q}\mathrm{Im}\ \delta\Phi(q + i0) \tag{27}$$

for $q > 0$. The expression (4), (10), (27) of the regularized free energy thus reads

$$F_{\mathrm{reg}} = \frac{2T}{\pi}\mathrm{Im}\int_0^{\infty+\mathrm{i}0} dk \frac{\delta\Phi(k) - \delta\Phi(\mathrm{i}0)}{k} \ln\left(2\mathrm{sh}\frac{\hbar ck}{2T}\right)\chi(k) . \tag{28}$$

The substraction of the real number $\delta\Phi(\mathrm{i}0)$, which is equal to $\delta\Phi(0)$ when the box Σ is finite, ensures the convergence of the integral; it corresponds to the fact that static fields associated with $k=0$ do not contribute to the Casimir effect.

In order to take advantage of the convergence of the expansions (23) and (25) for $\mathrm{Im}k > |\mathrm{Re}k|$ in (26), (28), we shall deform the integration contour in (28) towards the imaginary axis. We also *get rid of the spatial cutoff* Σ. We therefore introduce the function

$$\Psi(y) = \lim_{\Sigma\to\infty} \delta\Phi(\mathrm{i}y) \ . \tag{29}$$

When taking the limit $\Sigma\to\infty$, we note that G_0 defined by (15) and hence M and K defined by (14) and (18) decrease exponentially with the distance l as e^{-yl}. The function $\Phi(\mathrm{i}y)$ is represented, through its definition (26) and the multiple scattering expansions (23), (25), by *closed paths* that bounce an *even* number of times on S or Σ. All paths which involve only scatterings on Σ are compensated for, owing to the substraction in (26). The contributions of all the paths which involve at least one back and forth travel between S and Σ contain a factor e^{-2yL} where L is the minimum distance from S to Σ; they disappear when Σ is pushed away to infinity. The remaining paths which contribute to $\Psi(y)$ involve only scatterings on S, in even number; before integration over $r'=r$ of (26) they contain for large r a factor e^{-2yL}, where L is the shortest distance between r and S, so that the integral over r is convergent. We shall perform it explicitly below. Note that the poles on the real axis of $[\delta\Phi(k)-\delta\Phi(\mathrm{i}0)]/k$ with residues $\pm\frac{1}{2}$, above which the integral (28) runs, become dense in the limit $\Sigma\to\infty$ and are replaced by a cut.

It remains to *get rid of the ultraviolet cutoff* $\chi(k)$ which eliminates the high frequencies. In order to deform the contour of (28) towards the half line $k=\mathrm{i}y, y>0$ where the singularities of the logarithm lie, we take for $\chi(k)$ a meromorphic function of the form

$$\chi(k) = \sum_i \frac{a_i}{k^2-\mu_i^2} + \frac{a_i^*}{k^2-\mu_i^{*2}} \ . \tag{30}$$

The poles μ_i lie in the first quadrant and their residues satisfy

$$\sum_i \mathrm{Re}\ a_i = 0\ , \qquad -2\sum_i \mathrm{Re}(a_i/\mu_i^2) = 1\ . \tag{31}$$

The moduli $|\mu_i|$ have the same order of magnitude as a number Q that will tend to infinity, so that $\chi(q)$ is close to 1 as long as $q \ll Q$ and decreases as q^{-4} when $q \gg Q$. This behaviour ensures the convergence of (10) on account of the behaviour of the residual terms in (8).

Deforming the contour of (28) towards the imaginary axis produces terms associated with the residues at the points $k=\mu_i$. The properties of $\delta\Phi(k)$ then ensure that these residues vanish when $|\mu_i| \propto Q\to\infty$.

Indeed, for $k\to\infty$ in the first quadrant, it turns out that the multiple scattering expansion (23), (25) is dominated by its lowest, two-scattering term

$MK^2\mathbf{j}_0$, provided the surface S is twice differentiable. The result is found from a short-distance expansion as

$$\Psi(y) = \frac{1}{32\pi}\int_S d^2\alpha \left(\frac{1}{R_1R_2} - \frac{3}{R^2}\right) + \mathcal{O}\left(\frac{1}{y^2}\right) , \tag{32}$$

where R_1 and R_2 are the two principal curvature radii and $1/R$ the average curvature at the point α. Finally, near the origin, Ψ behaves as

$$\Psi(y) \approx -n(1 - Ay) + \mathcal{O}(y^2) , \tag{33}$$

where n is the genus of S, depending only on its topology ($n = 0$ for a sphere, $n = 1$ for a torus), and where $A > 0$.

6 The renormalized Casimir free energy

Altogether, using the above properties, we find the limit of (10) for $\Sigma \to \infty$ and $Q \to \infty$ or $\chi(q) \to 1$ as

$$F = \frac{\hbar c}{\pi}\int_0^\infty dy[\Psi(y) - \Psi(\infty)] + 2T\int_0^\infty \frac{dy}{y}[\Psi(y) - \Psi(+0)]g(y) , \tag{34}$$

where the *temperature* appears through the sawteeth function

$$g(y) = \frac{1}{2} - \frac{y}{\eta} + \sum_{n=1}^{\infty}\theta(y - n\eta) , \ \eta = \frac{2\pi T}{\hbar c} . \tag{35}$$

with $\theta(x) = 0$ for $x < 0$, $\theta(x) = 1$ for $x > 0$

A closed expression for the function $\Psi(y)$, which encapsulates the effect of the *geometry* of the boundaries S on the modes of the field, is found by integrating (26) on r in the whole space. We noted that its integrand is the same for all the regions v separated by S, so that this integral depends only on the first and last scattering points on S in the expansions (23), (25). The integration over $r = r'$ of the corresponding product of M and $K\mathbf{j}_0$ can be expressed only in terms of the kernel K on S, and it simply yields $\frac{1}{2}y\, dK/dy$. Hence we find

$$\Psi(y) = -\frac{y}{4}\frac{d}{dy}\mathrm{Tr}\ln(1 - K^2) , \tag{36}$$

where the trace Tr and the products stand for integration over S of a variable α and summation on the tensor indices of K. Expressed as function of y, the kernel K on the surface S defined by (15), (18) reads

$$K(\alpha, \beta; y) = \frac{1}{2\pi}n_\alpha \wedge \mathrm{curl}_\alpha \left[\frac{\mathrm{e}^{-y|\alpha-\beta|}}{|\alpha - \beta|}\mathbf{1}\right] . \tag{37}$$

It is real, decreases exponentially at large distances, and locally vanishes as the product of the distance and the curvature for $|\alpha - \beta| \to 0$. When S consists of several disconnected pieces, the normals n_α on each of them should be oriented compatibly; for instance, all of them should point towards the outermost region. Along the integration path of (34), the series obtained by expanding $\ln(1-K^2)$ in powers of K^2 converges. The values of $\Psi(y)$ at both ends of the integration path were given by (32) and (33).

The two terms of (34) correspond to the two different phenomena that we are studying. The first one is the *Casimir energy proper*, associated with the *variation of the zero-point energy* of the electromagnetic modes that is induced by the introduction of the perfectly conducting shells. It is the product of $\hbar c$ by a factor with dimension L^{-1} depending on the shape of S. The second one is the variation of the *free energy of the black body* due to the effect of the boundaries on the gas of *real photons*. It has no ultraviolet divergence. The temperature T is that of the walls, which carry random currents in equilibrium with the quantized field.

When there are several material bodies, the surface S involves several disconnected sheets. One can then classify the various terms of the expansion of (36) in powers of K according to the position on these sheets of the scattering points $\alpha, \beta, \cdots$. The terms for which all these points lie on the same connected sheet describe the free energy of *each separate body*; they do not contribute to the forces between indeformable bodies but determine internal constraints for thin conducting foils. Those for which some propagation $K(\alpha, \beta)$ occurs between two points α and β situated on different sheets describe the *interaction* free energy.

7 Conditions for the existence of the Casimir effect

We have proved above the existence of a limit F for the free energy of the field only, for perfectly conducting, thin walls, which defines the Casimir effect. This has been made possible under two conditions, which we now discuss.

On the one hand, the Casimir effect exists only owing to the *electromagnetic* nature of the field. Our proof made use of the cancellation of the one-scattering terms $\pm MK\mathbf{j}_0$ of the expansions (23) and (25), which eliminated a divergent surface contribution. Let us show that the *ultraviolet divergence* of (10) *cannot be removed for a scalar field*, so that the *specific features of electromagnetism are essential* to assign an energy to the field separately. Actually, the high-frequency expansion of the density of eigenmodes that replaces (8) for a scalar field includes, after a volume term $vq^2/2\pi^2$, an *area term* equal to $-sq/8\pi$ for Dirichlet boundary conditions (cancellation of the field at the wall), or to $+sq/8\pi$ for Neumann conditions (cancellation of the normal derivative), where s is the area of the boundary of the domain v. This term is associated with the occurence of single scattering in the expansion analogous to (23). In the evaluation of the regularized free energy (10) the contributions of the two sides of the walls add up. A contribution $\mp(\hbar cs/8\pi)\int_0^\infty dq\, q^2\chi(q)$ proportional to the area s of the walls S occurs, and its divergence when the ultraviolet cutoff Q tends to infinity is incurable. A divergent Casimir force, tending either to stretch or to shrink the boundaries, would thus appear for a scalar field. The properties of matter interacting with the field could not be disregarded.

The inexistence of a divergent area term for the electromagnetic field can be traced to the mixed boundary conditions $\mathbf{E}_{\mathrm{t}} = 0$, $(\mathrm{curl}\mathbf{E})_{\mathrm{n}} = 0$ and to the constraint $\mathrm{div}\mathbf{E} = 0$. The first condition is of the Dirichlet type for the two tangential components of $\mathbf{E}$; the constraint yields a Neumann boundary condition for the normal component. However the condition $(\mathrm{curl}\mathbf{E})_{\mathrm{n}} = 0$ relates the two tangential components to each other, so that altogether the electromagnetic field behaves as two scalar fields, one with Dirichlet the other with Neumann boundary conditions. The fact that the area terms are opposite for these two boundary conditions entails their compensation.

A second condition is also necessary for the renormalization of the total Casimir energy, namely the *smoothness of the boundaries* S. We have relied, in our elimination of the ultraviolet divergence, on the behaviour (32) of $\Psi(y)$ for large y, which itself requires a *finite curvature* of S. To understand the origin of this condition, let us return to the expansion (8) of the eigenmode density $\rho^{(v)}$ for a given volume v. Its first term was cancelled by the subtraction $\rho^{(\Sigma)}$ in (10). The next curvature term of (8) yields in $F^{(v)}$, given by (7), a $\int q dq$ divergence. Fortunately, at each point α of the boundary S, the curvatures are opposite on the two sides of this boundary. Hence, in the summation over the domains v of (10), the curvature terms cancel one another. Accordingly, the large y expansion (32) of $\Psi(y)$ begins with a second order curvature contribution.

If, however, the surface S has a *sharp fold*, for instance if it is a hollow thin cube, this cancellation of the divergences from both sides S no longer occurs. The third, wedge term of (8) gives rise to the same $\int q\, dq$ divergence as the curvature term for each $F^{(v)}$. Let us evaluate its coefficient. Going from one side of the surface S to the other changes the dihedral angle θ into $2\pi - \theta$. The sum of the wedge contributions to $\delta\rho$ from the two neighbouring domains is thus

$$\frac{1}{12\pi^2}\int ds\left[\frac{(\pi-\theta)(\pi-5\theta)}{\theta}+\frac{(\theta-\pi)(5\theta-9\pi)}{2\pi-\theta}\right]=\frac{1}{6\pi}\int ds\frac{(\pi-\theta)^2}{\theta(2\pi-\theta)}\,, \tag{38}$$

and it is a positive number as soon as $\theta \neq \pi$. Hence F_{reg} is dominated by a positive, divergent term proportional to (38), and the field exerts in this idealized model an infinite constraint on the considered foil, which tends to *flatten* its dihedron.

The same divergence occurs for an open conducting foil. Its *edge* is equivalent to a dihedral angle $\theta = 2\pi$, the contribution of which to $\rho^{(v)}$ is $(3/8\pi)\int ds$. The divergence of F implies, for instance, that the zero-point energy of the field produces a strong attractive force which tends to join the two halves of a thin conducting foil that is cut along a line.

We also find a divergence if S involves several adjacent dihedra, for instance, if it is made of three half-planes joined along their common edge as in a *honeycomb*.

In this case, for $\theta = 2\pi/3$, we find a contribution to $\delta\rho$ equal to $(-7/24\pi)\int ds$ which is *negative*. Such a configuration would thus be particularly stable if the experiment could be realized.

The fact that one cannot define a finite Casimir energy for thin conducting foils with creases is related to the singularities of electromagnetic fields near sharp edges.

Nevertheless the above treatment can easily be adapted to Casimir forces between perfectly conducting *rigid bodies, even if they have sharp angles.* Consider, for instance, two bulky wedges. The free energy of each of them cannot be renormalized, because the kernel $K(\alpha,\beta)$ is singular as $1/|\alpha-\beta|^2$ for two neighbouring points α and β located on the different sides of the edge; the resulting contribution $\mathrm{Tr}K^2$ to (36), in particular, is seen to generate a divergence in (34). If, however, according to the remark at the end of section 6, we focus on the interaction between the two indeformable wedges, letting aside their own energies, such divergent contributions are irrelevant. The interaction free energy of the two wedges is thus obtained by keeping, in the expansion of (36) in powers of K, only those terms which involve scatterings on both wedges (with an even number of factors K). Their contribution to (34) is expected to be finite in spite of the short-distance singularity of K near the edges of the wedges, and the Casimir force between them is therefore well defined. For example, for two wedges with the same dihedral angle 2θ facing each other perpendicularly at a distance l, the two-scattering approximation expressed below by eq.(50) provides at $T = 0$ an attractive Casimir interaction energy equal to $-\hbar c\ \mathrm{tg}^2\theta/4\pi^2 l$.

We now review some applications of the general expression (34) for the Casimir free energy. More details can be found in [2].

8 Parallel plates

For two parallel plates with area S, lying at a distance l from each other, the function $\Psi(y)$ associated with the three regions separated by these plates is

$$\Psi(y) = \frac{Sy^2}{2\pi}\ln(1-\mathrm{e}^{-2yl})\ , \tag{39}$$

wherefrom we get the elementary Casimir effect at $T = 0$. The low temperature expansion of the free energy (34),

$$F(T) = -\frac{S\pi^2\hbar c}{720\ l^3} - \frac{ST^3\zeta(3)}{2\pi\ \hbar^2c^2} + \frac{S\pi^2 lT^4}{45\hbar^3c^3} + \mathcal{O}(T^2\mathrm{e}^{-\pi\hbar c/lT})\ , \tag{40}$$

shows that the Casimir attraction $-\partial F/\partial l$ increases with the temperature.

The comparison of (28), which includes the factor $\ln(1-\mathrm{e}^{-\hbar cq/T})$, with (34), (39), exhibits *a duality between high and low temperatures,*

$$F(T) - F(0) = \left(\frac{2lT}{\hbar c}\right)^4\left[F\left(\frac{\hbar^2c^2}{4l^2T}\right) - F(0)\right]\ . \tag{41}$$

At high temperature, we have

$$F(T) = -\frac{ST\ \zeta(3)}{8\pi l^2} + \mathcal{O}(T^2 \mathrm{e}^{-4\pi lT/\hbar c}) , \tag{42}$$

yielding again an attraction due to radiation pressure.

The *entropy* $-\partial F/\partial T$ rises at low temperatures as $3ST^2\zeta(3)/2\pi\hbar^2c^2$, independently of the distance between the plates, and tends to a *finite* limit $S\zeta(3)/8\pi l^2$ at high temperature.

9 Low temperatures

Owing to the more and more rapid oscillations of $g(y)$ when $T \to 0$, we can evaluate the second term of (34) for low temperatures by expanding $\Psi(y)$ around $y = 0$. This yields

$$F(T) - F(0) = \frac{\pi T^2}{3\hbar c}\Psi'(0) - \frac{\pi^3 T^4}{135\hbar^3 c^3}\Psi'''(0) + \mathcal{O}(T^6) . \tag{43}$$

This behaviour is related to the *topology* of the boundaries S, since according to (33) $\Psi'(0)$ vanishes for a singly connected surface, and equals nA for a multiply connected surface. Accordingly, the low-temperature *entropy* arising from (43) behaves for $n = 0$ as T^3 (like the entropy of the black body), but is large as $-2\pi\ n\ AT/3\hbar c$ for $n \neq 0$. This negative sign looks paradoxical. It is related to the fact that for torus-like topologies, permanent supercurrents can generate *static magnetic fields*. The occurence of a number $2n$ of such modes with $q = 0$, which do not contribute to the Casimir effect, entails a depletion in the distribution $\rho^{(v)}(q)$ for $q \neq 0$. In fact, we see from (27), (29), (33) that $\delta\rho(q)$ for $\Sigma \to \infty$ tends to $-2nA/\pi$ as $q \to 0$; it is this negative sign which is reflected in that of the Casimir entropy. However the total entropy of a quantum system must be positive. In the present case this results from the positivity of the total density of eigenmodes $\sum_v \rho^{(v)} = \rho^{(\Sigma)} + \delta\rho$ of the field, ensured by the fact that $\rho^{(\Sigma)} \sim Vq^2/\pi^2$ is infinite in the large Σ limit considered here.

The dimensionless parameter of the expansion (43) is $LT/\hbar c$ where L is the typical size of the system S. The lowest order contributions should become experimentally accessible since this parameter is 0,5 for $T = 300\mathrm{K}$ and $L = 3\mu\mathrm{m}$.

The different behaviour of (40) and (43) arises from the fact that (43) holds only for a finite system, whereas for an infinite system like a pair of parallel plates $\Psi'''(y)$ diverges when $y \to 0$ as shown by (39).

10 High temperatures

At high temperature the second term of (34) is dominated by the first sawtooth of $g(y)$. The corresponding calculation yields

$$F = -CT\ln(T/\hbar c Q) + \mathcal{O}(T^{-1}) , \tag{44}$$

$$\mathcal{C} = \Psi(+0) - \Psi(\infty) = \frac{1}{32\pi} \int_S d^2\alpha \left(\frac{3}{R^2} - \frac{1}{R_1 R_2} \right) - n \ , \qquad (45)$$

$$\ln \mathcal{Q} = -\frac{1}{\mathcal{C}} \int_0^\infty dy \ln y \Psi'(y) \ . \qquad (46)$$

Here the dimensionless parameter $\hbar c/TR$ of the expansion (44) is governed by a length R of the order $\mathcal{Q}^{-1}$ associated with the short-range behaviour of the kernel K. This characteristic length is therefore a typical curvature radius R of S. The high-temperature limit might become experimentally relevant for crippled foils with small R.

The dominant term of (44) is formally the same as the free energy (5) of a number $\mathcal{C}$ of *classical harmonic oscillators* with average frequency $c\mathcal{Q}/2\pi$. Contrary to what happens for the black body which requires a quantum treatment at any temperature, the Casimir contribution that we have calculated, which describes the change in the free energy of photons brought in by the boundaries S, takes a *classical* form for $T \gg \hbar c/R$. This is possible here because the modes that contribute to the Casimir effect have bounded frequencies, whereas the modes with $h\nu \gg T$ crucially contribute to the black-body radiation.

The internal energy $U \sim \mathcal{C}T$ arising from (44) expresses the classical *equipartition*, and as usual in the classical limit the entropy $\mathcal{C} \ln(\mathrm{e}T/\hbar c\mathcal{Q})$ depends on Planck's constant in its additive constant. The number $\mathcal{C}$, positive or negative, is interpreted as the *average number of modes* with finite frequency *added by the introduction of the boundaries S*. Here again the *topology* of S enters the expression (45) of $\mathcal{C}$ through the genus n and the integer $\int d^2\alpha/4\pi R_1 R_2$. For parallel plates, we have $\mathcal{C} = 0$ but $\mathcal{C} \ln \mathcal{Q} = -S\zeta(3)/8\pi l^2$.

The high-temperature Casimir *constraints* on the conductors S describe the effects of *radiation pressure*. To dominant order they behave as $T \ln T$ and are obtained by studying how $\mathcal{C}$ varies when the conductors are displaced or deformed. They tend to let F decrease, thus to let $\mathcal{C}$ increase. The only non-topologic part of $\mathcal{C}$, $3 \int d^2\alpha/(32\pi R^2)$, does not depend on the relative position of the conductors. Hence, to the dominant order in $T \ln T$, there are no forces between different conductors induced by the field. Moreover, since $\mathcal{C}$ is dimensionless and scale-invariant, there are no there forces tending to dilate or contract hollow conducting shells. However $\mathcal{C}$ increases with the average curvature $1/R$ of S, so that the Casimir effect tends at high temperature to *let conducting foils undulate*. This tendency is limited by the next terms of the expansion (44), and the curvature $|R|^{-1}$ tends to rise up to values of order $T/\hbar c$.

The next order contributions to the constraints, of order T, arise from the contribution $\mathcal{C}T \ln \mathcal{Q}$ to F. Since $\mathcal{Q}^{-1}$ is proportional to the size of S, these Casimir forces of order T would tend to contract S if $\mathcal{C}$ is negative, to expand it if $\mathcal{C}$ is positive.

11 The wrinkling effect

The existence of constraints that tend to wrinkle conducting surfaces at high temperature is confirmed by the study of *small deformations of a thin foil.* The evaluation of (34) for a weakly deformed conducting plane S is achieved by means of a two-dimensional Fourier analysis. It is found that the constraints created by the field tend to create ripples with wavelengths larger than $2,9\hbar c/T$, and to restore flatness for smaller wavelengths. In particular, the Casimir effect proper, at zero temperature, tends to suppress the curvatures. Thus a *conducting plane foil* is *stable at* $T = 0$, but *unstable at* $T \neq 0$ under small deformations.

This phenomena is confirmed by the study of the *distribution in space of the free energy* of the electromagnetic field. The density of Casimir free energy $f(r)$ is obtained in the same way as the total free energy (34), except for the integration over the point (26). In fact, taking into acocunt the spectral representation (19) of the magnetic Green function $\mathbf{\Gamma}$ and the similar one for the electric function $\mathbf{\Xi}$, the contribution of each pole $k = q_m$ of $\Phi^{(v)}(k)$ is then weighted by $\frac{1}{2}[\mathbf{B}_m^2(r) + \mathbf{E}_m^2(r)]$. (The coefficients μ_0^{-1} and ϵ_0 of (1) are recovered when the physical fields entering (3) are expressed in terms of the real and normalized functions entering $\mathbf{\Gamma}$ and $\mathbf{\Xi}$.) The density of free energy $f(r)$ is thus found as a series describing a wave starting from r, scattering an even number of times on S and returning to r. The two-scattering term is sufficient to provide the free energy density at a point r located near S, at a distance d from it much shorter than the local curvature radii. We find at low temperature

$$f(r) = -\frac{\hbar c}{30\pi^2 R d^3} + \frac{T^3 \zeta(3)}{2\pi R \hbar^2 c^2} + \cdots , \tag{47}$$

and at high temperature

$$f(r) = \frac{T}{16\pi R d^2}\left[\ln(2dT/\hbar c) + C - \frac{1}{4}\right] + \cdots , \tag{48}$$

where C is Euler's constant.

This density is not bounded when $d \to 0$, and its divergence is non-integrable. Hence, even after renormalization by subtraction of the free energy of the vacuum without boundaries, the free energy $F^{(v)}$ associated with *each region* is *divergent.* The total free energy $F = \sum_v F^{(v)}$ is finite because on the two sides of S the average curvatures R^{-1} are opposite at each point, so that the divergences from (47) or from (48) cancel each other.

The sign in (47) shows that the presence of a perfectly conduting foil produces the transfer of an *infinite* amount of *zero-point energy from the concave to the convex side*, whereas, according to (48), the energy of *real photons* is transferred *from the convex to the concave* side. These opposite signs are consistent with the stability or unstability of a plane foil against deformations, depending on the temperature.

12 Other examples

The Casimir forces between conductors lying *far apart* can be evaluated by means of the free energy (34) of the vacuum separating them. We noted that it may also be attributed to the *random currents* that circulate on their surface and produce the field. Such forces have thus the same nature as *van der Waals forces* of mutual induction, except for the retarded character of the interaction. We find from (34) that two conductors at a large distance l apart *attract* each other as $1/l^8$ at zero temperature, as $1/l^7T$ at high temperature. *Torques* are also found for anisotropic bodies. The same results hold for a small conducting body facing a miror, which is attracted by it.

For two *neighbouring conductors*, the multiple scattering expansion of (36) can be used to justify *Derjagin's approximation* (see in this volume the contribution of Duplantier). Consider for instance the Casimir force between a plane and a sphere with radius R, and denote by l their shortest distance. The trace in (36) can be replaced by an integration over a point α of the plane:

$$\Psi(y) = -\frac{1}{4}\int d^2\alpha \; y\frac{d}{dy}\left[\mathrm{tr}\ln(1-K^2)\right]_{\alpha\alpha} , \tag{49}$$

where the trace tr is meant on the tensor index only. Suppose $l \ll R$. Owing to the exponential decrease of K, the integral (49) is dominated by the contributions such that α lies at a distance x from the sphere of order l, and such that all successive scattering points also lie at a distance of order l from α. Thus, for each α, the integrand of (49) is approximately the same as for two parallel plates lying at a distance x apart, which according to (39) is given by $\pi^{-1}y^2\ln(1-e^{-2yx})$. This is just Derjagin's approximation. Corrections can be obtained from (49).

Another useful approximation is the *two-scattering approximation*, for which we retain for $\Psi(y)$ only the lowest order term $\frac{1}{2}\mathrm{Tr}KydK/dy$ of (36). More explicitly, this yields

$$\begin{aligned}\Psi^{(2)}(y) &= 2y\frac{d}{dy}\int d^2\alpha d^2\beta \frac{dG_0(|\alpha-\beta|)}{dn_\alpha}\frac{dG_0(|\alpha-\beta|)}{dn_\beta} \\ &= -\frac{y^2}{8\pi^2}\int d^2\alpha d^2\beta (n_\alpha.\rho)(n_\beta.\rho)\frac{1}{\rho}\frac{d}{d\rho}\frac{\mathrm{e}^{-2y\rho}}{\rho^2}\end{aligned} \tag{50}$$

where ρ is the vector $\alpha-\beta$. Numerical tests show that this approximation should be fairly good; for example it yields for the Casimir force at $T=0$ between two parallel plates the correct result times $90/\pi^4$, an error of 8%.

Using this approximation, we can evaluate the free energy of interaction between a plane and a sphere. The integrals in (50) can be completely worked out and yield

$$\Psi^{(2)}(y) = \left(-\frac{1}{2}Ry+\frac{1}{4}\right)\mathrm{e}^{-2yl} - \left(\frac{1}{2}Ry+\frac{1}{4}\right)\mathrm{e}^{-2yl-4yR} . \tag{51}$$

Hence we find at $T = 0$, for $l \ll R$,

$$E^{(2)} \approx -\frac{\hbar c R}{8\pi l^2} + \frac{\hbar c}{8\pi l} , \tag{52}$$

which exhibits a correction in l/R to the two-scattering Derjagin contribution.

The study of a *spherical shell* S with radius R shows that the Casimir energy at $T = 0$ behaves as

$$E = 0,046 \hbar c/R . \tag{53}$$

At high temperature, we find in agreement with (44)

$$F = -\frac{T}{4}[\ln(TR/\hbar c) + 0,769] - \left(\frac{\hbar c}{R}\right)^2 \frac{1}{3840T} + \mathcal{O}\left(\frac{1}{T^3}\right) . \tag{54}$$

In the previous examples we had found attractive Casimir forces. In contrast, these forces tend here at any temperature to *expand the sphere.* They increase with T. The radiation pressure exerted from inside thus exceeds that exerted from outside, contrary to what happens for parallel plates.

An intermediate geometry between a sphere and parallel plates is that of a *cylinder.* We evaluate the Casimir energy of a hollow cylinder S by means of the two-scattering approximation (50). For a cylinder with radius R and length $L \gg R$, we find at low temperature

$$F^{(2)} \sim -\frac{2\pi^3 \, L \, R^2 \, T^4}{45\hbar^3 \, c^3} , \tag{55}$$

and at high temperature

$$F^{(2)} \sim -\frac{3LT}{64 \, R} \ln \frac{4,56TR}{\hbar c} , \tag{56}$$

in agreement with (44). The cylinder tends to shrink at high temperature. At zero temperature, the zero-point energy *vanishes* in the considered approximation and its exact value is very small, an intermediate situation between the parallel plates (40) and the sphere (53).

Unfortunately all these Casimir constraints on thin conducting foils, including the wrinkling effect, are presently difficult to detect experimentally, because of their weakness compared to the binding forces that ensure the cohesion of the metallic sheets.

References

[1] R. Balian and B. Duplantier, Electromagnetic waves near perfect conductors, I. Multiple scattering expansions and distribution of modes, *Ann. Phys. NY* **104**, 300–335 (1977).

[2] *idem*, II. Casimir effect, *Ann. Phys. NY* **112**, 165–208 (1978).

Other important references are given in the other articles published in the present volume and in the book by K.A. Milton, The Casimir effect : physical manifestations of zero-point energy (World Scientific, 2001).

Roger Balian
Service de Physique Théorique
CEA/DSM/SPhT
Unité de recherche associée au CNRS
CEA, Saclay
F-91191 Gif-sur-Yvette Cedex

Poincaré Seminar 2002, 93 – 108

Poincaré Seminar 2002

Measurement of the Atom–Wall Interaction: from London to Casimir–Polder

Alain Aspect* and Jean Dalibard†

Abstract. We first present the Casimir–Polder result, giving the interaction potential between a ground state atom and a mirror. This result, obtained within the framework of quantum electrodynamics, is valid for any separation z between the atom and the mirror, provided the electronic cloud does not overlap with the mirror. For large z, this interaction potential varies as $U_{\rm CP}(z) \propto z^{-4}$. This results from the modification of vacuum fluctuations by the mirror and this is quite different from the simple electrostatic result obtained by neglecting any retardation effect, $U(z) \propto z^{-3}$. We also indicate how the Casimir–Polder potential is modified when the mirror is replaced by a dielectric (Lifshitz theory). We then describe three recent experiments which give a clear evidence for the existence of retardation terms in the atom-wall problem, and which are in good agreement with the Casimir–Polder prediction.

1 Introduction

The fact that the electromagnetic vacuum can interact with atomic particles and produce a measurable effect is certainly one of the most striking features of Quantum Mechanics. The name of the Dutch physicist H.B.G. Casimir is attached to some very spectacular manifestations of this interaction. In 1948 he predicted his famous result concerning the attractive force between two perfectly conducting plates [1]. The review of the current experimental state-of-the art for this problem will be done in the next presentation by Reynaud. The same year, Casimir made another essential contribution, together with his colleague Polder [2]. They addressed the following problem: what is the asymptotic behavior of the long range interaction between two atoms, or between an atom and its mirror image?

The existence of long range forces, acting when the constituents are separated by more than a typical atomic size, was predicted by van der Waals in 1881. The first quantitative estimate of these forces was performed by London [3], using an analysis based on classical electrodynamics. The question raised by Casimir and Polder, and that we would like to address here, is the existence of sizeable effects, originating from the quantization of the electromagnetic field, in the long range interaction between an atom prepared in its ground electronic state and a mirror.

*Unité mixte de recherche du CNRS.

†Unité de recherche de l'Ecole normale supérieure et de l'Université Pierre et Marie Curie, associée au CNRS.

2 The Casimir–Polder problem

2.1 The (relatively) short range result

When a static electric dipole $\boldsymbol{d}$ is placed in front of an ideally conducting wall, it interacts with its mirror image and the corresponding energy is

$$U(z) = -\frac{d_x^2 + d_y^2 + 2d_z^2}{64\pi\epsilon_0 z^3} \tag{1}$$

The $-$ sign means that the corresponding interaction is attractive. Here Oz denotes the axis normal to the plane and z is the distance between the atom and the plane. Consider now an atom in its internal ground state $|0\rangle$, placed in front of such a wall. A similar effect may occur, as long as the distance z is notably larger than the atom size, to avoid the overlap between the electron cloud of the atom and the wall itself. The reason for this attraction is clear: although the atom possesses no electric dipole moment in its ground state ($\langle 0|d_i|0\rangle = 0$, for $i = x, y, z$), the average values $\langle d_i^2\rangle$ are strictly positive. A simple picture then emerges in which a fluctuating dipole is associated to the atom, which polarizes the conducting charges of the wall; the induced charge distribution then interacts with the initial atomic dipole. This effect, predicted by Lennard-Jones in 1932 [4], leads to the following interaction potential:

$$U_{\mathrm{LJ}}(z) = -\frac{\langle 0|d_x^2 + d_y^2 + 2d_z^2|0\rangle}{64\pi\epsilon_0 z^3} \tag{2}$$

The passage from a truly static to a time-dependent atomic dipole introduces a time scale τ in the problem and hence, due to the finite speed of light, a length scale $c\tau$. The z^{-3} dependence of the electric field created by the dipole is valid only at distances smaller than $c\tau$. For larger distances, a new approach is needed to account for *retardation effects*, as pointed out in 1941 by J. A. Wheeler [5]. For instance, if one deals with a classical oscillating dipole $\boldsymbol{d}e^{-i\omega t}$, it is well known that the electromagnetic field which is radiated at long distances varies like the inverse of the distance to the dipole. The leading term in the interaction energy between the oscillating dipole and the conducting wall is then [6]:

$$\text{Large } z: \qquad U(z) \sim \frac{(d_x^2 + d_y^2)k^2}{32\pi\epsilon_0 z} \cos 2kz \ , \tag{3}$$

where we set $k = \omega/c$.

For an atom or a molecule, several questions now emerge. What is the relevant time scale τ? Is the picture of an oscillating dipole valid? Does the physics depend on the internal level (ground or excited) of the atom?

2.2 Retardation effects in the atom-wall problem

2.2.1 Atom in its ground electronic state

The problem for a ground state atom was solved in 1948 in a brilliant manner by Casimir and Polder, using the formalism of quantum electrodynamics [2, 7]. Their results show that the retardation effect anticipated above is indeed essential and that it leads to the replacement of the Lennard-Jones z^{-3} variation of the interaction energy by a z^{-4} variation. The length scale on which the transition between the z^{-3} and the z^{-4} regimes occurs is c/ω, where ω is a typical Bohr frequency of the atom. We shall not give the exact result of Casimir and Polder (denoted hereafter as $U_{\mathrm{CP}}(z)$). We simply recall that it is valid for any distance z, and that it coincides with $U_{\mathrm{LJ}}(z)$ for short distances. We now comment on its asymptotic form for large z. In this case, considering an atom with a single valence electron prepared in its ground state $|0\rangle$, one obtains:

$$\text{Large } z: \qquad U_{\mathrm{CP}}(z) \sim -\frac{3}{32\pi^2\epsilon_0}\frac{\hbar c\alpha}{z^4}\,. \tag{4}$$

Here α denotes the static polarisability of the atom in the state $|0\rangle$, of energy E_0:

$$\alpha = \frac{2q^2}{3}\sum_{n\neq 0}\frac{|\langle n|\hat{\boldsymbol{r}}_{\mathrm{e}}|0\rangle|^2}{E_n - E_0} \tag{5}$$

where the sum runs over all the atomic excited states n of energy E_n, and where $\hat{\boldsymbol{r}}_{\mathrm{e}}$ is the position operator of the electron with respect to the atom center-of-mass. Note that for hydrogen and for alkali atoms, the largest contribution to the sum (5) comes from the resonance line (1s $\leftrightarrow$ 2p and ns $\leftrightarrow$ np respectively).

The question now arises to interpret this result in terms either of vacuum fluctuations (modification of the atomic electron dynamics by the quantized electromagnetic field) or radiation reaction (action of the field radiated by the atom upon itself). Such a separation is possible in an unambiguous manner when one expresses the measurable physical quantities in terms of the correlation functions and linear susceptibilities of the two interacting systems, the atom and the electromagnetic field [8]. Using this formalism, the authors of [9] have shown that the result (4) is entirely due to vacuum fluctuations.

Actually one can recover (4) within a numerical factor by the following simple reasoning [10] (see also [11]). The physical origin of (4) is similar to that of the Casimir d^{-4} force between two perfectly conducting walls separated by a distance d (for a review, see the contributions of Balian and Duplantier, and of Reynaud in the same issue). At a distance z from the mirror, the modes of the electromagnetic field which are strongly modified by the presence of the conducting wall are those with a frequency ω such that $\omega \leq c/z$. The electric field associated with each mode is $\mathcal{E}_\omega = \left(\hbar\omega/2\epsilon_0 L^3\right)^{1/2}$, where L^3 is an arbitrarily large quantization volume. The contribution of each mode to the Lamb shift of the ground state of the atom is

$-\alpha\mathcal{E}_\omega^2/2$. Here we use the static polarizability α; indeed we assume that the atom is far enough from the wall so that all the considered modes have a frequency ω much lower than any atomic Bohr frequency. This crude estimation of the modification of the Lamb shift of the atomic ground state, due to the presence of the wall, then gives

$$U(z) \simeq - \sum_{\omega<c/z} 2 \times \alpha\mathcal{E}_\omega^2/2 = -\frac{\alpha\hbar}{4\pi^2\epsilon_0 c^3}\int_0^{c/z} \omega^3 \, d\omega = -\frac{1}{16\pi^2\epsilon_0}\frac{\hbar c\alpha}{z^4} \, ,$$

where the multiplicative factor 2 accounts for the two polarizations basis states for a given wave vector. This is a remarkably good approximation of the exact asymptotic result (4).

2.2.2 Atom in an excited state

The result that we just obtained for a ground state atom is very different from the one obtained for an atom prepared in an excited electronic state $|n\rangle$. In this case, one can show indeed that the leading term is [9, 10]:

$$\text{Large } z: \qquad U(z) \sim \frac{q^2}{8\pi\epsilon_0 z}\sum_{n'<n} k_{nn'}^2 \left(|\langle n|\hat{x}_e|n'\rangle|^2 + |\langle n|\hat{y}_e|n'\rangle|^2 \right) \cos(2k_{nn'}z) \tag{6}$$

where we have set $k_{nn'} = (E_n - E_{n'})/(\hbar c)$ and where the sum over n' runs only on levels with an energy $E_{n'}$ lower than E_n. Here we recover the $\cos(2kz)/z$ behavior characteristic of a classical oscillating dipole (3). As shown in [9], vacuum fluctuations and radiation reaction contribute equally to this result.

2.3 The Lifshitz approach

A few years after the work of Casimir and Polder, Lifshitz also addressed the problem of long range interactions between atomic particles and a macroscopic body [12]. He did not consider a metallic surface, but a bulk dielectric material characterized by a linear susceptibility $\epsilon(\omega)$. We shall not review Lifshitz theory in detail, and we simply give the long range potential for an atom in its ground electronic state, assuming that one electronic transition at frequency ω_A is dominating (for a review, see e.g. [13, 14, 15]):

$$\text{Large } z: \qquad U(z) \sim -\frac{3}{32\pi^2\epsilon_0}\frac{\hbar c\alpha}{z^4}\frac{\epsilon(\omega_A)-1}{\epsilon(\omega_A)+1}\Phi(\epsilon) \, . \tag{7}$$

The function $\Phi(\epsilon)$ is nearly constant and equal to 0.77 when the index of refraction $n = \sqrt{\epsilon}$ varies between 1 and 2, which accounts for most glasses. Note that one recovers the case of a perfectly conducting plate by taking the limit $\epsilon \to \infty$, in which case $\Phi(\epsilon) \to 1$.

We conclude this brief section on Lifshitz theory by noting that the use of a dielectric opens new perspectives with respect to a perfectly conducting wall. One can arrange the response function $\epsilon(\omega)$ of the dielectric to be resonant with some particular Bohr frequencies of the atoms. It is then possible to enhance or decrease the contribution of some atomic transitions to the interaction potential. For example, one can modify the coefficient appearing in front of the short range z^{-3} variation (2). It is even possible to change the sign of the interaction energy if the atom is prepared in an excited state, so that the Lennard-Jones attractive potential is turned into a repulsive one [16, 17].

3 Experimental results

3.1 A brief review of the experimental status

The main motivation of this presentation is to discuss the experimental tests of the Casimir–Polder–Lifshitz prediction, i.e. the long range z^{-4} interaction energy for an atom in front of a conducting wall or a dielectric material. We shall not address here the results obtained recently in cavity quantum electrodynamics, where the atom is surrounded by a cavity with a large quality factor, so that it couples resonantly to only one (or a few) of the cavity modes. We refer the reader to [6, 10], where these experiments are discussed in detail. We shall not discuss either the possible manifestations of long range forces inside an atom. These can occur for instance within a Rydberg helium atom, for which the outer electron sees a field which can be significantly different from the Coulomb field from the nucleus+inner electron. We refer the reader to [18], which present several interesting contributions on this topic.

Before addressing the Casimir–Polder z^{-4} prediction, we shall say a few words on experimental studies in the short range regime, where the Lennard-Jones z^{-3} variation dominates. This regime has first been studied in [19], for an atom or a molecule in front of a conducting material. The idea is to send an atomic or molecular beam very close to a metal cylinder and to look for the deflection of the beam. A deviation is actually detected, but it is difficult to extract quantitative conclusions from these experiments, the main reason being that the impact parameters are uniformly distributed over all possible values. The effect of the atom-wall attraction on the deflected beam is then strongly dominated by the atoms having the smallest impact parameter, where retardation effects play no role, and only the Lennard-Jones potential can be tested. Similar experiments are reported in [20, 21]. Note that the results of these experiments were only in qualitative agreement with the theoretical prediction (2). One can also prepare the atoms in a highly excited Rydberg state, so that the corresponding dipole is much larger, which allows for a more precise study of the Lennard-Jones prediction [22].

The Lifshitz prediction has been tested using liquid-helium films on cleaved surfaces of alkaline-earth fluoride crystals [23]. By varying the thickness of the film between 1 and 25 nm, the authors could obtain a test of Lifshitz theory over 5 orders of magnitude for the potential strength. These experiments showed a first evidence for the deviation from the z^{-3} law at long distances (i.e. thick films).

High resolution spectroscopy experiments can also reveal a position-dependent frequency shift of the atomic energy levels in the relatively short range (z^{-3}) regime. These methods have been used to test the Lennard-Jones prediction for excited atoms [24, 25, 26], and the atom-wall repulsion resulting from a well chosen dielectric response of the wall has been observed [27].

We now turn to three experiments where the Casimir–Polder retardation effect for an atom in its ground state has been observed and studied. We note that this observation cannot be performed using a spectroscopic measurement. Indeed one measures in this case an energy difference between the ground state and an excited state. Since the shift for any excited state (6) is much larger than the shift of the ground state (4), one would only access in this way to the excited level physics, and not to the ground state one. Clearly, one has to rely on a measurement dealing only with the ground state to test this prediction. This leaves several possibilities opened, as pointed out in [6]. One could use atomic interferometry to measure the shift [28, 29, 30]. One could also measure a differential shift between various sublevels of the ground state, in case the non-scalar part of the static susceptibility is significant. Finally, as done in the three experiments described below, one can look for a mechanical effect of the Casimir–Polder potential [31, 33, 41].

3.2 Atom metal force: the Yale experiment [31]

This remarkable experiment constitutes to our knowledge the first quantitative study of retardation effects in the interaction between an isolated ground state atom and a conducting wall. This experiment is precise enough to clearly discriminate between the Casimir–Polder value of the interaction energy and the Lennard-Jones result, in which the interaction is modelled by the instantaneous electrostatic interaction between the atomic dipole and its image in the metal. Figure 1 is a sketch of the experiment. A beam of sodium atoms travels inside a cavity formed by two almost parallel, gold coated, plates. The distance L between the plates can be varied between 0.7 μm and 8 μm. The length of the cavity is $D = 8$ mm.

The experiment consists in measuring the transmission T (or rather the opacity $1/T$) as a function of the separation L. For $L > 3$ μm, the transmission is found equal (within error bars) to the "geometrical" expectation. This geometrical expectation is determined using a Monte–Carlo simulation, in which one neglects any interaction between the atoms and the walls of the cavity. The straight (classical) atomic trajectories are determined by the initial Maxwell–Boltzmann distribution, and only atoms that do not hit the walls are transmitted.

For smaller separations L, the contribution of the atom-wall interaction to the opacity becomes appreciable, and the measured transmission is smaller than the geometrical one (see Figure 2). This reduction can be easily understood if one remembers that the atom-wall interaction is attractive, both for short and long

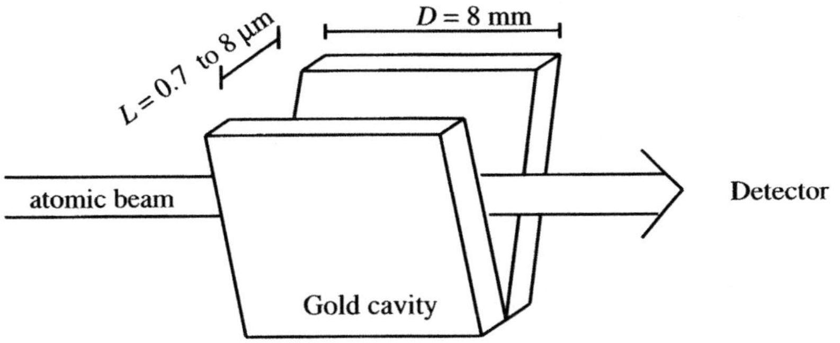

Figure 1: Yale experiment: an atomic beam is sent through a cavity made of two gold coated plates making a small wedge. The number of transmitted atoms is measured as a function of the distance L between the plates.

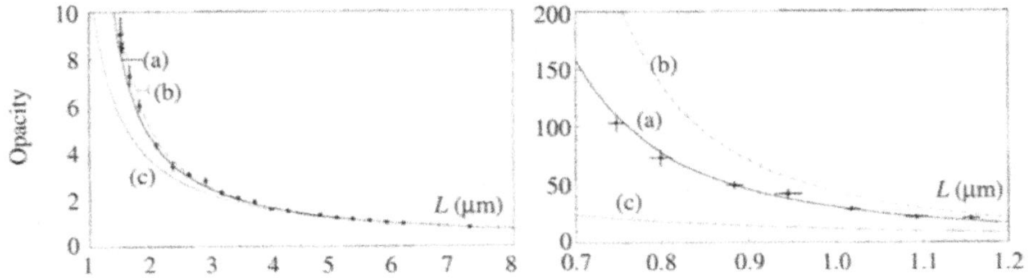

Figure 2: Inverse of the measured transmission (opacity) as a function of the plate separation L. Curves (a), (b), and (c) result from a Monte–Carlo calculation, assuming various atom-cavity interaction potentials: (a) Casimir–Polder interaction (exact); (b) Lennard-Jones interaction (no retardation); (c) no interaction (geometrical model).

distances. When an atom comes close enough to one of the walls, its trajectory is bent towards this wall. Therefore the number of atoms hitting the walls is larger than what is given by the geometrical analysis, and the effective aperture of the cavity is thus smaller than the geometrical aperture.

To obtain an order of magnitude of the critical wall spacing L_c for which the losses due to the atom–wall interaction become significant, we can compare the maximal transverse kinetic energy $E_{K\perp} \sim k_B T \, (L/2D)^2$ of an atom transmitted by the cavity (within the geometrical analysis), and the atom-wall interaction energy $U_{\text{cav}}(L/2)$ for an atom located at the center of the cavity. For simplicity, we evaluate $U_{\text{cav}}(L/2)$ using the short distance approximation (2). We notice that the value of $\langle d_i^2 \rangle$ essentially results from the contribution of the sodium resonance line 3s$\leftrightarrow$3p at $\lambda_{\text{res}} = 589$ nm, so that (2) can be written $U_{LJ}(z) = (3/16)\, \hbar\Gamma/(kz)^3$, where Γ is the radiative lifetime of the 3p level and $k = 2\pi/\lambda_{\text{res}}$ [10]. A back-of-the-envelope calculation then yields $L_c \sim 1$ μm, in good agreement with the observed value of the separation below which the measured transmission becomes significantly smaller than the geometrical one.

These experimental results constitute more than a mere evidence of the dramatic role of the atom-wall interaction at a distance $z \sim \lambda_{res}$. They allow a precise comparison with the exact Casimir–Polder result [32] and they clearly rule out a model which would simply extend the Lennard-Jones prediction (2) to any distance. Note that for the relevant atom-wall distances in this experiment ($z \sim \lambda_{res}$), the Casimir–Polder result significantly differs from the simple asymptotic form (4). One must use the exact Casimir–Polder potential $U_{CP}(z)$, which connects the short and long distance asymptotic forms. Now, if one fits the experimental data using the potential $\xi\, U_{CP}(z)$, where ξ is an adjustable parameter, one finds $\xi = 1$, within an uncertainty factor of 10% (at 1 standard deviation). To our knowledge, this is the most precise measurement of the interaction energy of a ground state atom and a metal wall at a distance sensitive to the retardation effects.

A close look at the data of Fig. 2 reveals an a priori paradoxical fact. The discrimination between the theoretical expressions with and without retardation effects is more dramatic for the smallest values of the cavity width. In fact, only the atoms travelling close to the center of the cavity are transmitted, and it is only for these atoms that the precise form of the interaction energy is important. Now, even for the smallest value $L = 0.7$ μm used in this experiment, the atom wall distance (0.35 μm) is not small compared to the wavelength λ_{res} of the dominant transition. It is therefore not surprising that the retardation effects play a significant role in this case. The fact that the relevant atoms are travelling at the center of the cavity is important in another respect. For these atoms, the wavelengths of the modes of the electromagnetic field which are affected by the walls are larger than L, that is 0.7 μm. At these wavelengths, gold behaves as an almost perfect conductor. It would not be so for shorter wavelengths, i.e. for smaller atom-wall distances.

3.3 Atom dielectric force: the Orsay experiment [33]

A key ingredient for the success of the experiment above is the fact that, for a small plate separation, the detected atoms are at a well defined distance from the attracting plates, since they travel close to the center of the cavity (atoms departing from this symmetry plane are attracted and stick to the plates). With a well defined impact parameter, it is possible to test the interaction energy law with a good accuracy.

With the advent of methods for laser cooling and manipulating atoms [34, 35, 36], it has become possible to accurately control atomic trajectories, and this offers new possibilities to define precisely the impact parameter. As suggested in [37], atomic mirrors allow to control the distance of minimum approach to a dielectric wall, and to measure the interaction energy. Figure 3 sketches an experiment recently performed in Orsay in this purpose. Laser cooled and trapped rubidium (^{87}Rb) atoms, at a temperature of about 10 μK (i.e. a *r.m.s.* velocity

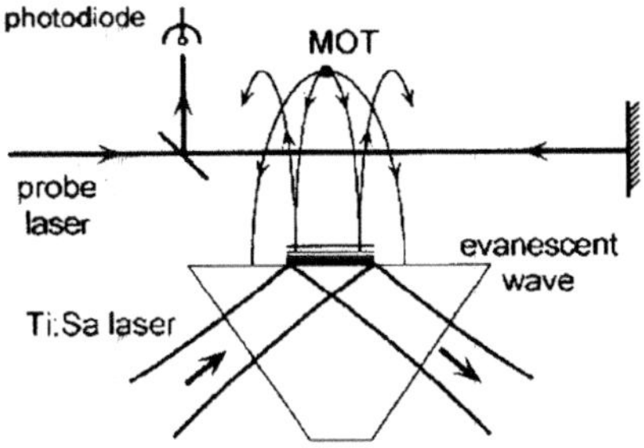

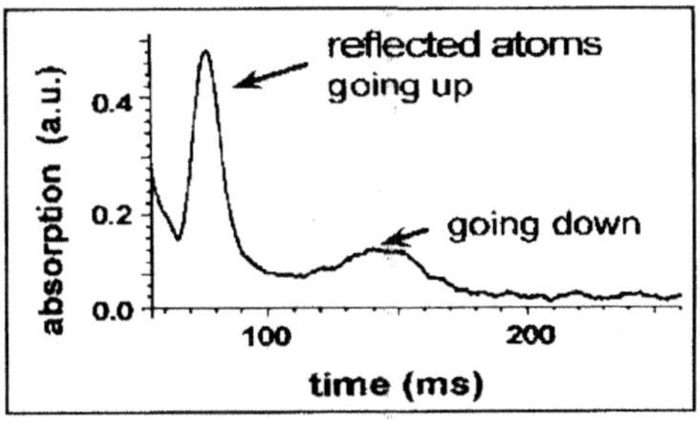

Figure 3: Orsay experiment: trapped cold atoms at 10 μK are released on an evanescent wave atomic mirror located 15 mm below. The number of reflected atoms is measured by monitoring the absorption of a resonant probe laser above the atomic mirror.

of 4 cm/s), are released on an atomic mirror located 15 mm below. The incident atoms, with a kinetic energy dispersion less than 1%, are reflected from the quasi resonant evanescent wave resulting from the total internal reflection of a laser beam in the prism. The reflecting potential is due to the interaction between the evanescent wave electric field and the atomic electric dipole induced by this field. This *dipole potential* is proportional to the square of the electric field (intensity) in the evanescent wave, and therefore decays exponentially as a function of the distance to the surface [38]:

$$U_{\text{dip}}(z) = \frac{\hbar\Gamma}{8}\,\frac{I}{I_{\text{sat}}}\,\frac{\Gamma}{\delta}\,e^{-2\kappa z} = \hbar\Lambda\,e^{-2\kappa z}\ , \tag{8}$$

where I is the light intensity at the surface of the prism, and $\delta = \omega_L - \omega_A$ is the detuning between the laser frequency ω_L and the atomic resonance frequency ω_A ($\lambda_{\text{res}} = 2\pi c/\omega_A = 780$ nm). The quantity $I_{\text{sat}} = 16$ W/m^2 is the saturation intensity of the atomic transition and $\Gamma = 3.7 \times 10^7$ s^{-1} is the radiative width of the relevant excited state. The decay length κ^{-1} is of the order of $\lambda_{\text{res}}/2\pi$, the exact value depending on the laser direction (in this experiment, $\kappa^{-1} = 114$ nm).

The reflecting potential is repulsive, in contrast to the Casimir–Polder potential which is attractive at all distances. For the choice of parameters of the experiment, the Casimir–Polder potential, which varies with z as a sum of power laws, dominates at short and very large distances, but there is an intermediate range of position z for which the dipole potential dominates. In this case, a clear maximum of the total potential exists (Figure 4a). The height of this potential barrier depends univocally on the ratio I/δ, and the experiment consists in decreasing this parameter to find the threshold value $(I/\delta)_{\text{T}}$ below which the atoms are no longer reflected. This measured value can then be compared to the value predicted with different expressions of the atom-dielectric potential, by stating that the potential barrier height is exactly equal to the kinetic energy of the incident

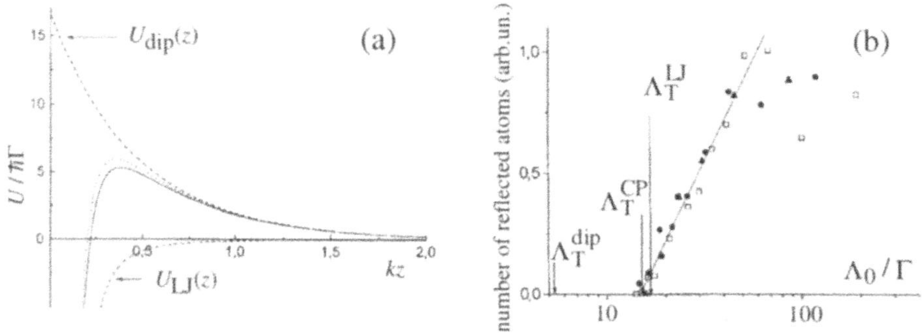

Figure 4: (a) Atoms incident on the atomic mirror experience a total potential which is the sum of the evanescent wave reflecting potential $U_{\rm dip}(z)$ and of the atom-dielectric interaction $U(z)$. The height of the resulting potential barrier is controlled by changing the parameter I/δ of the evanescent wave and it depends on the mathematical form assumed for $U(z)$. Solid line: total potential neglecting retardation in the atom-dielectric interaction $U(z)$ (Lennard-Jones). Dotted line: total potential with the Casimir–Polder–Lifshitz expression for the atom-dielectric interaction.
(b) Number of reflected atoms as a function of $\ln(I_0/\delta)$ (expressed in suitable units, hence the notation Λ_0). The various symbols correspond to different laser intensities. The results can be fitted by a straight line, whose extrapolation to 0 gives the measured value of the threshold, to be compared to the values calculated with the various potentials of Figure 4a, and indicated by arrows. $\Lambda_{\rm T}^{\rm dip}$: no atom-dielectric interaction; $\Lambda_{\rm T}^{\rm LJ}$: Lennard-Jones form of the atom-dielectric potential (no retardation); $\Lambda_{\rm T}^{\rm CP}$: Casimir–Polder–Lifshitz potential.

atoms.

One may, at this point, raise the experimental problem of having a perfectly uniform evanescent wave intensity, in order to have an abrupt threshold. This is so difficult that the authors of [37] renounced to make a precise measurement. The Orsay group has circumvented the difficulty by keeping the standard gaussian transverse profile of the laser beam, and by noticing the following fact. When one changes the parameter I_0/δ (where I_0 is now the intensity *at the center* of the laser beam), the number $N_{\rm R}$ of reflected atoms varies as $\ln((I_0/\delta)/(I_0/\delta)_{\rm T})$. Indeed the effective mirror — *i.e.* the location where the potential barrier height is larger than the kinetic energy of the incident atoms — is an ellipse of area proportional to that quantity.

We have plotted in Figure 4b the number of reflected atoms as a function of $\ln(I_0/\delta)$. One clearly sees that the experimental points are aligned. A fit to a straight line then yields the measured threshold value $(I_0/\delta)_{\rm T}$. We have indicated on the x axis the various threshold values corresponding to the various potentials shown in Figure 4a. The threshold $\Lambda_{\rm T}^{\rm dip}$ is calculated for the dipole potential alone, without any atom-dielectric interaction. It differs from the observed value by a factor of 3, clearly showing the dramatic effect of this atom-dielectric interaction.

The threshold Λ_T^{LJ} is calculated with the non retarded Lennard-Jones potential (2). Here we take into account that we deal with a dielectric prism and not an ideal mirror; we assume a dielectric constant ϵ independent of the frequency [39] so that:

$$U_{LJ}(z) = -\frac{\epsilon - 1}{\epsilon + 1}\frac{\langle d^2 \rangle}{48\pi\epsilon_0 z^3} = -A\frac{\hbar\Gamma}{(2kz)^3}\ , \tag{9}$$

where we used the fact that the dipole is isotropic: $\langle d^2 \rangle = 3\langle d_i^2 \rangle, i = x, y, z$. We take the value of ϵ at the wavelength λ_{res} of the resonant transition, that completely dominates the dipole fluctuations. Using known atomic data, we calculate the square of the atomic electric dipole in the ground state and find $A = 0.88$ with an accuracy of 1%.

We see on Figure 4b that the threshold Λ_T^{LJ} slightly exceeds the experimental value. Actually, the difference is of the order of our estimation of the uncertainty, which is dominated by the uncertainty on the absolute value of the laser intensity. Therefore the agreement between our result and a model using the Lennard-Jones potential is only marginal. On the other hand, it is clear that the threshold Λ_T^{CP} agrees better with the experimental result. To calculate this threshold, we have used an expression of the Casimir–Polder potential given by [40]. As for the case of the Yale experiment, this measurement is done at an intermediate distance, where one cannot use the asymptotic form of equation (4). More precisely, at the position of the potential barrier — *i.e.* at 48 nm from the wall, see Figure 4a — the correction to (9) due to retardation is 30%.

3.4 Quantum reflection by a Casimir–Polder potential: the Tokyo experiment [41]

We start by explaining briefly the concept of quantum reflection. For $z > 0$, consider a potential $U(z) < 0$ which tends to zero at infinity. We assume that this potential is attractive ($dU/dz > 0$) and we consider incident atoms with an energy E_i at $z = +\infty$. Quantum reflection is predicted to occur for atoms with a low incident kinetic energy E_i, if the potential changes rapidly enough. In this case, the atoms are reflected well before reaching the minimum of the potential $U(z)$ located in $z = 0$, so that the presence probability of the atoms remains vanishingly small around this minimum (Fig. 5a).

More precisely, the condition for quantum reflection is

$$\frac{d\lambda_{dB}}{dz} \geq 1\ , \tag{10}$$

where $\lambda_{dB}(z)$ is the local de Broglie wavelength of the particle at a distance z, calculated in a semi-classical analysis ($\lambda_{dB}(z) = h/\sqrt{2m(E_i - U(z))}$). This condition can be seen as a breakdown of the validity of the semi-classical (WKB) approximation, which would imply that an incident particle always reaches $z = 0$, whatever

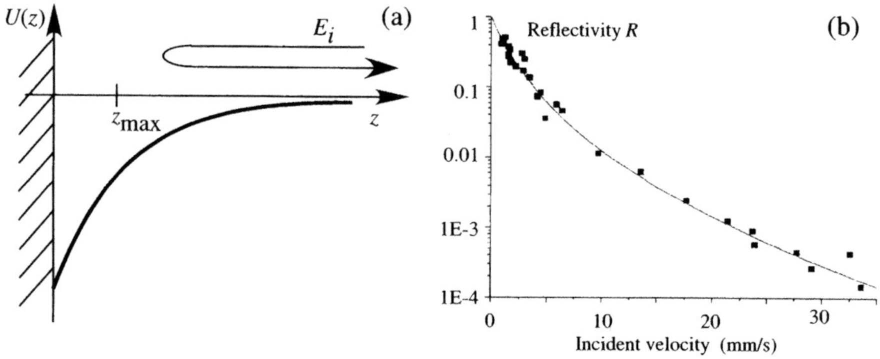

Figure 5: Tokyo experiment. (a) Quantum reflection of a particle with incident energy E_i on a purely attractive potential $U(z)$. (b) Reflectivity vs. velocity for metastable neon atoms impinging on a silicon surface. The solid curve is the reflectivity calculated using the model potential (11) with $C_4 = 6.8\ 10^{-56}$ J/m^4 and $z_0 = 64$ nm.

its initial energy. For a power law potential $U(z) = -C_n/z^n$ with $n > 2$, and for particles with a sufficiently low incident energy E_i, the condition (10) is fulfilled over some range of distances z. Indeed the maximum of $\phi(z) = d\lambda_{\rm dB}/dz$ is found in $z_{\rm max} = ((n-2)C_n/(2(n+1)E_i))^{1/n}$ and this maximum scales as $E_i^{(2-n)/(2n)}$. Both quantities $z_{\rm max}$ and $\phi(z_{\rm max})$ tend to infinity as E_i tends to zero.

Suppose now that the potential $U(z)$ is created by a bulk material located in the domain $z < 0$. In the case of quantum reflection on the surface of the material, the particle is reflected before it reaches the immediate vicinity of the material, where it could stick. One therefore expects an elastic reflection coefficient $R(E_i)$ which tends to 1 when the incident energy E_i goes to zero.

This phenomenon has first been observed for the reflection of helium and hydrogen atoms on a liquid helium surface [42, 43, 44, 45]. In the experiment that we wish to describe here [41], quantum reflection has been demonstrated for a solid surface: very slow metastable Neon atoms bounce elastically and specularly on the purely attractive potential created by a piece of silicon (semi conductor) or glass (dielectric). The idea is then to extract information on the potential $U(z)$ from the measurement of $R(E_i)$.

As shown in Fig. 5b, obtained with silicon, the author measures the reflectivity for a range of incident velocities between 1 mm/s and 30 mm/s. As expected, he finds that the reflectivity increases when the velocity decreases. The largest reflectivity is $R = 0.5$ at 1 mm/s, and the data are consistent with the extrapolated value $R = 1$ at zero velocity.

The data are fitted by a simple theoretical model, which consists in connecting the asymptotic behaviors of the semi-classical atom wave function for short and long distances z. In this model, one assumes that a particle which can reach the location $z = 0$ sticks to the surface (absorptive boundary conditions). The atom-bulk silicon potential is modelled by

$$U(z) = -\frac{C_4}{(z + z_0)z^3}\,, \tag{11}$$

which gives an account for the behavior seen above for both short (z^{-3}) and long (z^{-4}) distances. The C_4 coefficient deduced from the fit is in agreement with the one expected from the Casimir–Polder theory: $C_4/C_4^{\mathrm{CP}} = 0.7 \pm 0.4$. The value of λ is $z_0 = 0.06\ \mu$m, with a range within σ confidence of $0 - 0.7\mu$m. This value for z_0 is much smaller than the distance between the turning point and the surface that one derives from the above considerations (typically 1 μm). This shows that, in this experiment, one is sensitive mostly to the retarded z^{-4} Casimir–Polder potential.

4 Conclusion

Thanks to experimental results obtained during the last ten years, there is now a clear evidence for retardation effects in the interaction between a ground state atom and a wall, either a metal, a semi-conductor, or a dielectric. The experiments clearly rule out a pure Lennard-Jones z^{-3} potential, which would exist in absence of retardation.

For the three experiments that we have described, the typical minimum distance between the atom and the surface varies between 0.05 μm to 1 μm. The most accurate fit to the Casimir–Polder potential is obtained in the Yale experiment, where an agreement between theory and experiment is found at the 10 % level. There is a strong hope that the theoretical predictions can be tested with an improved accuracy when coherent atom sources, emerging from Bose-Einstein condensates, will be easily available. With these *atom lasers* [46], one will be able to have a better control of the parameters of the atomic beam incident on the surface. Together with the possibility of performing an interferometric measurement of the shift induced by the atom-wall potential, this should allow for an extension of the range of distances over which the Casimir–Polder potential is tested.

References

[1] H.B.G. Casimir, *Proc. K. Ned. Akad. Wet* **60**, 793 (1948).

[2] H.B.G. Casimir and D. Polder, *Phys. Rev.* **73**, 360 (1948).

[3] F. London, *Z. Physik* **63**, 245 (1930).

[4] J.E. Lennard-Jones, *Trans. Faraday Soc.* **28**, 334 (1932).

[5] J.A. Wheeler, *Phys. Rev.* **59**, 928 (1941).

[6] E.A. Hinds, *Adv. At. Mol. Opt. Phys.*, sup. 2, 1 (1994).

[7] The same paper contains a second calculation, addressing the long-range force between two atoms. Due to the lack of space, we only address the first part of the Casimir-Polder paper.

[8] J. Dalibard, J. Dupont-Roc and C. Cohen-Tannoudji, *J. Phys.* (Paris) **43**, 1617 (1982).

[9] D. Meschede, W. Yhe and E. Hinds, *Phys. Rev. A* **41**, 1587 (1990).

[10] S. Haroche, in *Fundamental Systems in Quantum Optics*, Les Houches LIII, J. Dalibard, J.-M. Raimond and J. Zinn-Justin, Eds., North-Holland (Amsterdam, 1992).

[11] L. Spruch, in *Long Range Casimir Forces*, F.S. Levin and D.A. Micha, eds., Plenum Press (New York).

[12] E.M. Lifshitz, *Sov. Phys. JETP* **2**, 73 (1956).

[13] E. Elizalde and A. Romeo, *Am. J. Phys.* **59**, 711 (1991).

[14] Y. Tikochinsky and L. Spruch, *Phys. Rev. A* **48**, 4223 (1993).

[15] A. Landragin, Thèse d'université (Orsay, 1997).

[16] J.M. Wylie and J.E. Sipe, *Phys. Rev. A* **32**, 2030 (1985).

[17] M. Fichet, F. Schuller, D. Bloch, and M. Ducloy, *Phys. Rev. A* **51**, 1553 (1995).

[18] F.S. Levin and D.A. Micha, eds., *Long Range Casimir Forces*, Plenum Press (New York).

[19] D. Raskin and P. Kusch, *Phys. Rev.* **179**, 712 (1969).

[20] A. Shih and V.A. Parsigian, *Phys. Rev. A* **12**, 835 (1975).

[21] M.J. Mehl and W.L. Schaich, *Phys. Rev. A* **21**, 1177 (1980).

[22] A. Anderson, S. Haroche, E.A. Hinds, W. Jhe and D. Meschede, *Phys. Rev. A* **37**, 3594 (1988).

[23] E.S. Sabisky and C.H. Anderson, *Phys. Rev. A* **7**, 790 (1973).

[24] M. Oria, M. Chevrollier, D. Bloch, M. Fichet and M. Ducloy, *Europhys. Lett.* **14**, 527 (1991).

[25] V. Sandoghdar, C.I. Sukenik, E.A. Hinds, and S. Haroche, *Phys. Rev. Lett.* **68**, 3432 (1992).

[26] M. Boustimi, B. Viaris de Lesegno, J. Baudon, J. Robert, and M. Ducloy, *Phys. Rev. Lett.* **86**, 2766 (2001) and refs. in.

[27] H. Failache, S. Saltiel, M. Fichet, D. Bloch, and M. Ducloy, *Phys. Rev. Lett.* **83**, 5467 (1999).

[28] P. Szriftgiser, D. Guéry-Odelin, M. Arndt, and J. Dalibard, *Phys. Rev. Lett.* **77**, 4 (1996).

[29] L. Cognet, V. Savalli, G.Z.K. Horvath, D. Holleville, R. Marani, N. Westbrook, C.I. Westbrook, and A. Aspect, *Phys. Rev. Lett.* **81**, 5044 (1998).

[30] M. Gorlicki, S. Feron, V. Lorent, and M. Ducloy, *Phys. Rev. A* **61**, 013603 (2001).

[31] C.I. Sukenik, M.G. Boshier, D. Cho, V. Sandoghdar, and E.A. Hinds, *Phys. Rev. Lett.* **70**, 560 (1993).

[32] For the generalisation of the Casimir–Polder result to the case where an atom interacts with the two walls of a cavity, see G. Barton, *Proc. Roy. Soc. London A* **410**, 175 (1987).

[33] A. Landragin, J.-Y. Courtois, G. Labeyrie, N. Vansteenkiste, C. I. Westbrook, and A. Aspect, *Phys. Rev. Lett.* **77**, 1464 (1996).

[34] S. Chu, *Rev. Mod. Phys.* **70**, 685 (1998).

[35] C. Cohen-Tannoudji, *Rev. Mod. Phys.* **70**, 707 (1998).

[36] W.D. Phillips, *Rev. Mod. Phys.* **70**, 721 (1998).

[37] M. Kasevich, K. Moler, E. Riis, E. Sunderman, D. Weis, and S. Chu, in Atomic Physics 12, ed. by J. C. Zorn and R. R. Lewis, AIP Conf. Proc. No 233 (AIP, New York 1991), p.47.

[38] R.J. Cook and R.K. Hill, *Opt. Commun.* **43**, 258 (1982); *Phys. Rev. A* **32**, 2030 (1985).

[39] Since the dipole fluctuations are totally dominated by the first resonance transitions at 780 nm and 795 nm, it is legitimate to take ϵ constant.

[40] J. M. Wylie and J.E. Sipe, *Phys. Rev. A* **30**, 1185 (1984).

[41] F. Shimizu, *Phys. Rev. Lett.* **86**, 987 (2001).

[42] V. U. Nayak, D. O. Edwards, and N. Masuhara, *Phys. Rev. Lett.* **50**, 990 (1983).

[43] J. J. Berkhout et al., *Phys. Rev. Lett.* **63**, 1689 (1989).

[44] J. M. Doyle et al., *Phys. Rev. Lett.* **67**, 603 (1991).

[45] I. A. Yu et al., *Phys. Rev. Lett.* **71**, 1589 (1993).

[46] K. Helmerson *et al.*, Physics World, August 1999, p. 31.

Alain Aspect
Laboratoire Charles Fabry de l'Institut d'Optique
B.P. 147
F-91403 Orsay Cedex, France

Jean Dalibard
Laboratoire Kastler Brossel
24, rue Lhomond
F-75005 Paris, France

Poincaré Seminar 2002, 109 – 126

Poincaré Seminar 2002

Recent Experiments on the Casimir Effect: Description and Analysis

Astrid Lambrecht and Serge Reynaud

1 Motivations

After its prediction in 1948 [1], the Casimir force has been observed in a number of 'historic' experiments which confirmed its existence and main properties [2, 3, 4, 5]. The Casimir force has recently been measured with a largely improved experimental precision [6] which allows for an accurate comparison between measured values of the force and theoretical predictions. This comparison is interesting for various reasons.

The Casimir force is the most accessible effect of vacuum fluctuations in the macroscopic world. As the existence of vacuum energy raises difficulties at the interface between the theories of quantum and gravitational phenomena, it is worth testing this effect with the greatest care and highest accuracy [7, 8]. But the comparison between theory and experiment should take into account the important differences between the real experimental conditions and the ideal situation considered by Casimir.

Casimir calculated the force between a pair of perfectly smooth, flat and parallel plates in the limit of zero temperature and perfect reflection. He found an expression for the force F_{Cas} and the corresponding energy E_{Cas} which only depend on the distance L, the area A and two fundamental constants, the speed of light c and Planck constant $\hbar$

$$\begin{aligned} F_{\mathrm{Cas}} &= \frac{\hbar c \pi^2 A}{240 L^4} = -\frac{\mathrm{d}E_{\mathrm{Cas}}}{\mathrm{d}L} \\ E_{\mathrm{Cas}} &= \frac{\hbar c \pi^2 A}{720 L^3} \end{aligned} \tag{1}$$

Each transverse dimension of the plates has been supposed to be much larger than L. Conventions of sign are chosen so that F_{Cas} and E_{Cas} are positive. They correspond to an attractive force ($\sim 0.1\mu$N for $A = 1\mathrm{cm}^2$ and $L = 1\mu$m) and a binding energy.

Most experiments have been performed in a sphere-plane geometry which differs from the plane-plane geometry considered by Casimir. In the former geometry, the force is derived from the Deriagin approximation [9], often called in a somewhat improper manner the proximity force theorem. With this approximation, the force is obtained as the integral of force contributions corresponding to the various inter-plate distances as if these contributions were independent. In the plane-sphere geometry, the force is thus determined by the radius R of the sphere and by the Casimir energy as evaluated in the plane-plane configuration. The Deriagin approximation is also used to evaluate the surface roughness corrections.

The fact that the Casimir force (1) only depends on fundamental constants and geometrical features is remarkable. In particular it is independent of the fine structure constant which appears in the expression of the atomic Van der Waals forces. This universality property is related to the assumption of perfect reflection used by Casimir in his derivation. Perfect mirrors correspond to a saturated response to the fields since they reflect 100 % of the incoming light. This explains why the Casimir effect, though it has its microscopic origin in the interaction of electrons with electromagnetic fields, does not depend on the fine structure constant. Now, real mirrors are not perfect reflectors. The most precise experiments are performed with metallic mirrors which behave as nearly perfect reflectors at frequencies smaller than a characteristic plasma frequency but become poor reflectors at higher frequencies. Hence the Casimir expression has to be modified to account for the effect of finite conductivity. At the same time, experiments are performed at room temperature whereas the Casimir formula (1) only holds in vacuum, that is at zero temperature.

A precise knowledge of the Casimir force is a key point in many accurate force measurements for distances ranging from nanometer to millimeter. These experiments are motivated either by tests of Newtonian gravity at millimetric distances [10, 11, 12, 13] or by searches for new weak forces predicted in theoretical unification models with nanometric to millimetric ranges [14, 15, 16, 17, 18]. Basically, they aim at putting limits on deviations of experimental results from present standard theory. The Casimir force is the dominant force between two neutral nonmagnetic objects in the range of interest so that any new force would appear as a difference between experimental measurements and theoretical expectations of the Casimir force.

As far as the aim of a theory-experiment comparison is concerned, the accuracy of theory is as crucial as the precision of experiments. If a given accuracy, say at the 1% level, is aimed at in the comparison, then the theoretical and experimental accuracy have to be mastered at this level independently from each other. Since the various corrections to the Casimir formula which have already been alluded to may have a magnitude much larger than the 1% level, a high-accuracy comparison necessarily requires a precise analysis of the differences between the ideal case considered by Casimir and real situations studied in experiments.

2 Experiments before 1997

We first review some of the experiments performed before 1997.

The first experiment to measure the Casimir force between two metals was carried out by Spaarnay in 1958 [19]. A force balance based on a spring balance was used to measure the force between two flat neutral plates for distances between 0.5 and 2μm. Measurements were carried out for Al-Al, Cr-Cr and Cr-steel plates through electromechanical techniques. Spaarnay discussed the major difficulties of the experiments, in particular the control of the parallelism of the two plates, the determination of the distance between them, and the control of the neutrality of the two metal plates which is delicate since the Casimir force can easily be masked by electrostatic forces. The experiment gave evidence of an attractive force between the two plates and Sparnaay cautiously reported that "the observed attractions do not contradict Casimir's theoretical prediction". For the sake of comparison with experiments described below, an error bar of the order of 100 % may be attributed to this experiment.

Probably the first unambiguous measurement of the Casimir force between metallic surfaces was performed by van Blokland and Overbeek in 1978 [20]. The force was measured with the help of a spring balance between a lens and a flat plate, both coated with 50-100nm thick chromium layers, for distances from 132 to 670nm, measured by determining the capacitance of the system. The use of a lens instead of a second flat plate simplified the control of the geometry by suppressing the problem of parallelism. The force in this configuration was evaluated with the help of Deriagin's approximation discussed in more detail below. The investigators compared their experimental results to theoretical calculations using the Lifshitz theory for chromium and concluded to an agreement between the measured and calculated force values, confirming for the first time the effect of finite conductivity. For this experiment, one may estimate the accuracy to be of the order of 25%.

The Casimir force has been observed in a number of other experiments, in particular [21, 22, 23, 24]. More detailed or systematic reviews may be found in [2, 3, 4, 5, 6].

3 Recent experiments

Recently new measurement techniques were used to measure the Casimir effect with improved accuracy. Quite a number of experiments have been carried out in the last years and we will describe some of them which seem to be the most significant ones.

In 1997 Steve Lamoreaux measured the Casimir force by using a torsion pendulum [25]. The force was measured between a metallized sphere and a flat metallic plate with controlled but unequal electrostatic potential. Since the electrostatic and Casimir forces were acting simultaneously, it was necessary to substract precisely the effect of the electrostatic force in order to deduce the value of the Casimir force. This measurement was made for distances between 0.6 and 6 microns. Comparison between the experimental results and the theoretical predictions was reported to be in agreement at the level of 5 %.

After the correction of inaccuracies in the initial report [26, 27, 28], the results of this experiment can be summarized as follows: the force has been measured, probably with an error bar of the order of 10 % at the shortest distances where the effect of finite conductivity of the Au and Cu metallic layers used in the experiments was unambiguously observed; the error bar was certainly much larger at distances larger than a few μm where the magnitude of the force is much weaker; this probably explains why the temperature correction has not been seen though it should have been seen at the largest distance $\sim 6\mu$m explored in the experiment (see below). It is difficult to be more affirmative on this topic, in particular because this experiment was stopped by the relocation of Steve Lamoreaux.

Shortly after this publication, a second measurement was reported by Umar Mohideen [29] followed by several reports with an improved precision [30, 31]. This experiment is based on the use of an atomic force microscope (AFM). A metallized sphere is fixed on the cantilever of the microscope and brought to the close vicinity of a flat metallic plate, at a distance between 0.1 and 0.9μm. Both surfaces are put at the same electrostatic potential and the Casimir force is measured by the deflection of a laser beam on the top of the cantilever, as shown on Figure 1.

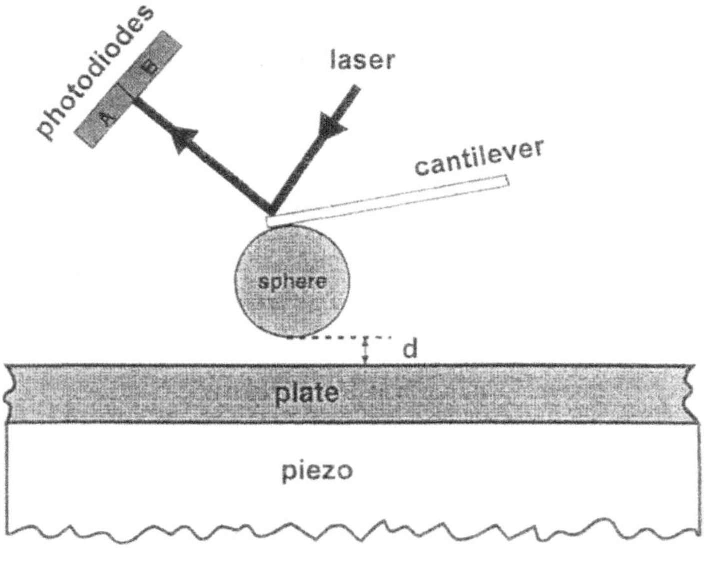

Figure 1: Experimental setup of the Casimir force measurement in [29, 30, 31]. The force is measured between the sphere and the plate with the distance of closest approach d (denoted L in the present report). The sphere is fixed on the cantilever of an AFM and its position measured by the deflection of a laser beam on the top of the cantilever. With kind courtesy of Umar Mohideen.

The comparison between experimental results and theoretical predictions has been performed for Al and Au coated surfaces. A typical experimental accuracy at the level of 1% is obtained with a comparable agreement with theory, as depicted on Figure 2. Theoretical points are based on the methods described below: they take into account the large effects of conductivity and plane-sphere geometry as well as the effects of temperature and roughness which have a smaller influence ($< 1\%$) for this experiment. The same group has also studied the effect of sinusoidal corrugations on the properties of the Casimir force [32] and has observed the lateral Casimir force between corrugated plates [33].

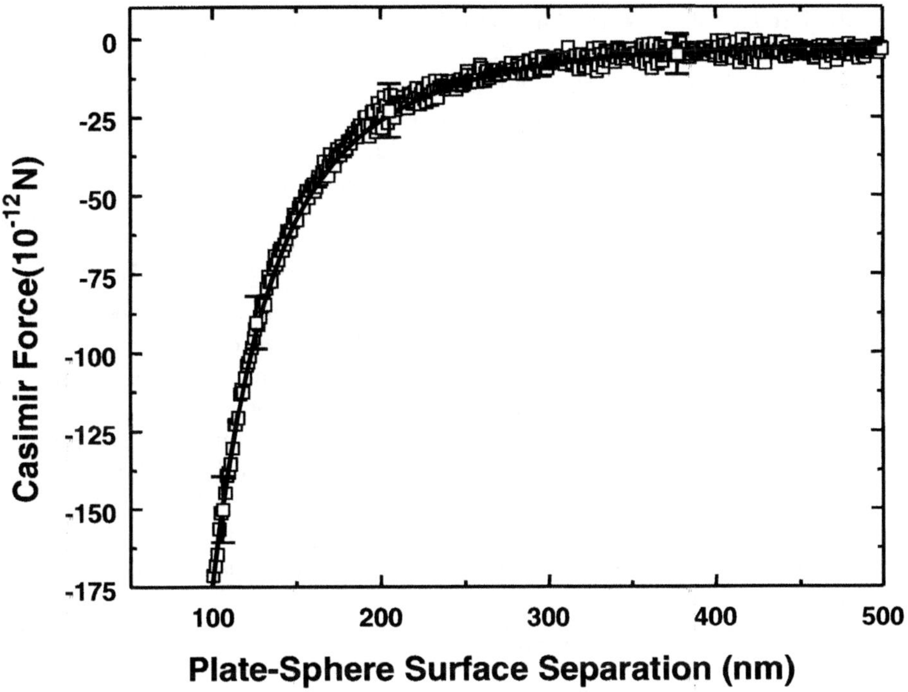

Figure 2: Comparison between experimentally measured values and theoretical predictions of the Casimir force, as reported in [30]; the squares and bars represent experimental points and errors bars for a few of them; the solid line represents theoretical predictions. With kind courtesy of Umar Mohideen.

An independent measurement has been published in 2000 by Thomas Ederth [34] who also used an AFM. The force was measured between two neutral metallic crossed cylinders (curvature 10mm) at short distances ranging from 20 to 100nm. Great efforts allowed Ederth to reduce the surface roughness, which was a necessity at these short distances. After a careful error analysis, Ederth concluded to an accuracy at the level of several %.

The experiments done by Federico Capasso and his group at Bell Labs, using microelectromechanical systems (MEMS), also deserve a special mention [35, 36]. MEMS are movable structures fabricated on a semiconductor wafer through integrated circuit technology and they are used as a new generation of sensors and actuators working in the micrometer or submicrometer distance range. The Casimir force is measured between a polystyrene sphere and a polysilicon plate with metallic coatings. The plate is suspended so that it can rotate around an axis. The variation of the plate rotation angle when the sphere is approached to a distance between 100nm and 1μm reveals the Casimir force with a magnitude agreeing with theory. When the plate is set into oscillation, frequency shifts, histeretic behavior and bistability are observed, again in agreement with the effect of the Casimir force predicted by the theory. The main interest of these experiments is to show that the Casimir force plays a significant role in systems of technological interest like the MEMS. It is indeed the dominant force in the micrometer range and this experiment shows that mechanical effects of quantum vacuum fluctuations have to be taken into account in micro- or nanotechnology [37].

Experiments described in the present section up to this point use a sphere-plane geometry or a crossed cylinders geometry. Their analysis relies on the accuracy of the Deriagin approximation which is not precisely known. This is not the case for the experiments performed in the initial Casimir geometry with two parallel flat plates. A measurement in this geometry has recently been reported on by Bressi, Carugno, Onofrio and Ruoso [38, 39]. The force is observed between two parallel flat plates coated with chromium, one of which is mounted on a silicon cantilever while the other one is fixed on a rigid piezoelectric stack. The plate fixed on the piezoelectric stack is set into oscillatory motion and this induces a varying Casimir force onto the plate mounted on the cantilever. The motion of the latter is then monitored by using a tunneling electromechanical transducer. The measurement has been performed for distances between 0.5 and 3μm and the result has been found to agree with theory at the 15% precision level.

4 The effect of imperfect reflection

As explained in the introduction, a precise theory-experiment comparison requires not only a detailed control of the experiments but also a careful estimation of the theoretical expectation of the force in the real conditions of the experiments. We begin here by the more spectacular "correction" to the ideal Casimir formula (1) which is associated with imperfect reflection of mirrors.

No real mirror can be considered as a perfect reflector at all field frequencies. In particular, the most precise experiments are performed with metallic mirrors which show perfect reflection only at frequencies smaller than a characteristic plasma frequency ω_{P} which depends on the properties of conduction electrons in the metal. Hence the Casimir force between metal plates does fit the ideal Casimir formula (1) only at distances L much larger than the plasma wavelength

$$\lambda_{\mathrm{P}} = \frac{2\pi c}{\omega_{\mathrm{P}}} \tag{2}$$

For metals used in the recent experiments, this wavelength lies in the 0.1μm range

(107nm for Al and 136nm for Cu and Au). At distances smaller than or of the order of the plasma wavelength, the finite conductivity of the metal has a significant effect on the force. The idea has been known since a long time [40, 41] but the investigation of the effect of imperfect reflection has been systematically developed only recently.

We first consider the initial Casimir geometry with perfectly plane, flat and parallel plates at zero temperature. We thus restrict our attention on the effect of the reflection properties of the mirrors described by scattering amplitudes which depend on the frequency of the incoming field. Assuming that these amplitudes obey general properties of unitarity, high-frequency transparency and causality, one derives a regular expression of Casimir force which is free from the divergences usually associated with the infiniteness of vacuum energy. The cavity formed by the two mirrors can be dealt with by using the Fabry–Pérot theory. Vacuum field fluctuations impinging the cavity have their energy either enhanced or decreased inside the cavity, depending on whether their frequency is resonant or not with a cavity mode. The radiation pressure associated with these fluctuations exerts a force on the mirrors which is directed either inwards or outwards respectively. It is the balance between the inward and outward contributions, when they are integrated over the field frequencies and incidence angles, which gives the net Casimir force [42].

The techniques of analytical continuation of the response functions in the complex plane then allow one to write the Casimir force as an integral over imaginary frequencies $\omega = i\xi$ with ξ real

$$F = \frac{\hbar A}{\pi} \sum_p \int \frac{\mathrm{d}^2\mathbf{k}}{4\pi^2} \int_0^\infty \mathrm{d}\xi \frac{\kappa r_1^p\left[i\xi,\mathbf{k}\right] r_2^p\left[i\xi,\mathbf{k}\right]}{e^{2\kappa L} - r_1^p\left[i\xi,\mathbf{k}\right] r_2^p\left[i\xi,\mathbf{k}\right]}$$

$$\kappa \equiv \sqrt{\mathbf{k}^2 + \frac{\xi^2}{c^2}} \tag{3}$$

$r_j^p\left[\omega,\mathbf{k}\right]$ is the reflection amplitude for the two mirrors $j = 1, 2$ and the field mode characterized by a frequency ω, a tranverse wavevector $\mathbf{k}$ (transverse means orthogonal to the main direction of the cavity, that is also parallel to the plane of the plates) and a polarization p. The amplitudes appear in expression (3) at imaginary frequencies $\omega = i\xi$ where they have real and positive values. The fraction appearing in (3) represents the difference between the radiation pressures on outer and inner sides of the cavity after the continuation to the imaginary axis. It is determined by the product of the reflection amplitudes of the two mirrors and by an exponential factor $e^{2\kappa L}$ representing the propagation dephasing for the field after a roundtrip in the cavity, that is a propagation length $2L$. Expression (3) includes the contribution of the modes freely propagating inside and outside the cavity but also the contribution of evanescent waves confined to the vicinity of the mirrors.

Equation (3) is a convergent integral for any couple of mirrors described by scattering amplitudes obeying the properties of causality, passivity and high frequency transparency. This means that the potential divergence associated with the infiniteness of vacuum energy has been cured by using the physical properties of scattering amplitudes, that is also by describing mirrors just as opticians do. Furthermore expression (3) does not depend on any particular microscopic model but may be applied to any reflection amplitude obeying the general properties already discussed. Itcan be used for lossy as well as lossless mirrors [43].

The ideal Casimir result is recovered at the limit where mirrors may be considered as perfect over the frequency range of interest, that is essentially over the first few resonance frequencies of the cavity [42]. This can be considered as an alternative demonstration of the Casimir formula without any reference to a renormalization or regularization technique. For real mirrors, the effect of imperfect reflection is described by a reduction factor η_{F} which multiplies the ideal Casimir expression (1) to give the force F

$$F = \eta_{\mathrm{F}} F_{\mathrm{Cas}} \tag{4}$$

In order to go further, we have to specialize the general expression (3) to a model of mirrors. The commonly used model corresponds to reflection on bulk mirrors with an optical response described by a dielectric function $\varepsilon(\omega)$. The reflection amplitudes corresponding to the two polarizations $p = \mathrm{TE}, \mathrm{TM}$ are thus given by the Fresnel formulas for each mirror

$$\begin{aligned} r_j^{\mathrm{TE}}[i\xi, \mathbf{k}] &= -\frac{\sqrt{\xi^2 \varepsilon(i\xi) + c^2\mathbf{k}^2} - c\kappa}{\sqrt{\xi^2 \varepsilon(i\xi) + c^2\mathbf{k}^2} + c\kappa} \\ r_j^{\mathrm{TM}}[i\xi, \mathbf{k}] &= \frac{\sqrt{\xi^2 \varepsilon(i\xi) + c^2\mathbf{k}^2} - c\kappa\varepsilon(i\xi)}{\sqrt{\xi^2 \varepsilon(i\xi) + c^2\mathbf{k}^2} + c\kappa\varepsilon(i\xi)} \end{aligned} \tag{5}$$

Taken together, relations (3,5) reproduce the Lifshitz expression for the Casimir force at zero temperature [40]. It is worth stressing again that relations (3) have a wider domain of validity since, as already discussed, they allow one to deal with more general scattering amplitudes than (5).

The optical response of conduction electrons in metals is approximately described by a plasma model, that is by a dielectric function

$$\varepsilon(\omega) = 1 - \frac{\omega_{\mathrm{P}}^2}{\omega^2} \tag{6}$$

A better description is given by the Drude model which accounts for the relaxation of conduction electrons

$$\varepsilon(\omega) = 1 - \frac{\omega_{\mathrm{P}}^2}{\omega(\omega + i\gamma)} \tag{7}$$

Since the ratio $\frac{\gamma}{\omega_{\mathrm{P}}}$ is much smaller than unity, the relaxation parameter γ has a significant effect on ε only at frequencies where the latter is much larger than unity and where, accordingly, the mirror is nearly perfectly reflecting. It follows that relaxation has a limited influence on the value of the Casimir force [44].

In contrast, the modification of the dielectric constant due to interband transitions has an observable effect on the Casimir force measured at distances of the order of the plasma wavelength [44]. This appears on the results of numerically integrated values of the reduction factor η_F shown on Figure 3. The solid line represents the factor calculated for two identical Au mirrors described by the plasma model with the plasma wavelength $\lambda_P = 136$nm corresponding to Au. The dashed line represents the factor calculated by using the tabulated optical data [44].

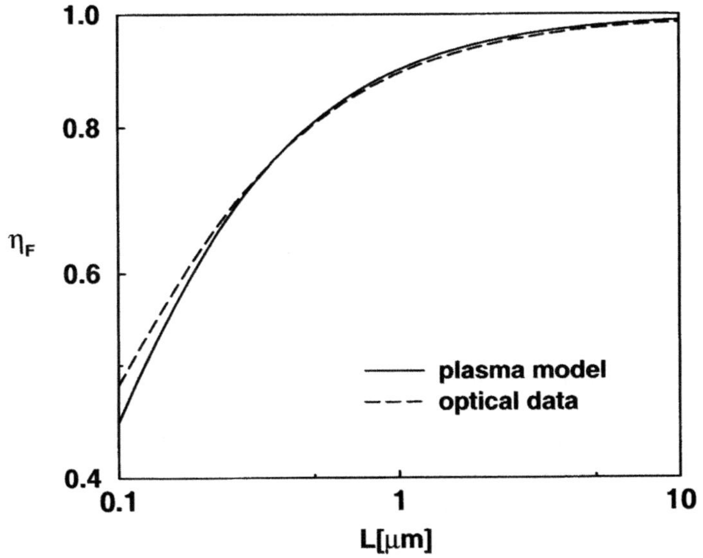

Figure 3: Reduction factor η_F for the Casimir force between two identical Au mirrors at zero temperature as a function of the distance L. The solid and dashed lines correspond to evaluations based respectively on the plasma model with $\lambda_P = 136$nm and on tabulated optical data [44].

This figure clearly shows that the effect of imperfect reflection is important at the smallest distances explored in the experiments: the reduction factor is of the order of 50% for Au mirrors at a distance around 0.1μm. It also appears that a careful description of the optical properties of metals is necessary to obtain a precise estimation of the force: in particular, the description of metals by the plasma model is not sufficient if an accuracy in the 1% range is aimed at.

5 The effect of temperature

The preceding estimations were corresponding to experiments at zero temperature. But all experiments to date have been performed at room temperature and the radiation pressure of thermal field fluctuations has a significant contribution to the force at distances larger than or of the order of a thermal wavelength [45, 46]

$$\lambda_{\mathrm{T}} = \frac{\hbar c}{k_{\mathrm{B}} T} \tag{8}$$

with $\lambda_{\mathrm{T}} \sim 7\mu\mathrm{m}$ at room temperature.

It is in principle quite simple to describe the effect of thermal field fluctuations which are superimposed to vacuum fluctuations. At zero temperature indeed, the field energy per mode is simply the vacuum contribution $\frac{1}{2}\hbar\omega$. At a non zero temperature, the field energy is the sum of this vacuum contribution and of the energy of the mean number n of photons per mode given by Planck law

$$\frac{1}{2}\hbar\omega \quad \longrightarrow \quad \left(\frac{1}{2} + n\right)\hbar\omega \tag{9}$$

This means that the contribution of a mode of frequency ω to the Casimir force has to be multiplied by a factor

$$1 + 2n\left(\omega\right) \quad = \quad \frac{1}{\tanh \frac{\hbar\omega}{2k_{\mathrm{B}}T}} \tag{10}$$

After the analytical continuation to the imaginary axis, expression (3) has to be modified by inserting a factor $1 + 2n\left(i\xi\right)$ in the integrand. This factor has poles at the Matsubara frequencies $\xi_m = m\frac{2\pi k_{\mathrm{B}}T}{\hbar}$ (m integer). The first of these poles lies at zero frequency where the metallic response functions also diverge and it must therefore be treated with great care. This delicate point has recently given rise to a burst of controversial results for the evaluation of the Casimir force between real dissipative mirrors at a non zero temperature [47, 48, 49] (see also [50, 51, 52, 53, 54, 55, 56]).

Here we use equation (7) of [57] as the starting point of numerical integration of the correction factor η_{F}. This equation is based on a uniform expansion of the terms to be integrated to obtain the Casimir force and it is valid for all the optical models of real mirrors. As far as the recent controversy is concerned, the evaluations deduced in this manner are in agreement with the results of [49] and at variance with the conclusions of [47, 48].

The resulting correction factor is drawn on Figure 4 as a function of the distance L. Here, we have chosen to consider two identical Al mirrors described by a plasma model with the plasma wavelength $\lambda_{\mathrm{P}} = 107\mathrm{nm}$. The solid line represents the correction factor η_{F} in such a configuration at room temperature $T = 300\mathrm{K}$. For the sake of comparison, we have also represented, as the dashed line, the plasma correction $\eta_{\mathrm{F}}^{\mathrm{P}}$ evaluated with the same mirrors at zero temperature and, as the dotted-dashed line, the thermal correction η_F^T evaluated with perfect reflectors at room temperature.

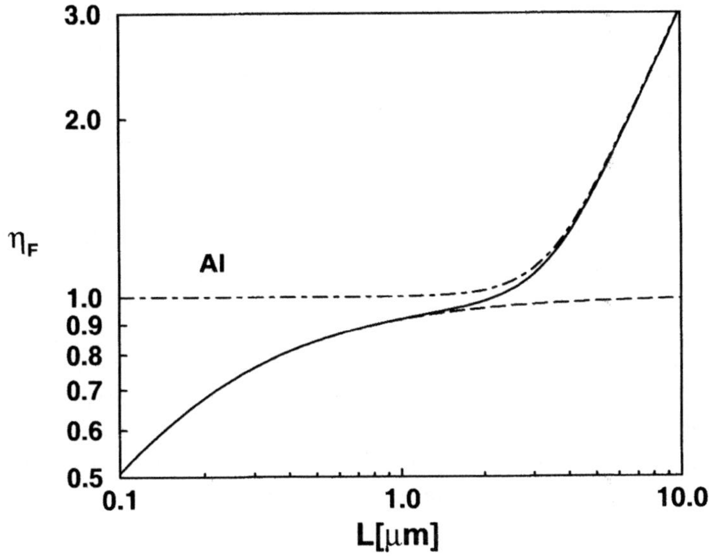

Figure 4: Correction factors for the Casimir force between two identical Al mirrors described by a plasma model with $\lambda_P = 107$nm at room temperature $T = 300$K as functions of the distance L. The solid, dashed and dotted-dashed lines represent respectively the whole correction factor η_F, the plasma correction factor η_F^P describing only the effect of imperfect reflection and the thermal correction factor η_F^T describing only the effect of temperature.

The plasma correction factor η_F^P describes only the effect of imperfect reflection and corresponds to the reduction of the force discussed in the preceding section. Meanwhile the thermal correction factor η_F^T describes only the effect of temperature: it is computed for perfect reflection and corresponds to an increase of the force. The two factors are appreciable respectively at distances smaller than 1μm and larger than 1μm. It follows that the whole correction η_F giving the force F when both effects are simultaneously accounted for is nearly equal to the product of the plasma and thermal correction factors. This is however an approximation the accuracy of which has to be carefully discussed when a precise evaluation is aimed at.

In order to evaluate the quality of this approximation, it is worth writing the whole correction factor as

$$\eta_F = \eta_F^P \eta_F^T (1 + \delta_F) \tag{11}$$

A null value for δ_F would mean that the whole correction factor may effectively be evaluated as the product of the plasma and thermal corrections computed independently from each other. In contrast, a non null value represents a correlation of the plasma and thermal corrections.

The correlation factor δ_F has been discussed in a detailed manner in [57, 58]. It should be taken into account when an accuracy at or beyond the 1% level is needed. This stems from the fact that the correlation scales as the ratio $\frac{\lambda_P}{\lambda_T}$ of the two wavelengths which characterize respectively the plasma and thermal effects and is of the order of 10^{-2} for ordinary metals at room temperature. The correlation factor is appreciable at distances larger than 1μm where the plasma model is known to be a good effective description of the metallic optical response. This justifies the use of this model in [57, 58]. At short distances, say around 0.1-0.5μm, a more complete description of the metallic optical response is needed but the temperature correction is negligible in this distance range. Note also that an analytical approximation of the correlation factor has been given in [57] through a perturbative development of the force to first order in $\frac{\lambda_P}{\lambda_T}$. The resulting expression is found to fit well the results of the complete numerical integration, with an accuracy much better than the 1% level. It provides one with a simple method for getting an accurate theoretical expectation of the Casimir force throughout the whole distance range explored in the experiments.

6 Effect of the geometry

It now remains to describe how the effect of geometry is included in the theoretical estimations of the Casimir force.

As already discussed, most experiments are performed in a sphere-plane geometry which differs from the plane-plane geometry for which exact expressions are available. The force in the former geometry is derived from the Deriagin approximation [9] which basically amounts to sum up the contributions corresponding to various inter-plate distances as if these contributions were independent. In the plane-sphere geometry, the result is simply determined by the radius R of the sphere and by the Casimir energy as evaluated in the plane-plane configuration

$$\begin{aligned} F_{\text{sphere-plane}} &= \frac{2\pi R}{A} E_{\text{plane-plane}} \\ E_{\text{plane-plane}} &= \int_L^\infty \mathrm{d}x \; F_{\text{plane-plane}}(x) = \eta_E E_{\text{Cas}} \end{aligned} \tag{12}$$

We have introduced a correction factor η_E for the Casimir energy, evaluated for the plane-plane geometry in the same manner as η_F for the Casimir force in (4).

Collecting these results leads to the final expression of the Casimir force in the sphere-plane geometry

$$F_{\text{sphere-plane}} = \frac{\hbar c \pi^3 R}{360 L^3} \eta_E \tag{13}$$

We have shown on Figure 5 the numerically integrated values of the reduction factor η_E for two identical Au mirrors at zero temperature. As on Figure 3, the solid line represents the factor calculated for mirrors described by the plasma model with $\lambda_P = 136$nm whereas the dashed line represents the factor deduced from the tabulated optical data for Au [44].

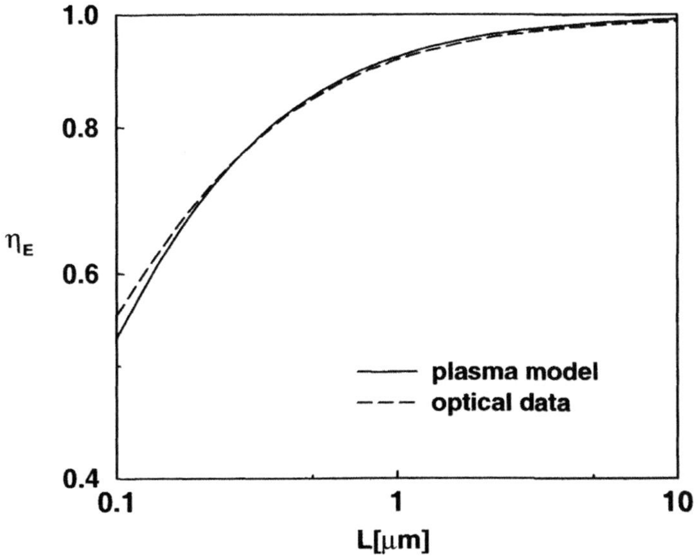

Figure 5: Reduction factor η_E for the Casimir energy between two identical Au mirrors at zero temperature as a function of the distance L; same conventions as on Figure 3.

We have considered here the case of a null temperature so that the evaluation is correct only at distances smaller than 1μm which corresponds to the most precise results. At longer distances, the temperature correction has to be taken into account by following the method presented in the preceding section.

At short distances, surface roughness corrections are also significant. They are included by using again the Deriagin approximation [59], which amounts to average the value of the Casimir forces on the various values of the inter-plate distances. A recent publication [60] opens the route to more precise evaluations of the plate corrugation and, potentially, of the surface roughness. As it could be expected, the effect of corrugation is found to depend on the wavelength of the surface perturbation and not only on its amplitude. In this new evaluation, the result of the Deriagin approximation is recovered only at the limit of large wavelengths or, equivalently, small wavevectors of the surface perturbation.

At this point, it is worth noting that the problem is in fact a more general deficiency of the Deriagin approximation. This approximation amounts to add the contributions corresponding to different distances but we know with certainty that the Casimir force is not additive (see a recent detailed discussion in [61]). As a result, the Deriagin method, though often called the proximity force theorem, can not be exact. A few results are available for the plane-sphere geometry which suggest that the approximation leads to correct results when the radius of the sphere is much larger than the distance of closest approach [62, 63, 64]. But the accuracy of the approximation is not known in a more general case or for an application to the evaluation of roughness effects.

Summary

It is clear that the Casimir effect has now been unambiguously observed: the experimental precision is already at the 1% level and it will certainly be improved in the future. This precision has allowed the experiments to observe the effect of imperfect reflection. However, the effect of temperature has not been seen at the largest distances explored in the experiments although it should have been. This is probably due to an insufficient precision at these distances.

An accurate theory-experiment comparison requires not only precise measurements but also accurate and reliable theoretical estimations. Important advances have been recently reported for the estimation of the effects of imperfect reflection and non null temperature. Efforts are still needed for the effects of geometry and surface roughness. It is worth keeping in mind that not only the accuracy of the approximations used to treat these effects should be carefully studied for perfect mirrors in vacuum but also that the corrections due to these effects are probably correlated to the effects of imperfect reflection and temperature in the same manner as the two latter effects are now known to be correlated to each other.

An attractive alternative is to come back to the initial plane-plane geometry but experiments in this geometry have not been able so far to reach the precision of sphere-plane experiments.

New advances are expected to occur quite soon in this domain, both on the experimental and theoretical sides. These new results will probably allow one to progress towards an improvement of the precision of the theory-experiment comparison. Any such improvement, at the 1% level or beyond, is important, since it either confirms a central prediction of Quantum Field Theory or otherwise reveals surprising new results in the domain of forces with nanometric to millimetric ranges.

Acknowledgments
The writing of this report has greatly benefited of discussions and correspondance with C. Genet, M.T. Jaekel, G. Barton, F. Capasso, E. Fischbach, S. Lamoreaux, J. Long, U. Mohideen, R. Onofrio and C. Speake.

References

[1] H.B.G. Casimir, *Proc. K. Ned. Akad. Wet.* **B51** (1948) 793.

[2] M.J. Sparnaay, in *Physics in the Making*, eds. Sarlemijn A. and Sparnaay M.J. (North-Holland, 1989) 235 and references therein.

[3] P.W. Milonni, *The quantum vacuum* (Academic, 1994).

[4] V.M. Mostepanenko and N.N. Trunov, *The Casimir effect and its applications* (Clarendon, 1997).

[5] S.K. Lamoreaux, Resource Letter in *Am. J. Phys.* **67** (1999) 850.

[6] M. Bordag, U. Mohideen and V.M. Mostepanenko, *Phys. Reports* **353** (2001) 1 and references therein.

[7] S. Reynaud, A. Lambrecht, C. Genet and M.T. Jaekel, *C. R. Acad. Sci. Paris* **IV-2** (2001) 1287 and references therein.

[8] C. Genet, A. Lambrecht and S. Reynaud, preprint (2002) arXiv:quant-ph/0210173.

[9] B.V. Deriagin, I.I. Abrikosova and E.M. Lifshitz, *Quart. Rev.* **10** (1968) 295.

[10] E. Fischbach and C. Talmadge, *The Search for Non Newtonian Gravity* (AIP Press/Springer Verlag, 1998).

[11] C.D. Hoyle, U. Schmidt, B.R. Heckel, E.G. Adelberger, J.H. Grundlach, D.J. Kapner and H.E. Swanson, *Phys. Rev. Lett.* **86** (2001) 1418.

[12] E.G. Adelberger et al, preprint (2002) arXiv:hep-ex/0202008.

[13] J.C. Long et al, preprint (2002) arXiv:hep-ph/0210004.

[14] G. Carugno, Z. Fontana, R. Onofrio and C. Rizzo, *Phys. Rev.* **D55** (1997) 6591.

[15] M. Bordag, B. Geyer, G.L. Klimchitskaya and V.M. Mostepanenko, *Phys. Rev.* **D60** (1999) 055004.

[16] E. Fischbach and D.E. Krause, *Phys. Rev. Lett.* **82** (1999) 4753.

[17] J.C. Long, H.W. Chan and J.C. Price, *Nucl. Phys.* **B539** (1999) 23.

[18] E. Fischbach, D.E. Krause, V.M. Mostepanenko and M. Novello, *Phys. Rev.* **D64** (2001) 075010.

[19] M.J. Spaarnay, *Physica* **24** (1958) 751.

[20] P.H.G.M. van Blokland and J.Th. Overbeek, *J. Chem. Soc. Faraday Trans.* **I74** (1978) 2637.

[21] B.V. Deriagin and I.I. Abrikosova, *Sov. Phys. JETP* **3** (1957) 819.

[22] D. Tabor and R.H.S. Winterton, *Nature* **219** (1968) 1120.

[23] W. Black, J.G.V. De Jong, J.Th.G. Overbeek and M.J. Sparnaay, *Trans. Faraday Soc.* **56** (1968) 1597.

[24] E.S. Sabisky and C.H. Anderson, *Phys. Rev.* **A7** (1973) 790.

[25] S.K. Lamoreaux, *Phys. Rev. Lett.* **78** (1997) 5.

[26] S.K. Lamoreaux, Erratum on [25] in *Phys. Rev. Lett.* **81** (1998) 5475.

[27] A. Lambrecht and S. Reynaud, Comment on [25] in *Phys. Rev. Lett.* **84** (2000) 5672.

[28] S.K. Lamoreaux, Reply to [27] in *Phys. Rev. Lett.* **84** (2000) 5673.

[29] U. Mohideen and A. Roy, *Phys. Rev. Lett.* **81** (1998) 4549.

[30] A. Roy, Ch. Lin and U. Mohideen, *Phys. Rev.* **D60** (1999) 111101.

[31] B.W. Harris, F. Chen and U. Mohideen, *Phys. Rev.* **A62** (2000) 052109.

[32] A. Roy and U. Mohideen, *Phys. Rev. Lett.* **82** (1999) 4380.

[33] F. Chen, U. Mohideen, G.L. Klimchitskaya and V.M. Mostepanenko, *Phys. Rev. Lett.* **88** (2002) 101801.

[34] Th. Ederth, *Phys. Rev.* **A62** (2000) 062104.

[35] H.B. Chan, V.A. Aksyuk, R.N. Kleiman, D.J. Bishop and F. Capasso, *Science* **291** (2001) 1941.

[36] H.B. Chan, V.A. Aksyuk, R.N. Kleiman, D.J. Bishop and F. Capasso, *Phys. Rev. Lett.* **87** (2001) 211801.

[37] D. Bishop, P. Gammel and C. Randy Giles, *Phys. Today* (October 2001) 38.

[38] G. Bressi, G. Carugno, A. Galvani, R. Onofrio, G. Ruoso and F. Veronese, *Class. Quant. Grav.* **18** (2001) 3943.

[39] G. Bressi, G. Carugno, R. Onofrio and G. Ruoso, *Phys. Rev. Lett.* **88** (2002) 041804.

[40] E.M. Lifshitz, *Sov. Phys. JETP* **2** (1956) 73.

[41] J. Schwinger, L.L. de Raad Jr. and K.A. Milton, *Ann. Physics* **115** (1978) 1.

[42] M.T. Jaekel and S. Reynaud, *J. Physique* **I-1** (1991) 1395.

[43] C. Genet, A. Lambrecht and S. Reynaud, preprint (2002) arXiv:quant-ph/0210174.

[44] A. Lambrecht and S. Reynaud, *Eur. Phys. J.* **D8** (2000) 309.

[45] J. Mehra, *Physica* **57** (1967) 147.

[46] L.S. Brown and G.J. Maclay, *Phys. Rev.* **184** (1969) 1272.

[47] M. Boström and Bo E. Sernelius, *Phys. Rev. Lett.* **84** (2000) 4757.

[48] V.B. Svetovoy and M.V. Lokhanin, *Mod. Phys. Lett.* **A15** (2000) 1013 and 1437; *Phys. Lett.* **A280** (2001) 177.

[49] M. Bordag, B. Geyer, G.L. Klimchitskaya and V.M. Mostepanenko, *Phys. Rev. Lett.* **85** (2000) 503.

[50] S.K. Lamoreaux, Comment on [47], *Phys. Rev. Lett.* **87** (2001) 139101.

[51] Bo E. Sernelius, Reply to [50], *Phys. Rev. Lett.* **87** (2001) 139102.

[52] Bo E. Sernelius and M. Boström, Comment on [49], *Phys. Rev. Lett.* **87** (2001) 259101.

[53] M. Bordag, B. Geyer, G.L. Klimchitskaya and V.M. Mostepanenko, Reply to [52], *Phys. Rev. Lett.* **87** (2001) 259102.

[54] G.L. Klimchitskaya and V.M. Mostepanenko, *Phys. Rev.* **A63** (2001) 062108.

[55] V.B. Bezerra, G.L. Klimchitskaya and V.M. Mostepanenko, Phys. Rev. **A 65**, 052113 (2002).

[56] J.R. Torgerson and S.K. Lamoreaux, preprint (2002) arXiv:quant-ph/0208042.

[57] C. Genet, A. Lambrecht and S. Reynaud, *Phys. Rev.* **A62** (2000) 012110.

[58] C. Genet, A. Lambrecht and S. Reynaud, *Int. J. Mod. Phys.* **A17** (2002) 761.

[59] G.L. Klimchitskaya, A. Roy, U. Mohideen and V.M. Mostepanenko, *Phys. Rev.* **A60** (1999) 3487.

[60] T. Emig, A. Hanke, R. Golestanian and M. Kardar, *Phys. Rev. Lett.* **87** (2001) 260402.

[61] G. Barton, *J. Phys.* **A34** (2001) 4083.

[62] D. Langbein, *J. Phys. Chem. Solids* **32** (1971) 1657.

[63] J.E. Kiefer et al, *J. Colloid and Interface Sci.* **67** (1978) 140.

[64] R. Balian and B. Duplantier, *Annals of Phys.* **112** (1978) 165.

Astrid Lambrecht and Serge Reynaud *
Laboratoire Kastler Brossel †
UPMC case 74
Campus Jussieu
F-75252 Paris cedex 05, France

*mailto:reynaud@spectro.jussieu.fr ; http://www.spectro.jussieu.fr/Vacuum
†Laboratoire du CNRS, de l'ENS et de l'Université Pierre et Marie Curie

Poincaré Seminar 2002, 127 – 136

Poincaré Seminar 2002

Dark Energy and the Destiny of the Universe

Michael S. Turner

Abstract. Only 0.5% of the material in the Universe exists in the form of stars. The rest exists as dark matter (about 1/3) and as dark energy (about 2/3). While we have now determined that our universe is nearly spatially flat, the presence of dark energy breaks the simple and familiar relationship between the geometry of the Universe and its destiny: a positively curved universe recollapses and a flat or negatively universe expands forever. Our ignorance of the nature of the dark energy allows for three futures: continued accelerated expansion and a darkening of the sky in 150 billion years; eventual slowing, with the number of visible galaxies increasing with time; or even recollapse. I summarize what we presently know about dark energy and the prospects for getting at its nature with future cosmological measurements.

1 The New Cosmology

Cosmology is enjoying the most exciting period of discovery ever. Over the past three years a new, improved standard cosmology has emerged. It incorporates the highly successful standard hot big-bang cosmology [1] and extends our understanding of the Universe to times as early as 10^{-32} sec, when the largest structures in the Universe were still subatomic quantum fluctuations.

This "New Cosmology" is characterized by

- Spatially flat [2], accelerating Universe [3, 4]
- Early period of rapid expansion (inflation)
- Density inhomogeneities produced from quantum fluctuations during inflation
- Composition: 2/3 dark energy; 1/3 dark matter; 1/200 bright stars
- Matter content: $(29 \pm 4)\%$ cold dark matter; $(4 \pm 1)\%$ baryons; $\sim 0.3\%$ neutrinos [5]

The New Cosmology is certainly not as well established as the standard hot big bang. However, the evidence is mounting. One of the most striking features of the New Cosmology is the fact that 99.5% of the material in the Universe is dark, i.e., not in the form of stars.

By now, most scientists are familiar with dark matter, the name given by Zwicky to the undetected matter whose gravity holds together cosmic structures from galaxies to the great clusters of galaxies. It is currently believed that the bulk of the dark matter exists in a sea of slowly moving elementary particles ("cold dark matter") left over from the earliest moments. The two leading candidates for the CDM particle are the axion and the neutralino [6]. At present, there is no experimental evidence for the existence of either.

The CDM hypothesis is remarkable; it modestly holds that a new form of matter exists and accounts for the bulk of the matter in the Universe. It is being tested by experiments that seek to directly detect the dark matter particles that hold our own galaxy together and by accelerator experiments that seek to produced neutralinos, whose mass mass is expected to be some 100 times that of the proton [6].

Dark energy makes dark matter seems absolutely mundane! Dark energy is my term for the causative agent of the current epoch of accelerated expansion. According to the second Friedmann equation,

$$\frac{\ddot{R}}{R} = -\frac{4\pi G}{3}(\rho + 3p) \tag{1}$$

this stuff must have negative pressure, with magnitude comparable to its energy density, in order to produce accelerated expansion [recall $q = -(\ddot{R}/R)/H^2$; R is the cosmic scale factor]. Further, since this mysterious stuff does not show its presence in galaxies and clusters of galaxies, it must be relatively smoothly distributed.

That being said, dark energy has the following defining properties: (1) it emits no light; (2) it has large, negative pressure, $p_X \sim -\rho_X$; and (3) it is approximately homogeneous (more precisely, does not cluster significantly with matter on scales at least as large as clusters of galaxies). Because its pressure is comparable in magnitude to its energy density, it is more "energy-like" than "matter-like" (matter being characterized by $p \ll \rho$). Dark energy is qualitatively very different from dark matter.

It has been said that the sum total of progress in understanding the acceleration of the Universe is naming the causative agent. While not too far from the truth, there has been progress which I summarize below.

2 Dark Energy: Seven Lessons

2.1 Two lines of evidence for an accelerating Universe

Two lines of evidence point to an accelerating Universe. The first is the direct evidence based upon measurements of type Ia supernovae carried out by two groups, the Supernova Cosmology Project [3] and the High-z Supernova Team [4]. These two teams used different analysis techniques and different samples of high-z supernovae and came to the same conclusion: the Universe is speeding up, not slowing down.

The recent discovery of a supernovae at $z = 1.755$ bolsters the case significantly [7] and provides the first evidence for an early epoch of decelerated expansion [8]. SN 1997ff falls right on the accelerating Universe curve on the magnitude – redshift diagram, and is a magnitude brighter than expected in a dusty open Universe or an open Universe in which type Ia supernovae are systematically fainter at high-z.

The second, independent line of evidence for the accelerating Universe comes from measurements of the composition of the Universe, which point to a missing energy component with negative pressure. The argument goes like this. CMB anisotropy measurements indicate that the Universe is flat, $\Omega_0 = 1.0 \pm 0.04$ [2]. In a flat Universe, the matter density and energy density must sum to the critical density. However, matter only contributes about 1/3rd of the critical density, $\Omega_M = 0.33 \pm 0.04$ [5]. (This is based upon measurements of CMB anisotropy, of bulk flows, and of the baryonic fraction in clusters.) Thus, two thirds of the critical density is missing!

In order to have escaped detection this missing energy must be smoothly distributed. In order not to interfere with the formation of structure (by inhibiting the growth of density perturbations) the energy density in this component must change more slowly than matter (so that it was subdominant in the past). For example, if the missing 2/3rds of critical density were smoothly distributed matter ($p = 0$), then linear density perturbations would grow as $R^{1/2}$ rather than as R. The shortfall in growth since last scattering ($z \simeq 1100$) would be a factor of 30, far too little growth to produce the structure seen today.

The pressure associated with the missing energy component determines how it evolves:

$$\begin{aligned} \rho_X &\propto R^{-3(1+w)} \\ \rho_X/\rho_M &\propto (1+z)^{3w} \end{aligned} \tag{2}$$

where w is the ratio of the pressure of the missing energy component to its energy density (here assumed to be constant). Note, the more negative w, the faster the ratio of missing energy to matter goes to zero in the past. In order to grow the structure observed today from the density perturbations indicated by CMB anisotropy measurements, w must be more negative than about $-\frac{1}{2}$ [9].

For a flat Universe the deceleration parameter today is

$$q_0 = \frac{1}{2} + \frac{3}{2} w \Omega_X \sim \frac{1}{2} + w$$

Therefore, knowing $w < -\frac{1}{2}$ implies $q_0 < 0$ and accelerated expansion.

2.2 Gravity can be repulsive in Einstein's theory, but ...

In Newton's theory mass is the source of the gravitational field and gravity is always attractive. In general relativity, both energy and pressure source the gravitational field. This fact is reflected in Eq. (1). Sufficiently large negative pressure leads to repulsive gravity. Thus, accelerated expansion can be accommodated within Einstein's theory.

Of course, that does not preclude that the ultimate explanation for accelerated expansion lies in a fundamental modification of Einstein's theory.

Repulsive gravity is a stunning new feature of general relativity. It leads to a prediction every bit as revolutionary as black holes – the accelerating Universe. If the explanation for the accelerating Universe fits within the Einsteinian framework, it will be an important new triumph for general relativity.

2.3 The biggest embarrassment in theoretical physics

Einstein introduced the cosmological constant to balance the attractive gravity of matter. He quickly discarded the cosmological constant after the discovery of the expansion of the Universe. Whether or not Einstein appreciated that his theory predicted the possibility of repulsive gravity is unclear.

The advent of quantum field theory made consideration of the cosmological constant obligatory not optional: The only possible covariant form for the energy of the (quantum) vacuum,

$$T^{\mu\nu}_{\rm VAC} = \rho_{\rm VAC} g^{\mu\nu},$$

is mathematically equivalent to the cosmological constant. It takes the form for a perfect fluid with energy density $\rho_{\rm VAC}$ and isotropic pressure $p_{\rm VAC} = -\rho_{\rm VAC}$ (i.e., $w = -1$) and is precisely spatially uniform. Vacuum energy is almost the perfect candidate for dark energy.

Here is the rub: the contributions of well-understood physics (say up to the 100 GeV scale) to the quantum-vacuum energy add up to 10^{55} times the present critical density. (Put another way, if this were so, the Hubble time would be 10^{-10} sec, and the associated event horizon would be 3 cm!) This is the well known cosmological-constant problem [10, 11].

While string theory currently offers the best hope for a theory of everything, it has shed precious little light on the problem, other than to speak to the importance of the problem. Thomas has suggested that using the holographic principle to count the available number of states in our Hubble volume leads to an upper bound on the vacuum energy that is comparable to the energy density in matter + radiation [12]. While this reduces the magnitude of the cosmological-constant problem very significantly, it does not solve the dark energy problem: a vacuum energy that is always comparable to the matter + radiation energy density would strongly suppress the growth of structure.

The deSitter space associated with the accelerating Universe poses serious problems for the formulation of string theory [13]. Banks and Dine argue that all explanations for dark energy suggested thus far are incompatible with perturbative string theory [14]. At the very least there is high tension between accelerated expansion and string theory.

The cosmological constant problem leads to a fork in the dark-energy road: one path is to wait for theorists to get the "right answer" (i.e., $\Omega_X = 2/3$); the other path is to assume that even quantum nothingness weighs nothing and something else with negative pressure must be causing the Universe to speed up. Of course, theorists follow the advice of Yogi Berra: where you see a fork in the road, take it.

2.4 Parameterizing dark energy: for now, it's w

Theorists have been very busy suggesting all kinds of interesting possibilities for the dark energy: networks of topological defects, rolling or spinning scalar fields (quintessence and spintessence), influence of "the bulk", and the breakdown of the Friedmann equations [11, 16]. An intriguing recent paper suggests dark matter and dark energy are connected through axion physics [15].

In the absence of compelling theoretical guidance, there is a simple way to parameterize dark energy, by its equation-of-state w [9].

The uniformity of the CMB testifies to the near isotropy and homogeneity of the Universe. This implies that the stress-energy tensor for the Universe must take the perfect fluid form [1]. Since dark energy dominates the energy budget, its stress-energy tensor must, to a good approximation, take the form

$$T_X{}^{\mu}_{\nu} \approx \mathrm{diag}[\rho_X, -p_X, -p_X, -p_X] \tag{3}$$

where p_X is the isotropic pressure and the desired dark energy density is

$$\rho_X = 2.7 \times 10^{-47}\,\mathrm{GeV}^4$$

(for $h = 0.72$ and $\Omega_X = 0.66$). This corresponds to a tiny energy scale, $\rho_X^{1/4} = 2.3 \times 10^{-3}\,\mathrm{eV}$.

The pressure can be characterized by its ratio to the energy density (or equation-of-state):

$$w \equiv p_X/\rho_X$$

which need not be constant; e.g., it could be a function of ρ_X or an explicit function of time or redshift. (Note, w can always be rewritten as an implicit function of redshift.)

For vacuum energy $w = -1$; for a network of topological defects $w = -N/3$ where N is the dimensionality of the defects (1 for strings, 2 for walls, etc.). For a minimally coupled, rolling scalar field,

$$w = \frac{\frac{1}{2}\dot{\phi}^2 - V(\phi)}{\frac{1}{2}\dot{\phi}^2 + V(\phi)} \tag{4}$$

which is time dependent and can vary between -1 (when potential energy dominates) and $+1$ (when kinetic energy dominates). Here $V(\phi)$ is the potential for the scalar field.

I believe that for the foreseeable future getting at the dark energy will mean trying to measure its equation-of-state, $w(t)$.

2.5 The Universe: the lab for studying dark energy

Dark energy by its very nature is diffuse and a low-energy phenomenon. It probably cannot be produced at accelerators; it isn't found in galaxies or even clusters of galaxies. The Universe itself is the natural lab – perhaps the only lab – in which to study it.

The primary effect of dark energy on the Universe is on the expansion rate. The first Friedmann equation can be written as

$$H^2(z)/H_0^2 = \Omega_M(1+z)^3 + \Omega_X \exp\left[3\int_0^z [1+w(x)]d\ln(1+x)\right] \tag{5}$$

where Ω_M (Ω_X) is the fraction of critical density contributed by matter (dark energy) today, a flat Universe is assumed, and the dark-energy term follows from energy conservation, $d(\rho_X R^3) = -p_X dR^3$. For constant w the dark energy term is simply $\Omega_X(1+z)^{3(1+w)}$. Note that for a flat Universe $H(z)/H_0$ depends upon only two parameters: Ω_M and $w(z)$.

While $H(z)$ is probably not directly measurable (however see Ref. [17]), it does affect two observable quantities: the (comoving) distance to an object at redshift z,

$$r(z) = \int_0^z \frac{dz}{H(z)},$$

and the growth of (linear) density perturbations, governed by

$$\ddot{\delta}_k + 2H\dot{\delta}_k - 4\pi G\rho_M\delta_k = 0,$$

where δ_k is the Fourier component of comoving wavenumber k and overdot indicates d/dt.

The comoving distance $r(z)$ can be probed by standard candles (e.g., type Ia supernovae) through the classic cosmological observable, luminosity distance $d_L(z) = (1+z)r(z)$. It can also be probed by counting objects of a known intrinsic comoving number density, through the comoving volume element, $dV/dzd\Omega = r^2(z)/H(z)$.

Both galaxies and clusters of galaxies have been suggested as objects to count [18]. For each, their comoving number density evolves (in the case of clusters very significantly). However, it is believed that much, if not all, of the evolution can be modelled through numerical simulations and semi-analytical calculations in the CDM picture. In the case of clusters, evolution is so significant that the number count test probe is affected by dark energy through both $r(z)$ and the growth of perturbations, with the latter being the dominant effect.

The various cosmological approaches to ferreting out the nature of the dark energy have been studied extensively (see other articles in this *Yellow Book*). Based largely upon my work with Dragan Huterer [19], I summarize what we know about the efficacy of the cosmological probes of dark energy:

- Present cosmological observations prefer $w = -1$, with a 95% confidence limit $w < -0.6$ [21].
- Because dark energy was less important in the past, $\rho_X/\rho_M \propto (1+z)^{3w} \to 0$ as $z \to \infty$, and the Hubble flow at low redshift is insensitive to the composition of the Universe, the most sensitive redshift interval for probing dark energy is $z = 0.2 - 2$ [19].
- The CMB has limited power to probe w (e.g., the projected precision for Planck is $\sigma_w = 0.25$) and no power to probe its time variation [19].
- A high-quality sample of 2000 SNe distributed from $z = 0.2$ to $z = 1.7$ could measure w to a precision $\sigma_w = 0.05$ (assuming an irreducible error of 0.14 mag). If Ω_M is known independently to better than $\sigma_{\Omega_M} = 0.03$, σ_w improves by a factor of three and the rate of change of $w' = dw/dz$ can be measured to precision $\sigma_{w'} = 0.16$ [19].
- Counts of galaxies and of clusters of galaxies may have the same potential to probe w as SNe Ia. The critical issue is systematics (including the evolution of the intrinsic comoving number density, and the ability to identify galaxies or clusters of a fixed mass) [18].
- Measuring weak gravitational lensing by large-scale structure over a field of 1000 square degrees (or more) could have comparable sensitivity to w as type Ia supernovae. However, weak gravitational lensing does not appear to be a good method to probe the time variation of w [20]. The systematics associated with weak gravitational lensing have not yet been studied carefully and could limit its potential.
- Some methods do not look promising in their ability to probe w because of irreducible systematics (e.g., Alcock–Paczynski test and strong gravitational lensing of QSOs). However, both could provide important independent confirmation of accelerated expansion.

2.6 Why now?: the Nancy Kerrigan problem

A critical constraint on dark energy is that it not interfere with the formation of structure in the Universe. This means that dark energy must have been relatively unimportant in the past (at least back to the time of last scattering, $z \sim 1100$). *If* dark energy is characterized by constant w, not interfering with structure formation can be quantified as: $w \lesssim -\frac{1}{2}$ [9]. This means that the dark-energy density evolves more slowly than $R^{-3/2}$ (compared to R^{-3} for matter) and implies

$$\begin{aligned} \rho_X/\rho_M &\to 0 \qquad \text{for } t \to 0 \\ \rho_X/\rho_M &\to \infty \qquad \text{for } t \to \infty \end{aligned}$$

That is, in the past dark energy was unimportant and in the future it will be dominant! We just happen to live at the time when dark matter and dark energy have comparable densities. In the words of Olympic skater Nancy Kerrigan, "Why me? Why now?"

Perhaps this fact is an important clue to unraveling the nature of the dark energy. Perhaps not. And God forbid, it could be the basis of an anthropic explanation for the size of the cosmological constant.

2.7 Dark energy and destiny

Almost everyone is aware of the connection between the shape of the Universe and its destiny: positively curved recollapses, flat; negatively curved expand forever. The link between geometry and destiny depends upon a critical assumption: that matter dominates the energy budget (more precisely, that all components of matter/energy have equation of state $w > -\frac{1}{3}$). Dark energy does not satisfy this condition.

In a Universe with dark energy the connection between geometry and destiny is severed [22]. A flat Universe (like ours) can continue expanding exponentially forever with the number of visible galaxies diminishing to a few hundred (e.g., if the dark energy is a true cosmological constant); the expansion can slow to that of a matter-dominated model (e.g., if the dark energy dissipates and becomes sub dominant); or, it is even possible for the Universe to recollapse (e.g., if the dark energy decays revealing a negative cosmological constant). Because string theory prefers anti-deSitter space, the third possibility should not be forgotten.

Dark energy holds the key to understanding our destiny!

3 The Challenge

As a New Standard Cosmology emerges, a new set questions arises. (Assuming the Universe inflated) What is physics underlying inflation? What is the dark-matter particle? How was the baryon asymmetry produced? Why is the recipe for our Universe so complicated? What is the nature of the Dark Energy? All of these questions have two things in common: making sense of the New Standard Cosmology and the deep connections they reveal between fundamental physics and cosmology.

Of these new, profound cosmic questions, none is more important or further from resolution than the nature of the dark energy. Dark energy could well be the number one problem in all of physics and astronomy.

The big challenge for the New Cosmology is making sense of dark energy.

Because of its diffuse character, the Universe is likely the lab where dark energy can best be attacked (though one should not rule other approaches – e.g., if the dark energy involves a light scalar field, then there should be a new long-range force [23]).

While type Ia supernovae look particularly promising – they have a track record and can in principle be used to map out $r(z)$ – there are important open issues. Are they really standardizable candles? Have they evolved? Is the high-redshift population the same as the low-redshift population?

The dark-energy problem is important enough that pursuing complimentary approaches is both justified and prudent. Weak-gravitational lensing shows considerable promise. While beset by important issues involving number evolution and the determination of galaxy and cluster masses [18], counting galaxies and clusters of galaxies should also be pursued.

Two realistic goals for the next decade are the determination of w to 5% and looking for time variation. Achieving either has the potential to rule out a cosmological constant: For example, by measuring a significant time variation of w or by pinning w at 5σ away from -1. Such a development would be a remarkable, far reaching result.

After determining the equation-of-state of the dark energy, the next step is measuring its clustering properties. A cosmological constant is spatially constant; a rolling scalar field clusters slightly on very large scales [24]. Measuring its clustering properties will not be easy, but it provides an important, new window on dark energy.

We do live at a special time: There is still enough light in the Universe to illuminate its dark side.

Acknowledgments. This work was supported by the DoE (at Chicago and Fermilab) and by the NASA (at Fermilab by grant NAG 5-7092).

References

[1] See e.g., S. Weinberg, *Gravitation and Cosmology* (Wiley & Sons, NY, 1972); E.W. Kolb and M.S. Turner, *The Early Universe* (Addison-Wesley, Redwood City, CA, 1990); or P.J.E. Peebles, *Physical Cosmology* (Princeton University Press, Princeton, NJ, 1971.

[2] P. de Bernardis et al, *Nature* **404**, 955 (2000); S. Hanany et al, *Astrophys. J.* **545**, L5 (2000); C.B. Netterfield et al, astro-ph/0104460; C. Pryke et al, astro-ph/0104490.

[3] S. Perlmutter et al, *Astrophys. J.* **517**, 565 (1999).

[4] A. Riess et al, *Astron. J.* **116**, 1009 (1998).

[5] M.S. Turner, astro-ph/0106035.

[6] See e.g., M.S. Turner, *Physica Scripta* **T85**, 210 (2000); B. Sadoulet, *Rev. Mod. Phys.* **71**, S197 (1999) or K. Griest and M. Kamionkowski, *Phys. Rep.* **333-4**, 167 (2000).

[7] A. Riess et al, *Astrophys. J.*, in press (astro-ph/0104455).

[8] M.S. Turner and A. Riess, astro-ph/0106051 (submitted to *Astrophys. J.*).

[9] M.S. Turner and M. White, *Phys. Rev.* **56**, R4439 (1997).

[10] S. Weinberg, *Rev. Mod. Phys.* **61**, 1 (1989)

[11] S. Carroll, http://www.livingreviews.org/Articles/Volume4/2001-1carroll.

[12] S. Thomas, hep-th/0010145.

[13] E. Witten, hep-th/0106109.

[14] T. Banks and M. Dine, hep-th/0106276.

[15] S. Barr and D. Seckel, astro-ph/0106239.

[16] M.S. Turner, *Physica Scripta* **T85**, 210 (2000).

[17] A. Loeb, *Astrophys. J.* **499,** L111 (1998).

[18] See e.g., J. Newman and M. Davis, *Astrophys. J.* **534**, L11 (2000); G.P. Holder et al, *Astrophys. J.* **553**, 545 (2001); S. Podariu and B. Ratra, astro-ph/0106549.

[19] D. Huterer and M.S. Turner, astro-ph/0012510 (*Phys. Rev. D* in press).

[20] D. Huterer, astro-ph/0106399 (submitted to *Phys. Rev. D*).

[21] S. Perlmutter, M.S. Turner, and M. White, *Phys. Rev. Lett.* **83**, 670 (1999).

[22] L. Krauss and M.S. Turner, *Gen. Rel Grav.* **31**, 1453 (1999).

[23] S. Carroll, *Phys. Rev. Lett.* **81**, 3067 (1998).

[24] K. Coble, S. Dodelson, and J. Frieman, *Phys. Rev. D* **55**, 1851 (1997).

Michael S. Turner
Center for Cosmological Physics
Departments of Astronomy & Astrophysics and of Physics
Enrico Fermi Institute
The University of Chicago, Chicago, IL 60637-1433
and
NASA/Fermilab Astrophysics Center
Fermi National Accelerator Laboratory
Batavia, IL 60510-0500

Part II

The Renormalization

Poster of the second Seminar 2002

Poincaré Seminar 2002, 139 – 177

Poincaré Seminar 2002

An Introduction to Renormalization

Vincent Rivasseau

Abstract. We review the theory of perturbative renormalization, discuss its limitations, and give a brief introduction to the powerful point of view of the renormalization group, which is necessary to go beyond perturbation theory and to define renormalization in a constructive way.

1 Introduction

The precise quantitative formulation of physical laws usually requires to introduce particular parameters or constants. It was early recognized that interaction with a particular medium or substrate can change the effective value of these constants. For instance Descartes laws for the refraction of light require a medium dependent index n and later Gauss's and Ampère's law introduced electric or magnetic permittivities whose values ϵ and μ in a non-empty medium such as water or glass reflect in a complex way the interaction of light with the atoms of this medium.

Even more simply, Alain Connes's favorite example of physical renormalization (see his contribution to this volume) is that of a ping-pong ball immersed in water. The ball accelerates upwards because of Archimedes's law, but with an experimentally measured inertial mass (ratio of force to acceleration) much bigger than its bare mass, because of the complex interaction with surrounding water.

New effective constants for the often multiplicative laws of physics can be considered as new normalizations of these laws. This is probably the origin of the name "renormalization". But a crisis occurred when physicists of the XXth century realized that this change of constants due to interaction is apparently infinite in the case of quantum field theory. This is disturbing because quantum field theory, which combines quantum mechanics and special relativity, was at that time considered the ultimate framework for the fundamental experimental laws of nature at the microscopic level[1]. Its consistency is therefore a matter of principle, whose importance can hardly be overemphasized.

The way out of this great "renormalization crisis" is a long story which required the efforts of many theoretical and mathematical physicists over the second half of the XXth century. I shall roughly divide it into two main chapters.

First the structure of the infinities or "divergences" in physical quantum field theories such as electrodynamics was elucidated. A recursive process, due

[1] It is still today to a large extent, although string theory holds great promises for an even more fundamental theory that would encompass gravity and have a natural fundamental ultraviolet length scale, the Planck scale.

to Bogoliubov and followers, was found to hide these infinities into unobservable "bare" parameters that describe the fundamental laws of physics at experimentally inaccessible extremely short distances. Although technically very ingenious, this solution left many physicists and probably most mathematicians under the impression that a real difficulty had been just "pulled under the rug".

It would be unfortunate however to remain under this impression. Indeed the second chapter of the story, known under the curious and slightly inaccurate name of the "renormalization group" (RG), truly solved the difficulty. It was correctly recognized by Wilson and followers that in a quantum theory with many scales involved, the change of parameters from bare to renormalized values is a phenomenon too complex to be described in a single step. Just like the trajectory of a complicated dynamical system, it must be instead studied step by step through a local evolution rule.

The change of scale in the RG plays the role of time in dynamical systems. This analogy is deep. There is a natural arrow of time, related to the second principle of thermodynamics, and there is similarly a natural arrow for the RG evolution: microscopic laws are expected to determine macroscopic laws, not the converse. The RG erases unnecessary detailed short scale information or "irrelevant operators". Even cosmology made now everybody familiar with the idea that the passing of time and the change of scale in physics are intimately related.

Apart from these almost philosophical comments, the RG improved point of view lead also concretely to many applications in various domains, some of which are also reviewed here. What seems less known, still today, is that RG also solved in a better way the old problem of infinities in perturbation theory. In the RG, the infinitesimal or discrete evolution under change of scale is perfectly well defined and finite. The old infinities are recognized as artefacts, due to an incorrect interchange of limits. In fact in the non-Abelian gauge theories which are presently at the backbone of the Standard Model, infinities disappear completely. Even after integrating evolution over an infinite sequence of intermediary scales, the RG flow remains perfectly bounded. The bare coupling constant, the ultimate "rug" under which perturbative infinities where supposed to hide, is in fact zero, the most finite of all possible values!

It is this amazing story that I will try to summarize in this note. As a testimony to its central place in recent theoretical physics, let me simply recall the many Nobel prizes awarded for major works on renormalization or related subjects. In 1965, R. Feynman, J. Schwinger and S.-I. Tomonaga received the Nobel prize for their formulation of quantum electrodynamics, the first theory to require renormalization. S. Glashow, S. Weinberg and A. Salam received the 1979 prize for unifying electromagnetic and weak interactions, two renormalizable field theories. In 1999, G. 't Hooft and M. Veltman received the prize for achieving the proof of renormalizability of this electroweak theory and of non-Abelian gauge theories in general. In 1982 the Nobel prize was awarded to K. Wilson for his invention of the renormalization group and its application to critical phenomena. Finally, among other contributions, P.G. de Gennes received the prize in 1991 for applying RG results to polymer physics. Besides these Nobel-winning contributions there have been so many other important works on renormalization that it is truly impossible to give full justice to all of them. So let me apologize in advance and refer to books such as [1, 2, 3, 4, 5, 6, 7, 8, 9, 10] for more complete references.

2 Perturbative (Euclidean) Quantum Field Theory

2.1 Functional Integral and the ϕ^4 Model

Quantum Field Theory is the second quantized formalism appropriate to treat in particular the collision experiments of particle physics, in which particle number is not conserved. Cross sections contain the physical information of the theory. They are the matrix elements of the diffusion matrix $\mathcal{S}$. Under a suitable asymptotic condition, there are "reduction formulae" which express the matrix elements of $\mathcal{S}$ in terms of the Green functions G_N (or time ordered vacuum expectation values) of the field ϕ, which is operator valued and acts on the Fock space:

$$G_N(z_1, ..., z_N) =< \psi_0, T[\phi(z_1)...\phi(z_N)]\psi_0 > . \tag{2.1}$$

where ψ_0 is the vacuum state and T is an operator, called T-product, that orders a product of operators such as $\phi(z_1)...\phi(z_N)$ according to decreasing times.

Consider a Lagrangian field theory, and split the total Lagrangian as the sum of a free plus an interacting piece, $\mathcal{L} = \mathcal{L}_0 + \mathcal{L}_{int}$. The Gell–Mann–Low formula expresses the Green functions as vacuum expectation values of a similar product of free fields with an $e^{i\mathcal{L}_{int}}$ insertion:

$$G_N(z_1, ..., z_N) = \frac{< \psi_0, T\left[\phi(z_1)...\phi(z_N)e^{i\int dx\mathcal{L}_{int}(\phi(x))}\right]\psi_0 >}{< \psi_0, T(e^{i\int dx\mathcal{L}_{int}(\phi(x))})\psi_0 >}. \tag{2.2}$$

In the functional integral formalism proposed by Feynman [11], the Gell–Mann–Low formula is itself replaced by a functional integral in terms of an (ill-defined) "integral over histories" which is formally the product of Lebesgue measures over all space time. It is interesting to notice that the integrand appearing in this formalism contains the full Lagrangian $\mathcal{L} = \mathcal{L}_0 + \mathcal{L}_{int}$, not just the interacting one. The corresponding formula is the Feynman–Kac formula:

$$G_N(z_1, ..., z_N) == \frac{\int \prod_j \phi(z_j)e^{i\int \mathcal{L}(\phi(x))dx}D\phi}{\int e^{i\int \mathcal{L}(\phi(x))dx}D\phi}. \tag{2.3}$$

This functional integral has potentially many advantages. First the rules of Gaussian integration make perturbation theory very transparent as shown in the next subsection. The fact that the full Lagrangian appears in (2.3) is interesting when symmetries of the theory are present which are not separate symmetries of the free and interacting Lagrangians, as is the case for non-Abelian gauge theories. It is also well adapted to constrained quantization, and to the study of non-perturbative effects.

There is a deep analogy between the Feynman–Kac formula and the formula which expresses correlation functions in classical statistical mechanics. For instance, the correlation functions for a lattice Ising model are given by

$$\langle \prod_{i=1}^{n} \sigma_{x_i} \rangle = \frac{\sum_{\{\sigma_x = \pm 1\}} e^{-L(\sigma)} \prod_i \sigma_{x_i}}{\sum_{\{\sigma_x = \pm 1\}} e^{-L(\sigma)}}, \tag{2.4}$$

where x labels the discrete sites of the lattice, the sum is over configurations $\{\sigma_x = \pm 1\}$ which associate a "spin" with value +1 or -1 to each such site and $L(\sigma)$ contains usually nearest neighbor interactions and possibly a magnetic field h:

$$L(\sigma) = \sum_{x,y \text{ nearest neighbors}} J\sigma_x \sigma_y + \sum_x h\sigma_x. \tag{2.5}$$

By analytically continuing (2.3) to imaginary time, or Euclidean space, it is possible to complete the analogy with (2.4), hence to establish a firm contact with statistical mechanics [5, 6, 7]. This idea also allows to give much better meaning to the path integral, at least for a free bosonic field. Indeed the corresponding free Euclidean measure $Z^{-1} e^{-\int L_0(\phi(x))dx} D\phi$, where Z is a normalization factor, can be defined easily as a Gaussian measure. This is simply because L_0 is a quadratic form of positive type[2].

The Green functions continued to Euclidean points are called the Schwinger functions of the model, and are given by the Euclidean Feynman–Kac formula:

$$S_N(z_1, ..., z_N) = Z^{-1} \int \prod_{j=1}^{N} \phi(z_j) e^{-\int \mathcal{L}(\phi(x))dx} D\phi \tag{2.6}$$

$$Z = \int e^{-\int \mathcal{L}(\phi(x))dx} D\phi. \tag{2.7}$$

The simplest interacting field theory is the theory of a one component scalar bosonic field ϕ with quartic interaction $g\phi^4$ (ϕ^3 which is simpler is unstable). In $\mathbb{R}^d$ it is called the ϕ^4_d model. For $d = 2, 3$ the model is superrenormalizable and has been built by constructive field theory. For $d = 4$ it is renormalizable in perturbation theory. Although the model lacks asymptotic freedom and a non-perturbative

[2]However the functional space that supports this measure is not in general a space of smooth functions, but rather of distributions. This was already true for functional integrals such as those of Brownian motion, which are supported by continuous but not differentiable paths. Therefore "functional integrals" in quantum field theory should more appropriately be called "distributional integrals".

version may therefore not exist, it remains a valuable tool for a pedagogical introduction to perturbative renormalization theory.

Formally the Schwinger functions of the ϕ^4_d are the moments of the measure:

$$d\nu = \frac{1}{Z} e^{-(g/4!)\int \phi^4 - (m^2/2)\int \phi^2 - (a/2)\int (\partial_\mu \phi \partial^\mu \phi)} D\phi, \tag{2.8}$$

where

- g is the coupling constant, usually assumed positive or complex with positive real part;
- m is the mass; it fixes an energy scale for the theory;
- a is the wave function constant. We often assume it to be 1;
- Z is a normalization factor which makes (2.8) a probability measure;
- $D\phi$ is a formal product $\prod_{x\in\mathbb{R}^d} d\phi(x)$ of Lebesgue measures at every point of $\mathbb{R}^d$.

But such an infinite product of Lebesgue measures is mathematically ill-defined. So it is better to define first the Gaussian part of the measure

$$d\mu(\phi) = \frac{1}{Z_0} e^{-(m^2/2)\int \phi^2 - (a/2)\int (\partial_\mu \phi \partial^\mu \phi)} D\phi. \tag{2.9}$$

where Z_0 is again the normalization factor which makes (2.9) a probability measure.

More precisely if we consider the translation invariant propagator $C(x,y) \equiv C(x-y)$ (with slight abuse of notation), whose Fourier transform is

$$C(p) = \frac{1}{(2\pi)^d} \frac{1}{p^2+m^2}, \tag{2.10}$$

we can use Minlos theorem and the general theory of Gaussian processes to define $d\mu(\phi)$ as the centered Gaussian measure on the Schwartz space of tempered distributions $S'(\mathbb{R}^d)$ whose covariance is C. A Gaussian measure is uniquely defined by its moments, or the integral of polynomials of fields. Explicitly this integral is zero for a monomial of odd degree, and for $n = 2p$ even it is equal to

$$\int \phi(x_1)...\phi(x_n) d\mu(\phi) = \sum_\gamma \prod_{l\in\gamma} C(x_{i(l)}, x_{j(l)}), \tag{2.11}$$

where the sum runs over all the pairings γ of the $2p$ arguments into p disjoint pairs $l = (i(l), j(l))$.

Note that since for $d \geq 2$, $C(p)$ is not integrable, $C(x,y)$ must be understood as a distribution. It is therefore convenient to also introduce a regularized kernel, for instance

$$C_\kappa(p) = \frac{1}{(2\pi)^d} \frac{e^{-\kappa(p^2+m^2)}}{p^2+m^2} \tag{2.12}$$

whose Fourier transform $C_\kappa(x,y)$ is now a smooth function and not a distribution. Such a regularization is called an ultraviolet cutoff, and we have (in the distribution sense) $\lim_{\kappa\to 0} C_\kappa(x,y) = C(x,y)$. Remark that due to the non zero m^2 mass term, the kernel $C_\kappa(x,y)$ decays exponentially at large $|x-y|$ with rate m, that is for some constant K and $d > 2$ we have:

$$|C_\kappa|(x,y)| \leq K\kappa^{1-d/2}e^{-m|x-y|}. \tag{2.13}$$

It is a standard useful construction to build from the Schwinger functions another class of functions called the connected Schwinger functions (in statistical mechanics connected functions are called Ursell functions or cumulants). These connected Schwinger functions are given by:

$$C_N(z_1,...,z_N) = \sum_{P_1\cup...\cup P_k=\{1,...,N\};\ P_i\cap P_j=0} (-1)^{k+1}\prod_{i=1}^{k} S_{p_i}(z_{j_1},...,z_{j_{p_i}}), \tag{2.14}$$

where the sum is performed over all distinct partitions of $\{1,...,N\}$ into k subsets $P_1,...,P_k$, P_i being made of p_i elements called $j_1,...,j_{p_i}$. For instance the connected 4-point function, when all odd Schwinger functions vanish due to the unbroken $\phi \to -\phi$ symmetry, is simply given by:

$$\begin{aligned} C_4(z_1,...,z_4) &= S_4(z_1,...,z_4) - S_2(z_1,z_2)S_2(z_3,z_4) \\ &\quad -S_2(z_1,z_3)S_2(z_2,z_4) - S_2(z_1,z_4)S_2(z_2,z_3). \end{aligned} \tag{2.15}$$

2.2 Feynman Rules

The full interacting measure may now be defined as the multiplication of the Gaussian measure $d\mu(\phi)$ by the interaction factor:

$$d\nu = \frac{1}{Z}e^{-(g/4!)\int \phi^4(x)dx}d\mu(\phi) \tag{2.16}$$

and the Schwinger functions are the normalized moments of this measure:

$$S_N(z_1,...,z_N) = \int \phi(z_1)...\phi(z_N)d\nu(\phi). \tag{2.17}$$

This formula is especially convenient to derive the perturbative expansion and Feynman rules of the theory. Indeed, expanding the exponential as a power series in the coupling constant g, one obtains for the Schwinger functions:

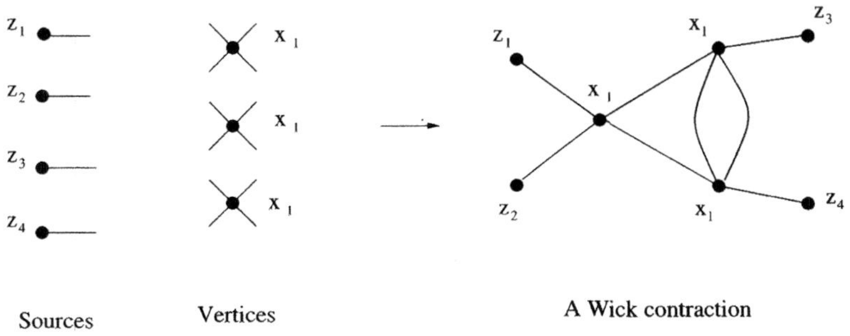

Figure 1: A contraction scheme

$$S_N(z_1,...,z_N) = \frac{1}{Z}\sum_{n=0}^{\infty}\frac{(-g)^n}{n!}\int\Big[\int\frac{\phi^4(x)dx}{4!}\Big]^n\phi(z_1)...\phi(z_N)d\mu(\phi) \tag{2.18}$$

It is now possible to perform explicitly the functional integral of the corresponding polynomial. The result gives at any order n a sum over "Wick contractions schemes $\mathcal{W}$", i.e. ways of pairing together $4n+N$ fields into $2n+N/2$ pairs. There are exactly $(4n+N-1)(4n+N-3)...5.3.1 = (4n+N-1)!!$ such contraction schemes.

Formally at order n the result of perturbation theory is therefore simply the sum over all these schemes $\mathcal{W}$ of the spatial integrals over $x_1,...,x_n$ of the integrand $\prod_{l\in\mathcal{W}} C(x_{i(l)}, x_{j(l)})$ times the factor $\frac{1}{n!}(\frac{-g}{4!})^n$. These integrals are then functions (in fact distributions) of the external positions $z_1,...,z_N$ But they may diverge either because they are integrals over all of $\mathbb{R}^4$ (no volume cutoff) or because of the singularities in the propagator C at coinciding points.

It is convenient to label the n dummy integration variables in (2.18) as $x_1,...,x_n$ and to draw a line for each contraction of two fields. Each position $x_1,...,x_n$ is then associated to a four-legged vertex and each external source z_i to a one-legged vertex, as shown in Figure 1.

For practical computations, it is obviously more convenient to gather all the contractions which lead to the same topological structure, hence the same integral. This leads to the notion of Feynman graphs. To any such graph is associated a contribution or amplitude, which is the sum of the contributions associated with the corresponding set of Wick contractions. The Feynman rules summarize how to compute this amplitude with its correct combinatoric factor.

We always use the following notations for a graph G:

- $n(G)$ or simply n is the number of internal vertices of G, or the order of the graph.

- $l(G)$ or l is the number of internal lines of G, i.e. lines hooked at both ends to an internal vertex of G.

- $N(G)$ or N is the number of external vertices of G; it corresponds to the order of the Schwinger function one is looking at. When $N = 0$ the graph is a vacuum graph, otherwise it is called an N-point graph.
- $c(G)$ or c is the number of connected components of G,
- $L(G)$ or L is the number of independent loops of G.

For a *regular* ϕ^4 graph, i.e. a graph which has no line hooked at both ends to external vertices, we have the relations:

$$l(G) = 2n(G) - N(G)/2, \tag{2.19}$$

$$L(G) = l(G) - n(G) + c(G) = n(G) + 1 - N(G)/2. \tag{2.20}$$

where in the last equality we assume connectedness of G, hence $c(G) = 1$. We like to define the superficial degree of divergence. For ϕ^4_d it is:

$$\omega(G) = dL(G) - 2l(G), \tag{2.21}$$

so that for a connected graph:

$$\omega(G) = (d-4)n(G) + d - \frac{d-2}{2}N(G). \tag{2.22}$$

It will be important also to define what we call a subgraph. This is not a completely straightforward notion. A *subgraph* F of a graph G is a subset of internal lines of G, together with the corresponding attached vertices. Hence there are exactly $2^{l(G)}$ subgraphs in G. We call the lines in the subset defining F the internal lines of F, and their number is simply $l(F)$, as before. Similarly all the vertices of G hooked to at least one of these internal lines of F are called the internal vertices of F and considered to be in F; their number by definition is $n(F)$. But remark that no external vertex of G can be of this kind. Precisely for this reason, the notion of external vertices does not generalize simply to subgraphs. Nevertheless for power counting we need at least to define a generalization of the number N for subgraphs. A good convention is to call external half-line of F every half-line of G which is not in F but which is hooked to a vertex of F; it is then the number of such external half-lines which we call $N(F)$. With this convention one has for ϕ^4 subgraphs the same relation (2.19) as for regular ϕ^4 graphs.

The definitions of c, L and ω then generalize to subgraphs in a straightforward way.

To compute the amplitude associated to a ϕ^4 graph, we have to add the contributions of the corresponding contraction schemes. This is summarized by the rules:

- To each line l_j with end vertices at positions x_j and y_j, associate a propagator $C(x_j, y_j)$.

- To each internal vertex, associate $(-g)/4!$.
- Count all the contraction schemes giving this diagram. The number should be of the form $(4!)^n n!/S(G)$ where $S(G)$ is an integer called the symmetry factor of the diagram. The 4! represents the permutation of the fields hooked to an internal vertex.
- Multiply all these factors, divide by $n!$ and sum over the position of all internal vertices.

The formula for the bare amplitude of a graph is therefore, as a distribution in $z_1, \dots . z_N$:

$$A_G(z_1, \dots, z_N) \equiv \int \prod_{i=1}^{n} dx_i \prod_{l \in G} C(x_l, y_l). \tag{2.23}$$

This is the "direct" or "x-space" representation of a Feynman integral. As stated above, this integral suffers of possible divergences. But the corresponding quantities with both volume cutoff and ultraviolet cutoff κ are well defined. They are:

$$A^{\kappa}_{G,\Lambda}(z_1, \dots, z_N) \equiv \int_{\Lambda^n} \prod_{i=1}^{n} dx_i \prod_{l \in G} C_\kappa(x_l, y_l). \tag{2.24}$$

The integrand is indeed bounded and the integration domain is a compact box Λ.

The *unnormalized* Schwinger functions are therefore formally given by the sum over all graphs with the right number of external lines of the corresponding Feynman amplitudes:

$$ZS_N = \sum_{\phi^4 \text{ graphs } G \text{ with } N(G)=N} \frac{(-g)^{n(G)}}{S(G)} A_G. \tag{2.25}$$

Z itself, the normalization, is given by the sum of all vacuum amplitudes:

$$Z = \sum_{\phi^4 \text{ graphs } G \text{ with } N(G)=0} \frac{(-g)^{n(G)}}{S(G)} A_G. \tag{2.26}$$

Let us remark that since the total number of Feynman graphs is $(4n+N)!!$, taking into account Stirling's formula and the symmetry factor $1/n!$ from the exponential we expect perturbation theory at large order to behave as $K^n n!$ for some constant K. Indeed at order n the amplitude of a Feynman graph is a 4n-dimensional integral. It is reasonable to expect that in average it should behave as c^n for some constant c. But this means that one should expect zero radius of convergence for the series (2.25). This is not too surprising. Even the one-dimensional integral

$$F(g) = \int_{-\infty}^{+\infty} e^{-x^2/2 - gx^4} dx \tag{2.27}$$

is well-defined only for $g \geq 0$. We cannot hope infinite dimensional functional integrals of the same kind to behave better than this one dimensional integral. In mathematically precise terms, F is not analytic near $g = 0$, but only Borel summable. A Borel summable function f can be entirely reconstructed from its asymptotic series $\sum_n a_n x^n$, but not by naively adding the terms in the series. One has rather to first define the Borel series

$$B(t) = \sum_n \frac{a_n}{n!} t^n \tag{2.28}$$

and to analytically continue this function B to a neighborhood of the real axis, then recover f through the integral formula

$$f(x) = \frac{1}{x} \int_0^\infty e^{-t/x} B(t) dt. \tag{2.29}$$

In the case of the function F, this process is guaranteed to converge (using the obvious analyticity of F for $\Re g > 0$, some uniform Taylor remainder estimates and Nevanlinna's theorem [12]). So we know the integral (2.29) can reconstruct F from the list of its Taylor coefficients, which in that particular case are nothing but

$$a_n = \frac{(-1)^n}{n!} \int_{-\infty}^{+\infty} x^{4n} e^{-x^2/2} dx = \frac{(-2\pi)^n (4n-1)!!}{n!}. \tag{2.30}$$

In general Bosonic functional integrals require some stability condition for the potential at large field (here e.g. $g \geq 0$), and their perturbation series do not converge. Borel summability is therefore the best we can hope for the ϕ^4 theory, and it has indeed been proved for the theory in dimensions 2 and 3 [13, 14].

From translation invariance, we do not expect $A^{\kappa}_{G,\Lambda}$ to have a limit as $\Lambda \to \infty$ if there are vacuum subgraphs in G. But we can remark that an amplitude factorizes as the product of the amplitudes of its connected components.

With simple combinatoric verification at the level of contraction schemes we can factorize the sum over all vacuum graphs in the expansion of unnormalized Schwinger functions, hence get for the normalized functions a formula analog to (2.25):

$$S_N = \sum_{\substack{\phi^4 \text{ graphs } G \text{ with } N(G)=N \\ G \text{ without any vacuum subgraph}}} \frac{(-g)^{n(G)}}{S(G)} A_G. \tag{2.31}$$

Now in (2.31) it is possible to pass to the thermodynamic limit (in the sense of formal power series) because using the exponential decrease of the propagator, each individual graph has a limit at fixed external arguments. There is of course no need to divide by the volume for that because each connected component in (2.31) is tied to at least one external source, and they provide the necessary breaking of translation invariance.

Finally one can determine the perturbative expansions for the connected Schwinger functions and the vertex functions. As expected the connected Schwinger functions are given by sums over connected amplitudes:

$$C_N = \sum_{\phi^4 \text{ connected graphs } G \text{ with } N(G)=N} \frac{(-g)^{n(G)}}{S(G)} A_G \tag{2.32}$$

and the vertex functions are the sums of the *amputated* amplitudes for proper graphs, also called one-particle-irreducible. They are the graphs which remain connected even after removal of any given internal line. The amputated amplitudes are defined in momentum space by omitting the Fourier transform of the propagators of the external lines. It is therefore convenient to write these amplitudes in the so-called momentum representation:

$$\Gamma_N(z_1, ..., z_N) = \sum_{\phi^4 \text{ proper graphs } G \text{ with } N(G)=N} \frac{(-g)^{n(G)}}{S(G)} A_G^T(z_1, ..., z_N), \tag{2.33}$$

$$A_G^T(z_1, ..., z_N) \equiv \frac{1}{(2\pi)^{dN/2}} \int dp_1 ... dp_N e^{i \sum p_i z_i} A_G(p_1, ..., p_N), \tag{2.34}$$

$$A_G(p_1, ..., p_N) = \int \prod_{l \text{ internal line of } G} \frac{d^d p_l}{p_l^2 + m^2} \prod_{v \in G} \delta(\sum_l \epsilon_{v,l} p_l). \tag{2.35}$$

Remark in (2.35) the δ functions which ensure momentum conservation at each internal vertex v; the sum inside is over both internal and external momenta; each internal line is oriented in an arbitrary way and each external line is oriented towards the inside of the graph. The incidence matrix $\epsilon(v, l)$ is 1 if the line l arrives at v, -1 if it starts from v and 0 otherwise. Remark also that there is an overall momentum conservation rule $\delta(p_1 + ... + p_N)$ hidden in (2.35). The drawback of the momentum representation lies in the necessity for practical computations to eliminate the δ functions by a "momentum routing" prescription, and there is no canonical choice for that.

2.3 Feynman representation

There are other convenient representations such as the "Feynman parametric representation" which do not need any non canonical choices. To define it we write the α or parametric representation of the propagator:

$$\hat{C}(p) = \frac{1}{(2\pi)^d} \int_0^\infty e^{-\alpha(p^2+m^2)} d\alpha, \tag{2.36}$$

$$\begin{aligned} C(x, y) &= \frac{1}{(2\pi)^d} \int_0^\infty d\alpha \int e^{ip\cdot(x-y)-\alpha(p^2+m^2)} d^d p \\ &= \frac{1}{(4\pi)^{d/2}} \int_0^\infty \frac{d\alpha}{\alpha^{d/2}} e^{-\alpha m^2 - |x-y|^2/(4\alpha)}. \end{aligned} \tag{2.37}$$

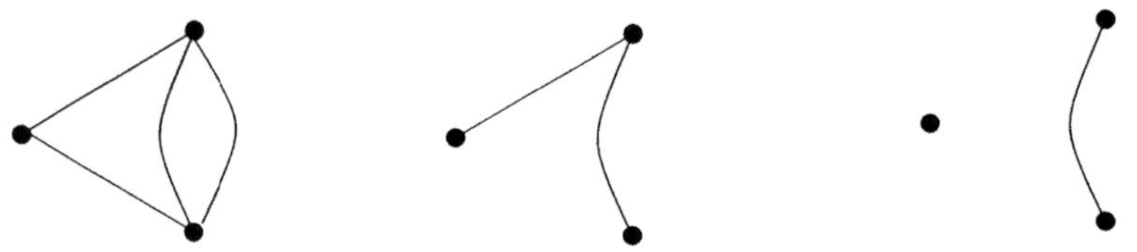

Figure 2: Spanning and two-trees

The x space or p space integrations can then be explicitly performed in any Feynman amplitude, since they are quadratic. The result is a compact formula with one scalar integration over a parameter α for each internal line of the graph:

$$A_G(p_1,...,p_N) = \delta(\sum_v P_v) \int_0^\infty \prod_l d\alpha_l \; e^{-\sum_l \alpha_l m^2 - V_G(\alpha,p)/U_G(\alpha)} \frac{1}{[U_G(\alpha)]^{d/2}} \tag{2.38}$$

where U_G and V_G are polynomials in α depending on the particular topology of the graph, called the Symanzik polynomials. Their explicit expression is:

$$U_G = \sum_S \prod_{l \text{ not in } S} \alpha_l, \tag{2.39}$$

$$V_G(p,\alpha) = (\sum_T \prod_{l \text{ not in } T} \alpha_l)(\sum_{a \in T_1} p_a)^2. \tag{2.40}$$

In (2.39) the sum runs over the spanning trees S of G. Such a spanning tree is a set of lines without loops connecting all the vertices of the graph. Similarly in (2.40), the sum runs over the two-trees T of G which separate G into two connected components, each containing a non empty set of external lines, one of which is T_1 (by overall momentum conservation, (2.40) does not change if T_1 is replaced by the set of external lines of the other connected component, which is the complementary of T_1) (see Figure 2 for an example).

In this elementary presentation we shall not reproduce the complete proof of these formulas (see [15] or [9]). They rely on a careful analysis of the quadratic form that one obtains in the exponential after rewriting all the propagators in α space. This quadratic form in turn can be deduced form the incidence matrix of the graph.

Remark that the parametric representation is not only "canonical" but also quite economical in large dimensions. In dimension 4, a four point subgraph of order n has $n-1$ loops hence the momentum integration is over a space of dimension $4n-4$; instead the parametric representation is over a space of dimension $l = 2(n-1)$, hence with only half as many scalar components to integrate. For

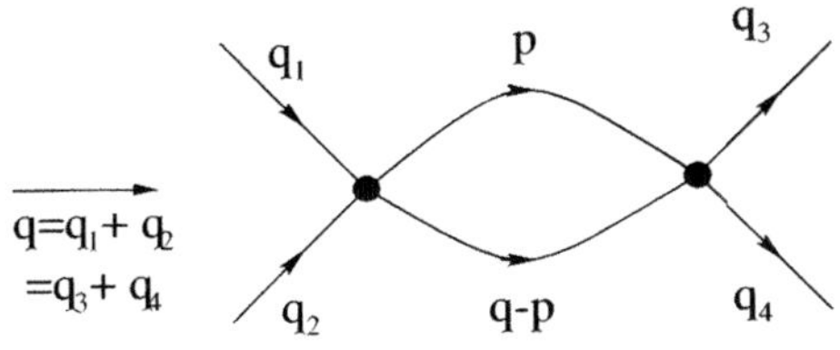

Figure 3: The graph G_0

instance the integral of the graph G_0 of Figure 3 involves only one total external momentum q and can be written formally as

$$\begin{aligned} A_{G_0}(q) &= \int d^4p \frac{1}{(p^2+m^2)((p-q)^2+m^2)} \\ &= \int_0^\infty \int_0^\infty \frac{d\alpha_1 d\alpha_2}{(\alpha_1+\alpha_2)^2} e^{-(\alpha_1+\alpha_2)m^2 - \frac{\alpha_1\alpha_2}{\alpha_1+\alpha_2}q^2}. \end{aligned} \tag{2.41}$$

However none of these two representations gives convergent integrals because of a divergence at large p or small α's. We return to the structure of these ultraviolet divergences in the next subsection.

The α-representation has also a fundamental interpretation in terms of Brownian motions [16]. In particular, the propagator (2.37) can be written as:

$$C(x,y) = \int_0^\infty d\alpha \exp(-m^2\alpha)\, P(x,y;\alpha) \tag{2.42}$$

where $P(x,y;\alpha) = (4\pi\alpha)^{-d/2}\exp(-|x-y|^2/4\alpha)$ is the Gaussian probability distribution of a Brownian path going from x to y in time α.

The Feynman diagrams can then be understood as made of Brownian paths interacting by Dirac distributions, as in the Edwards model for self-avoiding polymers [17]. This lead P.-G. de Gennes in 1972 to his famous relation between this polymer model and a $[(\phi)^2]^2$ field theory with $O(N)$ symmetry, in the $N \to 0$ limit [18]. This allowed RG results to be applied to polymer physics. A new development appeared when J. des Cloizeaux introduced a simple *direct* (dimensional) renormalization method for the Edwards model [19, 20], working explicitly in the α-representation.

2.4 Ultraviolet Divergences

The amputated amplitudes for a connected graph at finite external momenta are not always finite because of possible ultraviolet divergences. These divergences appear because the momentum integration over the loop variables in (2.35) may not always be absolutely convergent. This can be traced back to the distribution character of the propagator C in direct space, for $d \geq 2$, and the general impossibility to multiply distributions as should be done to define e.g. ϕ^4.

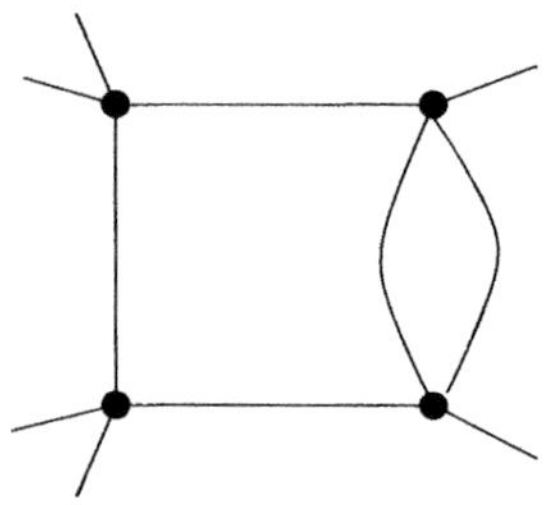

Figure 4: A 6-point subgraph with a divergent subgraph

This difficulty, also present in quantum electrodynamics, was the basic puzzle that the founding fathers of quantum field theory were confronted with. Let us explore it, increasing the dimension step by step. The naive global scaling of all internal momenta of the graph explains our definition of the superficial degree of divergence: it measures whether the integral over this global scaling parameter is convergent or not. Therefore graphs with $\omega(G) \geq 0$ are called primitively divergent.

– If $d = 2$, we find $\omega(G) = 2 - 2n$, so the only divergent graphs have $n = 1$, and $N = 0$ or $N = 2$. The only divergence is due to the "tadpole" loop $\int \frac{d^2p}{(p^2+m^2)}$ which is logarithmically divergent.

– If $d = 3$, we find $\omega(G) = 3 - n - N/2$, so the only divergent graphs have $n \leq 3$, $N = 0$, or $n \leq 2$ and $N = 2$. Such a theory with a finite number of "primitively divergent" subgraphs is called superrenormalizable.

– If $d = 4$, $\omega(G) = 4 - N$. Every two point graph is quadratically divergent and every four point graph is logarithmically divergent. This is in agreement with the superficial degree of these graphs being respectively 2 and 0. For instance the graph G_0 at zero momentum without ultraviolet cutoff is logarithmically divergent for large p:

$$A_{G_0}(0) = \int \frac{d^4p}{(p^2+m^2)^2} = +\infty \tag{2.43}$$

and the "*tadpole*" loop $\int \frac{d^dp}{p^2+m^2}$ is quadratically divergent. Theories in which the degree of divergence only depends on the number of external legs are called renormalizable.

– Finally for $d > 4$ we have infinitely many primitively divergent graphs with arbitrarily large number of external legs, and the theory is called non-renormalizable.

It was soon recognized that even graphs which have negative superficial degree of divergence, such as the 6-point subgraph of Figure 4 in $d = 4$, are not ultraviolet finite. Indeed they can contain divergent subgraphs, and the corresponding subintegrations do not converge.

The first progress on renormalization came in recognizing that for four-dimensional theories such as ϕ^4 or quantum electrodynamics, the superficially divergent graphs when suitably added to a local counterterm gives rise to a finite contribution. For instance in the case of the graph G_0 the "renormalized" amplitude

$$\begin{aligned} A^R_{G_0}(q) &= \int [\frac{1}{(p^2+m^2)((p+q)^2+m^2)} - \frac{1}{(p^2+m^2)^2}]\, d^4p \\ &= \int_0^\infty \int_0^\infty \frac{d\alpha_1 d\alpha_2 e^{-(\alpha_1+\alpha_2)m^2}}{(\alpha_1+\alpha_2)^2} \big[e^{-\frac{\alpha_1\alpha_2}{\alpha_1+\alpha_2}q^2} - 1\big]. \end{aligned} \quad (2.44)$$

is now finite.

Indeed let us prove finiteness of this amplitude. In the momentum representation, we reduce to the same denominator, and taking advantage of parity we obtain:

$$\begin{aligned} A^R_{G_0}(q) &= \int \frac{-2p.q - q^2}{(p^2+m^2)^2((p+q)^2+m^2)}\, d^4p \\ &= -\int \frac{q^2}{(p^2+m^2)^2((p+q)^2+m^2)}\, d^4p \end{aligned} \quad (2.45)$$

now an obviously convergent integral. In the parametric representation, using $|e^{-x} - 1| \leq x$ for positive x we can bound $A^R_{G_0}(q)$ by

$$\int_0^\infty \int_0^\infty d\alpha_1 d\alpha_2 e^{-(\alpha_1+\alpha_2)m^2} \frac{q^2\alpha_1\alpha_2}{(\alpha_1+\alpha_2)^3} \quad (2.46)$$

which is now a convergent integral. To be more precise, we should make additional remarks:

– the renormalized amplitude is negative

– it behaves as $c \log|q|$ as $|q| \to \infty$

– this large behavior at large q is solely due to the integral over the region $|p| \leq |q|$ of the *counterterm*. Indeed both

$$\int_{|p|\geq|q|} \frac{q^2}{(p^2+m^2)^2((p+q)^2+m^2)}\, d^4p \quad (2.47)$$

and

$$\int_{|p|\leq|q|} \frac{1}{(p^2+m^2)((p+q)^2+m^2)}\, d^4p \quad (2.48)$$

are well defined uniformly bounded integrals as $|q| \to \infty$.

Remark finally that the counterterm, when Fourier transformed, corresponds to a local ϕ^4 term, since the zero momentum value of the graph is nothing but the spatial integral over y of $C^2(x, y)$. This counterterm when added to the bare Lagrangian will renormalize G_0 not only as a primitive graph, but each time it appears as a subgraph in the expansion, since the combinatoric of inserting a ϕ^4 vertex or a G_0 subgraph at any place in a bigger diagram is clearly the same.

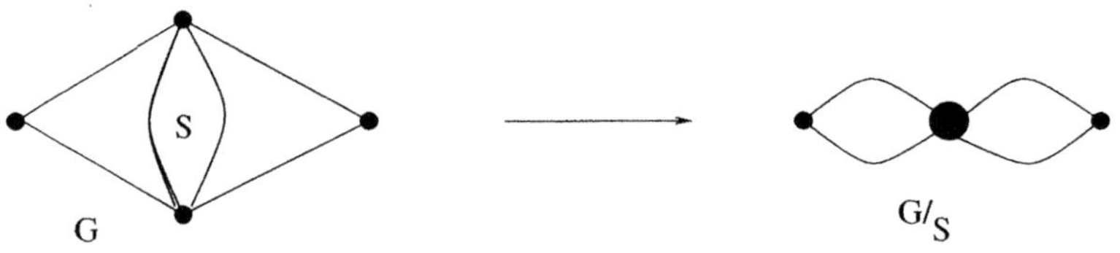

Figure 5: The reduction of a subgraph in a graph

In the same way local counterterms of the ϕ^4, ϕ^2 or $(\nabla\phi)^2$ type for any kind of primitively divergent graph, can be reabsorbed in the parameters of the Lagrangian of (2.8). Such an infinite redefinition which affects only the unobservable "bare" parameters of the theory hence it is not physically inconsistent.

But for a while it was not clear whether one could introduce a proper set of counterterms which is local and remove all the ultraviolet divergences of every graph, not only the main global primitive divergences but also all the divergences associated to subgraphs. This would make all particular submanifolds of the momentum integration convergent. The solution of this problem, by Bogoliubov, Parasiuk, Hepp and Zimmermann [21, 22, 23], and its extension to gauge theories by 'tHooft and Veltman [27] is a first great mathematical triumph of quantum field theory.

2.5 The Bogoliubov Recursion and Zimmermann's Solution

We have now to explain how to organize the set of all subtractions that should be performed in a renormalizable theory to make it ultraviolet finite in perturbation theory. When a local counterterm has been defined for a graph G_1 with N_1 external lines, the modified Lagrangian gives rise to a new vertex with N_1 lines. So for every graph G_2 which contains G_1 as a subgraph, to subtract the subintegration over G_1 corresponds to perform the sum

$$A_{G_1} + c_{G_1} A_{G_2/G_1} \tag{2.49}$$

where G_2/G_1 is the graph obtained by reducing G_1 to a single vertex in G_2 (see Figure 5 for an example). This reduction is an essential operation in renormalization theory. But remark already that if there are several divergent subgraphs in a graph G, we can define a reduced subgraph G/S only for families S of *disjoint* subgraphs S.

More generally, if G_2 itself is divergent, it seems clear that the counterterm for G_2 should be defined by taking the local part of (2.49), not of A_{G_1} itself. So the definition of counterterms is inductive, starting with the smaller graphs towards the bigger. This is after all the logic of perturbation theory. This induction was formalized by Bogoliubov. However since a graph G can contain overlapping divergent subgraphs S_1 and S_2 with non-trivial intersection S_3, such as in Figure 6, it is far from clear that this induction actually removes all ultraviolet divergences.

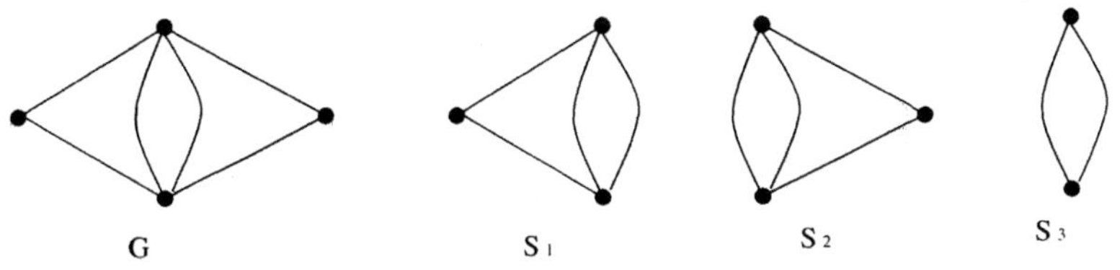

Figure 6: A graph with two overlapping divergent subgraphs

The first proof that Bogoliubov's induction actually leads to finite amplitudes is due to Hepp [22], and the first explicit solution of the induction, which involves the notion of "forests" is due to Zimmermann [23].

Suppose we have defined counterterms up to a given order n. Then for a graph G at order $n+1$ one defines a counterterm c_G and the renormalized amplitude A_G^R by

$$A_G^R = \sum_{\mathcal{S}} (A_{G/\mathcal{S}} \prod_{S \in \mathcal{S}} c_S) + c_G \tag{2.50}$$

where the sum is over all families $\mathcal{S}$ of disjoint primitively divergent subgraphs of G, including the empty one.

The exact definition of c_G contains some arbitrariness if the goal is to make renormalized amplitudes finite. In the BPHZ (Bogoliubov–Parasiuk–Hepp–Zimmermann) renormalization scheme, c_G is the local (or zero momentum part) of the sum in (2.50). More precisely, if we choose a system of loop momenta k for G and call p the external momenta we have

$$A_G(p) = \int dk I_G(p,k) \tag{2.51}$$

and for a primitively divergent graph one defines the counterterm c_G by a subtraction acting directly at the level of the integrand $I_G(p,k)$ in momentum space, to get

$$A_G^R(p) = \int dk(1 - \mathcal{T}^{d(G)}) I_G(p,k) \tag{2.52}$$

where $\mathcal{T}^{d(G)}$, the so-called Taylor "operator" selects the beginning of the Taylor expansion of $I_G(p,k)$ up to order $d(G)$ around the simple point $p=0$. This is in agreement with (2.44).

To generalize to graphs with divergent subgraphs one follows the Bogoliubov recursion. In fact renormalizing proper (i.e. connected one-particle-irreducible) subgraphs is enough, and the explicit solution of the Bogoliubov induction with this subtraction prescription is:

$$A_G^R = \int dk \mathcal{R} I_G(p,k) \tag{2.53}$$

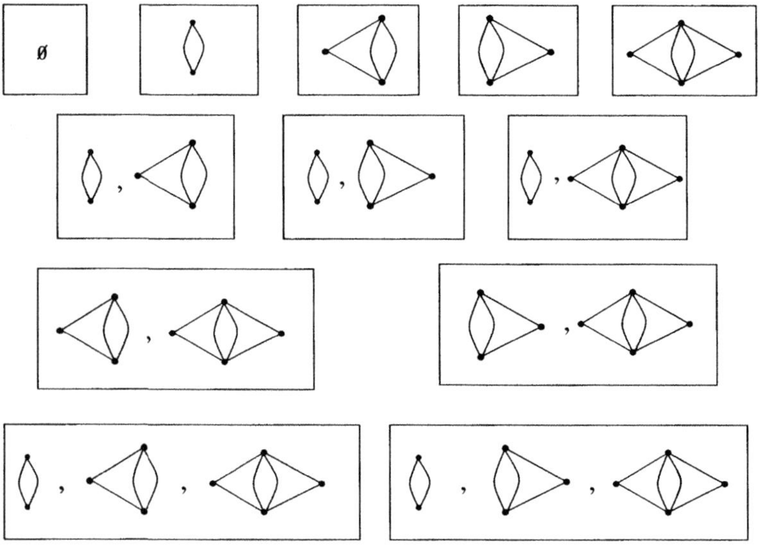

Figure 7: The twelve forests of G

$$\mathcal{R} = \sum_{\mathcal{F}} \prod_{S \in \mathcal{F}} (-\mathcal{T}^{d(S)}) \tag{2.54}$$

where the sum is over all forests of proper divergent subgraphs $S \subset G$, including the empty forest.

Definition 1 *A forest $\mathcal{F}$ is a subset of subgraphs such that for any pair S_1, S_2 of the forest, either $S_1 \subset S_2$ or $S_2 \subset S_1$ or S_1 and S_2 are disjoint.*

This definition ensures that the partial ordering by inclusion in a forest can indeed be pictured as a set of trees, hence the name "forest".

For example the graph G of Figure 5 which has 3 different divergent strict subgraphs, has 12 forests, namely

$$\{\emptyset\}, \{S_1\}, \{S_2\}, \{S_3\}, \{G\}, \{S_3, S_1\}, \{S_3, S_2\}, \{S_3, G\}, \{S_1, G\}, \{S_2, G\},$$
$$\{S_3, S_1, G\}, \{S_3, S_2, G\} \tag{2.55}$$

These 12 forests are shown in Figure 7 In formula (2.54) the product of the Taylor operators is taken following the partial ordering of the forest, that is from smaller to bigger graphs. Each Taylor operator selects the beginning of a Taylor expansion in the external momenta of a subgraph S, which can later become internal momenta for G. The definition of $\mathcal{R}$ may therefore depend on the choice of the momentum routing, hence of the loop momenta solving the δ functions in (2.35). This difficulty lead Zimmermann to define particular momentum routing called "admissible". For these choices, Zimmermann could then prove:

Theorem 2.1 *The integrals* (2.53) *do converge for any G and define amplitudes $A^R_G(p)$ which are tempered distributions when analytically continued to Minkowski space.*

The difficulty linked to momentum routing can be avoided completely by working instead in the parametric representation. It is indeed possible to define an $\mathcal{R}$ operator acting directly in the α-parametric space, equivalent to Zimmermann's operator, but bypassing completely the problem of admissible momentum routing [24, 25]. Then there exists a very explicit proof of finiteness of the renormalized amplitudes. One can divide, for any complete ordering of the parameters α_l, also called a "Hepp sector", the sum over all forests quite naturally into packets, so that each packet gives a finite contribution. The problem is that the packets themselves change when the Hepp sector changes! Nevertheless this method is then sufficiently explicit to not only prove finiteness but also to produce reasonable quantitative estimates of the size of renormalized perturbation theory at large order [26].

The definition of the packets is subtle, but let us try to sketch it. The number of forests in any packet is always a certain power of 2, that is is of the form 2^r for a certain integer r. Indeed the forests which compose any such packet are exactly those containing a fixed forest $\mathcal{F}_0$ and contained in another fixed forest $\mathcal{F}_0 \cup \mathcal{F}_1$. r is simply the number of elements in $\mathcal{F}_1$. So the forests in that packet are those $\mathcal{F}$ that satisfy $\mathcal{F}_0 \subset \mathcal{F} \subset \mathcal{F}_0 \cup \mathcal{F}_1$. Hence the sum of the Taylor subtractions for a given packet always reconstructs an operator

$$\prod_{S \in \mathcal{F}_0} (-\mathcal{T}_S) \prod_{S \in \mathcal{F}_1} (1 - \mathcal{T}_S). \tag{2.56}$$

In a given sector, there is exactly one packet for each forest $\mathcal{F}_0$ with a certain property, which roughly speaking says that $\mathcal{F}_0$ is made of subgraphs with some internal line α-parameter larger than some external line α-parameter in the ordering of the sector considered. Given such an $\mathcal{F}_0$, the forest $\mathcal{F}_1$ then is completely determined by $\mathcal{F}_0$ and the sector. It is made of the subgraphs with the opposite property, that is all α-parameters for the internal lines of these subgraphs of $\mathcal{F}_1$ are smaller than all α-parameters for their external lines in the ordering of the sector [3].

The factorization property (2.56) is what makes each packet finite. Indeed the defining property for the subgraphs of $\mathcal{F}_0$ means that they are not really divergent in the sector considered. This is because the smaller α-parameter for one of their external lines acts as a natural ultraviolet cutoff for the subgraph. In contrast the subgraphs of $\mathcal{F}_1$ are potentially divergent. But for these subgraphs the $1 - \mathcal{T}_S$ operators in (2.56) precisely provide the necessary subtractions! This is the basic mechanism which makes every packet finite.

[3]The true definition is a bit more complicated and inductive, because reduction by the elements of $\mathcal{F}_0$ (as shown in Figure 5) has to be taken into account, starting from the smallest subgraphs in $\mathcal{F}_0$ and working towards the largest.

2.6 Different renormalization schemes

To subtract the value of subgraphs at zero external momenta is obviously a natural but not a canonical choice. It may be even ill-defined if the theory contains massless particles, which is for instance the case of quantum electrodynamics. It is important therefore to have several different sets of renormalization schemes, and to understand how they are related to each other. Two different subsets of counterterms which both make the Feynman amplitudes finite must differ through finite counterterms. In practice one wants usually to fix some physical conditions such as the particular values of some Green functions at some given momenta, and to determine the renormalization scheme corresponding these conditions. It may require two steps: first to use a general scheme to get rid of infinities, then to adjust the scheme through finite counterterms so as to meet the physical conditions.

For instance the BPHZ scheme that we have considered for the massive Euclidean ϕ^4_4 theory corresponds to the following normalization conditions on the connected functions in momentum space:

$$C^4(0,0,0,0) = -g, \tag{2.57}$$

$$C^2(p^2=0) = \frac{1}{m^2}, \tag{2.58}$$

$$\frac{d}{dp^2} C^2|_{p^2=0} = -\frac{a}{m^4}. \tag{2.59}$$

Let us say a few words about another popular renormalization scheme, namely dimensional renormalization. The starting idea is that in the parametric representation (2.38) the dimension d can be considered as a complex parameter. The attentive reader can object that external momenta still live in $\mathbb{R}^4$. But since the amplitudes depend only on the Euclidean scalar invariants $(\sum_{a \in T_1} p_a)^2$ built on them (see (2.40)), this is not a major difficulty. Amplitudes such as I_{G_0} in (2.41) become meromorphic functions for $\Re d \leq 4$. They have then a pole at $d = 4$. It is therefore natural to define the finite part of the amplitude as the finite part of the corresponding Laurent series, hence to simply extract the pure pole with its correct residue at $d = 4$. When properly implemented according to Bogoliubov's induction this leads to the notion of dimensional renormalization.

This scheme has many advantages but one major drawback. The main advantage is that it preserves the symmetries of the theory such as gauge symmetries. Using it, 't Hooft and Veltman were able to show the renormalizability of the non-Abelian gauge theories at the core of the standard model [27]. For instance although the action $g^{-2}F_{\mu\nu}F^{\mu\nu}$ of a pure non-Abelian gauge theory contains terms of order 2, 3 and 4 in the field A_μ, it is possible with dimensional renormalization to preserve the basic relation between these three terms which makes the total Lagrangian a perfect square. In this way the theory remains of the same form after renormalization , but simply with a renormalized parameter g_{ren} instead of g. This success was extremely important to convince physicists to adopt non-Abelian gauge theories for particle physics. As other examples of use of this scheme, let us mention again the renormalization method for the Edwards model of polymers [19, 20] which has been shown to be equivalent to standard (dimensional) field-theoretic renormalization [28]. These works opened the way to the renormalization theory of interacting or self-avoiding crumpled membranes, where the Feynman diagrams are no longer made of lines but of extended surfaces (see, e.g., [29]). Dimensional renormalization is also at the core of the Riemann–Hilbert interpretation of renormalization [30].

But the big drawback of dimensional renormalization is that up to now it remains a purely perturbative technique. Nobody knows how to interpolate correctly in the space-time dimension d the infinite dimensional functional integrals (2.17) which are the basis for the non-perturbative or constructive version of quantum field theory. To solve this difficulty would certainly be a major progress.

2.7 What lies beyond perturbative renormalization?

The theory of perturbative renormalization is a brilliant piece of mathematical physics. The solution of the difficult "overlapping" divergence problem through Bogoliubov's recursion and Zimmermann's forests becomes particularly clear in the parametric representation using Hepp's sectors: in each sector there is a different classification of forests into packets so that each packet gives a finite integral.

Dimensional renormalization allows to preserve critical symmetries such as gauge symmetries, hence to prove renormalizability of four dimensional gauge theories, but does not seem adapted to non-perturbative theory. Note however that in this scheme the finite part of the Feynman amplitudes are related to ζ functions. This hints that this theory might be useful for mathematics, particularly number theory. The structure of the forests subtraction has been shown recently to be associated to a Hopf algebra and related to the Riemann–Hilbert problem in the works of Connes and Kreimer [31, 30].

But from the physical point of view we cannot conceal the fact that purely perturbative renormalization theory is also in some sense a conceptual maze. At least two facts already hint at a better theory which lies behind:

– The forest formula seems unnecessarily complicated, with too many terms. For instance if we examine closely the classification of forests into packets, we remark that in any given Hepp sector, only the particular packet corresponding to $\mathcal{F}_0 = \emptyset$ seems absolutely necessary to make the renormalized amplitude finite. The other packets, with non-empty $\mathcal{F}_0$ seem useless, a little bit like "junk DNA": they are there just because they are necessary for other sectors. This does not look optimal.

– The theory makes amplitudes finite, but at which cost! The size of some of these renormalized amplitudes becomes indeed unreasonably large as the size of the graph increases. This phenomenon is called the "renormalon problem". For instance it is easy to check that the renormalized amplitude (at 0 external momenta) of the graphs P_n with 6 external legs and $n + 2$ internal vertices in

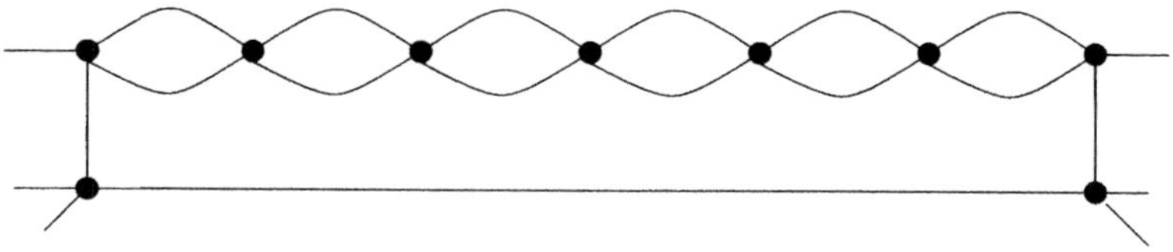

Figure 8: A family of graphs P_n producing a renormalon

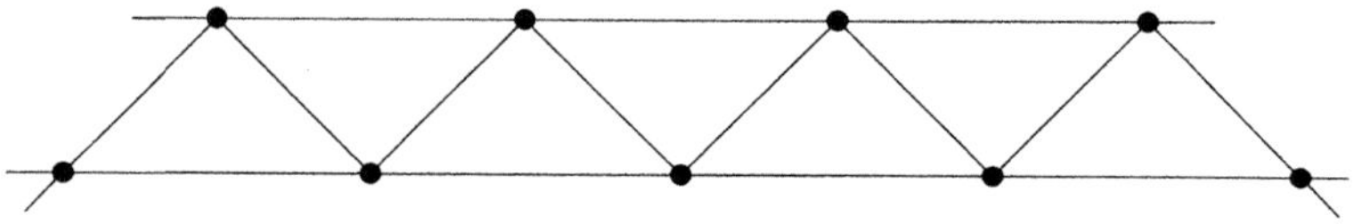

Figure 9: A family of convergent graphs Q_n, that do not produce any renormalon

Figure 8 becomes large as $c^n n!$ as $n \to \infty$. Indeed we remarked already that at large q the renormalized amplitude $A^R_{G_0}$ in (2.44) grows like $\log |q|$. Therefore the chain of n such graphs in Figure 8 behaves as $[\log |q|]^n$, and the total amplitude of P_n behaves as

$$\int [\log |q|]^n \frac{d^4 q}{[q^2 + m^2]^3} \simeq_{n \to \infty} c^n n! \tag{2.60}$$

So there are not only too many Feynman graphs to resum them, but some of them after renormalization also acquire so large values that the corresponding subfamilies of graphs cannot be resummed! These two hints are in fact linked. As their name indicates, renormalons are due to renormalization. Families of completely convergent graphs such as the graphs Q_n of Figure 9, are bounded by c^n, and produce no renormalons. But studying more carefully renormalization in the α parametric representation one can say more. One can check that renormalons are solely due to the forests packets with $\mathcal{F}_0 \neq \emptyset$ and in fact $\mathcal{F}_0$ large. A packet associated to a given $\mathcal{F}_0$ typically grows like $c^n |\mathcal{F}_0|!$ [26]. Recall that the forests $\mathcal{F}_0$ are made of those subgraphs which are not really divergent in the sector considered. So this renormalon analysis generalizes one of our previous remarks. Renormalons are due to subtractions that are not necessary to ensure convergence, just like the strange $\log |q|$ growth of $A^R_{G_0}$ at large q is solely due to the counterterm in the region where this counterterm is not necessary to make the amplitude finite.

We can therefore conclude that subtractions are not organized in an optimal way by the Bogoliubov recursion. The idea of renormalization itself is not wrong. But to use the size of the graph as the relevant parameter to organize Bogoliubov's induction is not the optimal idea. A better parameter to organize the induction was found in fact for other completely different reasons by Wilson and followers. It is not the size of the graph but rather the size of the line momenta in it that should be used to better organize the renormalization subtractions. This is the point of view of the renormalization group.

3 The Renormalization Group

The renormalization group is a strange name. It is in fact an (ill-defined) semi-group. Its discovery came in two steps: first by thinking about changing the renormalization scheme, field theorists such as Callan and Symanzik discovered a kind of "invariance" of the theory [32, 33]. Two renormalizable theories with two different sets of coupling constants but defined by subtracting at different scales can in fact be the same physical theory if the constants and scales are related through some "renormalization group" equations. It is in fact even possible to prove finiteness of perturbative renormalization, hence to bypass the BPHZ theorem by directly using these renormalization group equations [34].

Then came the conceptual breakthrough of Wilson and followers [36]: instead of renormalizing the theory at once, why not perform this difficult task in a sequence of steps? The evolution of the theory in this sequence of steps is then similar to the evolution of a dynamical system. In dynamical systems we know that it is usually easier (in particular numerically) to perform patiently a large sequence of local steps than to try to guess the global result, or to search for an analytic solution, which is very rare. The same is true in renormalization theory, in which some scale parameter plays the role of time.

Although this was not the historic path, it would have been perfectly possible to arrive also at the same renormalization group concept by simply trying to simplify Zimmermann's formula to get rid of renormalons. Indeed this is exactly what the RG also does!

This note is too short for a complete review of the renormalization group and in particular of its non-perturbative aspects. So we will sketch what it does on the simple example of ϕ^4_4.

3.1 Slicing

One needs first to separate the degrees of freedom of the theory, and to organize them into a sequence of slices, each slice corresponding to a given scale. It is convenient to choose this sequence of scales to form a geometric progression. The idea is then to perform the functional integral only over the modes of the field corresponding to momenta of a given scale and to compute an effective theory for the remaining scales. This should not be done in an arbitrary order: according to the usual scientific philosophy, microscopic laws should determine macroscopic behavior, not the converse [4]. So the "effective" field theory should emerge progressively from the bare theory like an effective picture progressively emerges from averaging the fine pixels in a detailed picture, or like thermodynamics with a few macroscopic parameters such as temperature or pressure should emerge from a very complicated and chaotic microscopic behavior governed by the laws of mechanics.

[4]This traditional philosophy is put in question by more holistic points of view such as those based on the dualities of string theory which exchange small and large distances. But in this note I will nevertheless stick to the old-fashioned standard wisdom!

In a theory such as ϕ_4^4, the mass fixes some particular scale beyond which no interesting physics happens because connected functions decay exponentially just as the propagator itself (2.13). So in this case the renormalization group will be used solely to treat the ultraviolet problem. One can slice the theory by dividing the Euclidean propagator into slices with an index $i \in \mathbb{N}$, and the slice i will correspond to momenta of order roughly M^i, where M is a fixed number, the ratio of the geometric progression (e.g. $M = 2$).

This can be done conveniently with the parametric representation, since α in this representation is roughly like $1/p^2$. So we can define the propagator within a slice as

$$C_i = \int_{M^{-2i}}^{M^{-2(i-1)}} e^{-m^2\alpha - \frac{|x-y|^2}{4\alpha}} \frac{d\alpha}{\alpha^{d/2}}. \tag{3.1}$$

We can intuitively imagine C_i as the piece of the field oscillating with Fourier components essentially only of size roughly M^i. In fact it is easy to prove the bound (for $d > 2$)

$$|C_i(x,y)| \leq K.M^{(d-2)i} e^{-M^i|x-y|} \tag{3.2}$$

where K is some constant.

For the first slice the formula is a little different because

$$C_0 = \int_1^{\infty} e^{-m^2\alpha - \frac{|x-y|^2}{4\alpha}} \frac{d\alpha}{\alpha^{d/2}}. \tag{3.3}$$

Now the full propagator with ultraviolet cutoff M^ρ, ρ being a large integer, may be viewed as a sum of slices:

$$C_{\leq\rho} = \sum_{i=0}^{\rho} C_i \tag{3.4}$$

Then the basic renormalization group step is made of two main operations:

- A functional integration
- The computation of a logarithm to define an effective action

Indeed decomposing a covariance in a Gaussian process corresponds to a decomposition of the field into independent Gaussian random variables ϕ^i, each distributed with a measure $d\mu_i$ of covariance C_i. Let us call

$$\Phi_i = \sum_{j=0}^{i} \phi_j. \tag{3.5}$$

This is the "low-momentum" field for all frequencies lower than i. The RG idea is that starting from scale ρ and performing $\rho - i$ steps, one arrives at an effective action for the remaining field Φ_i. Then writing $\Phi_i = \phi_i + \Phi_{i-1}$ splits the field into a "fluctuation" field ϕ_i and a "background" field Φ_{i-1}. The first step, functional integration, is performed solely on the fluctuation field, so it computes

$$Z_{i-1}(\Phi_{i-1}) = \int d\mu_i(\phi_i) e^{-S_i(\phi_i+\Phi_{i-1})}. \tag{3.6}$$

Then the second step rewrites this quantity as the exponential of an effective action, hence simply computes

$$S_{i-1}(\Phi_{i-1}) = -\log[Z_{i-1}(\Phi_{i-1})] \tag{3.7}$$

Now $Z_{i-1} = e^{-S_{i-1}}$ and one can iterate! The flow from the initial bare action $S = S_\rho$ for the full field to an effective renormalized action S_0 for the last "slowly varying" component ϕ_0 of the field is similar to the flow of a dynamical system. Its evolution is decomposed into a sequence of discrete steps from S_i to S_{i-1}.

Of course this program needs many modifications to become a mathematically correct (non-perturbative) prescription. But at least formally it has a non-perturbative potential because it is not formulated at the level of graphs. Integrating over a single "momentum slice" of the field is like computing a field theory with both ultraviolet and infrared cutoff, and should be much easier than a full-fledged ultraviolet or infrared problem.

A key feature of the standard presentation of the renormalization group has been also omitted. Usually one performs a third somewhat confusing operation in a RG step, which is a rescaling of all the lengths of the theory and of the field size. Here it would simply be

$$x \to M^{-1}x, \tag{3.8}$$

$$\phi \to M^{-(d-2)/2}\phi. \tag{3.9}$$

But this rescaling is made to compare more easily the effective action to the former one, just like a "reframing" of our averaged picture to always fit into a frame of fixed size. It is therefore some kind of analogue of changing the reference frame in a dynamical system, from the "laboratory frame" to a "moving frame". We prefer here not to introduce this rescaling, because in many situations the long distance behavior of a theory is not governed by a simple scaling around the point $p = 0$ in momentum space but by more complicated extended singularities. This phenomenon occurs in condensed matter, where the singularity is given by a so called Fermi surface, and in diffusion problems in Minkowski space, where the propagator is singular on a mass-shell. In these cases there is no single simple moving frame (but rather one different moving frame for each limit point of the extended singularity).

Of course there is lot of arbitrariness in the choice of the slicing for the RG. One can use for instance wavelets [37]. A very popular choice is "block-spinning", in which Φ_{i-1} is simply the average of Φ_i over a cube of side size M^{-i}. Again this is a choice which does not generalize easily to extended singularities (and also breaks the rotation invariance of the theory) so (when possible) slicing the covariance of the field seems the best technical tool.

It is clear that the RG strategy is not limited to the study of an ultraviolet problem in field theory. In fact since the renormalization group flows from ultraviolet scales to infrared ones, it is particularly well adapted to the study of critical phenomena in statistical mechanics [36, 5]. The bare critical action leading to an effective massless action corresponds to an initial point at some finite given spatial scale in a RG trajectory, for which a final condition (massless effective theory) is given at very long distance. Similarly "the ultraviolet limit" in field theory corresponds to a sequence of bare actions at smaller and smaller spatial scale which end up on the same renormalized theory at some given fixed spatial scale. So the two problems are very similar. Finally a massless field theory without ultraviolet cutoff is similar to a dynamical system with two boundary conditions one towards $t \to -\infty$ and one towards $t \to +\infty$.

3.2 The Flow

In this section we would like to sketch how the renormalization group deeply changes the way perturbation theory should be organized.

Naive field theory was formulated with a single set of coupling constants, and perturbatively renormalized field theory is formulated with two such sets, the bare and the renormalized constants. The bare couplings become infinite formal power series in the renormalized constants with coefficients which diverge when the ultraviolet cutoff is removed. But the correlation functions when expressed as power series in the renormalized coupling constant have perfectly finite ultraviolet limits order by order. This limit is the sum of the renormalized Feynman amplitudes given by the forest formulas. But in addition to the usual divergence of perturbation theory due to the large number of diagrams this perturbative renormalization theory suffers from a new non-perturbative disease, the renormalons generated by the anomalously large amplitudes of some families of graphs such as those of Figure 8.

How does this change with RG? RG tells us that we should neither use one nor two sets of coupling constants, but an infinite set, one for each scale. All these "running constants" are uniquely related to any one of them because they must lie on a single RG trajectory.

Clearly the RG philosophy means that we should neither compute the correlation functions as series in the bare coupling with diverging coefficients in the ultraviolet limit nor as renormalon-ill series in the renormalized coupling. We should compute them as multi-series in the infinite set of running constants.

Once this big change is accepted, everything falls into place.

The momentum slicing becomes the fundamental tool. The Feynman amplitudes are sliced into "assignments" $\mu = \{i_l\}$ with a slice index i_l for each line. There is also a vertex index i_v for each vertex, namely the highest line index flowing into that vertex. It is a natural convention to consider the true external lines of the graph as having index below all others, for instance here index -1. Then the amplitude for a graph is no longer proportional to the power of a single coupling but each vertex should be equipped with a running constant g_{i_v} corresponding to its scale in the assignment.

In this way we obtain the "effective expansion" for a given Schwinger function [35, 9]

$$S_N = \sum_{\substack{\phi^4 \text{ graphs } G \text{ with } N(G)=N \\ G \text{ without any vacuum subgraph}}} \sum_{\mu=\{i_l\}} \frac{1}{S(G)} [\prod_{v \in G} g_{i_v}] A^{R,eff}_{G,\mu}, \qquad (3.10)$$

where the effectively renormalized amplitude $A^{R,eff}_{G,\mu}$ contains only one subtraction packet, the one associated to $\mathcal{F}_0 = \emptyset$. More precisely the graph G and the assignment μ uniquely define a single "divergent forest" $\mathcal{F}(G, \mu)$ which is made of those divergent subgraphs in G for which the indices of internal lines are all greater than the indices of external lines. Then (for instance in the parametric representation)

$$A^{R,eff}_{G,\mu} = \int d\alpha \; [\prod_{S \in \mathcal{F}(G,\mu)} (1 - \mathcal{T}_S)] \; I_{G\mu}(\alpha). \qquad (3.11)$$

The Schwinger functions in this "effective expansion" are made of course of exactly the same pieces as the bare or the renormalized expansion. These pieces are simply reshuffled in a different way. Indeed in the effective expansion the subtractions associated to the additional packets responsible for all the complications of Zimmermann's formula have simply disappeared, exactly reabsorbed into the effective constants that equip the vertices. Since these packets were responsible for the renormalons, it is not surprising that the expansion (3.10) is free of the renormalon problem, as expressed by our next Lemma.

Remark that the subgraphs in $\mathcal{F}(G, \mu)$ are indeed exactly those divergent subgraphs which have short spatial scale compared to their external lines. Distances between internal vertices are then shorter than the typical oscillation lengths of the external legs. Since these legs are like sensors through which the subgraph communicates with the external world, subgraphs in $\mathcal{F}(G, \mu)$ look "quasi-point-like" when seen from the outside. It is therefore no surprise that subtracting a truly local counterterm for each such "quasi-local" subgraph, which is what $(1 - \mathcal{T}_S)$ does, leaves only a small remainder free of renormalons. More precisely one can prove (putting all external momenta to 0 to simplify):

Lemma 3.1 *There exists a constant K such that for any G*

$$\sum_{\mu} |A^{R,eff}_{G,\mu}| \leq K^{n(G)} \qquad (3.12)$$

One can conclude that although in the bare series the amplitudes were not subtracted at all, in the renormalized series they were subtracted too much because lots of useless forests gave rise to renormalons. By abandoning the idea of a single coupling constant, the effective expansion which lies between the bare and renormalized ones has exactly the right amount of subtractions, creating only small contributions.

Of course the attentive reader may object that the lemma has not too much meaning, because each piece $A^{R,eff}_{G,\mu}$ should be multiplied by a different factor $\prod_{v\in G} g_{i_v}$ before being summed over μ in the effective expansion. But let us suppose that all the running constants g_i remain bounded. In this case it is clear that the effective expansion is much better than the renormalized one from the point of view of resummation, since only the usual divergence linked to the large number of graphs remains. And bounded running constants are not uncommon: they occur in asymptotically free theories.

3.3 Asymptotic Freedom

In a just renormalizable theory like ϕ^4_4 the most interesting flow under the renormalization group is the one of the coupling constant. By a simple second order computation this flow is intimately linked to the sign of the graph G_0 of Figure 3. More precisely, we find that at second order the relation between g_i and g_{i-1} is

$$g_{i-1} \simeq g_i - \beta g_i^2 \tag{3.13}$$

(remember the minus sign in the exponential of the action), where β is a constant, namely the asymptotic value of $\sum_{j,j'/\,\mathrm{inf}(j,j')=i} \int d^4y C_j(x,y) C_{j'}(x,y)$ when $i \to \infty$. Clearly this constant is positive. So for the normal stable ϕ^4_4 theory, the relation (3.13) inverts into

$$g_i \simeq g_{i-1} + \beta g_{i-1}^2, \tag{3.14}$$

so that fixing the renormalized coupling seems to lead to a large, diverging bare coupling, incompatible with perturbation theory. This is the famous "Landau ghost" problem.

But in non-Abelian gauge theories an extra minus sign is created by the algebra of the Lie brackets. This surprising discovery has deep consequences. The flow relation becomes approximately

$$g_i \simeq g_{i-1} - \beta g_i g_{i-1}, \tag{3.15}$$

with $\beta > 0$, or, dividing by $g_i g_{i-1}$,

$$1/g_i \simeq 1/g_{i-1} + \beta, \tag{3.16}$$

with solution $g_i \simeq \frac{g_0}{1+g_0\beta i}$. A more precise computation to third order in fact leads to

$$g_i \simeq \frac{g_0}{1 + g_0(\beta i + \gamma \log i + O(1))}. \tag{3.17}$$

Such a theory is called asymptotically free (in the ultraviolet limit) because the effective coupling tends to 0 with the cutoff for a finite fixed small renormalized coupling. Physically the interaction is turned off at small distances. This theory is in agreement with scattering experiments which see a collection of almost free particles (quarks and gluons) inside the hadrons at very high energy. This was the main initial argument to adopt quantum chromodynamics, a non-Abelian gauge theory with $SU(3)$ gauge group, as the theory of strong interactions.

Remark that in such asymptotically free theories the flow and all running constants remain bounded (in fact by the renormalized coupling). The initial expectations that infinite Feynman diagrams should lead to infinite bare parameters are clearly wrong in this case since the bare coupling constant in fact tends to 0 with the ultraviolet cutoff!

Asymptotic freedom is not limited to the rather complicated non-Abelian gauge theories. As is well known, fermion diagrams have an extra minus sign per loop. The Gross–Neveu theory, a theory with quartic coupling and N species of Fermions in two dimensions, has the same power counting as ϕ^4_4, and is also asymptotically free in the ultraviolet limit. This is also the case for instance for the ϕ^4_4 theory with "wrong sign" of the coupling constant, which can be studied at least in the planar limit, which tames the natural instability due to that wrong sign. The "right sign" ϕ^4_4 is not asymptotically free in the ultraviolet but as a consequence it is asymptotically free in the infrared, which means that the corresponding massless critical theory (with fixed ultraviolet cutoff) is almost Gaussian in the long distance limit [36].

3.4 Some Comments on Constructive Renormalization

Constructive field theory has for ambitious goal to define the non-perturbative mathematically correct version of Lagrangian quantum field theory. This may be considered somewhat an academic problem for weakly coupled theories such as quantum electrodynamics, for which perturbative computations up to three loops seem sufficient. But there are strongly coupled theories such as quantum chromodynamics in which a non-perturbative approach is badly needed. Also it would be quite surprising if the patient analysis of the mathematical difficulties related to the summation of quantum perturbation theory did not lead to important new physical insights. After all the difficulties in resumming classical perturbation theory were very important for the modern understanding of dynamical systems [8].

For reviews of constructive theory we refer to [2, 9, 38, 39]. But here let us sketch how the RG has to be modified to become truly a non-perturbative tool, and review briefly the achievements of the theory.

The first difficulty if we try to resum perturbation theory has to do with the large number of Feynman graphs. Convergence of the functional integral itself, and the divergence of perturbation theory can be considered as "large field" problems, because they are related to the fact that a bosonic field is an unbounded variable. Physically a large field corresponds to a large number of excitations or particles being produced, and large field problems are generic in bosonic theories because bosons, in contrast with fermions, can pile up in large numbers at the same place.

In Fermionic theories the Pauli principle physically solves that problem: fermions cannot pile up at the same place. Mathematically the corresponding anticommuting functional integrals give rise to determinants. By Gram or Hadamard's inequalities an n by n determinant with elements bounded by 1 can never be of size $n!$ but at most $n^{n/2}$, so that fermionic perturbation theory converges, in sharp contrast with bosonic perturbation theory.

Clearly the RG as initially formulated by Wilson or summarized in (3.6)-(3.7) is not mathematically well-defined. In particular starting from any polynomial action it creates an effective action which is obviously no longer polynomial, and this even after a single step! Therefore the large field problem (integration on ϕ at large ϕ), appears! More precisely, even if the initial bare action is stable, i.e., bounded below, it is not clear that this remains true for $S_{eff}(\phi)$, even after a single RG step. Hence starting from a stable interaction, the second step of the RG may be already ill-defined. This point has to be stressed to physicists!

So constructive theory must modify carefully the two main operations in a RG step to make them well defined. The functional integral in a slice must be treated (in the bosonic case at least) with a tool called a *cluster expansion.* The idea of the cluster expansion is that since perturbation theory diverges we must keep most of it in the form of functional integrals. However one can test whether distant regions of space are joined or not by propagators. So one introduces a lattice of cubes of size comparable to the decay rate of the propagator (here M^{-i}) and one performs a battery of tests to know whether there are vertices or sources in different cubes joined by a propagator. This allows to rewrite the theory as a "polymer gas", the polymers being the sets of cubes joined together as the outcome of the cluster expansion. By construction this polymer gas has hardcore interactions: two connected components are always made of disjoint cubes. But when the coupling constant is small, the activities for the non-trivial polymers (containing more than one cube) are small. Hence the polymer gas is dilute and the statistical mechanics technique of the Mayer expansion, a tool which compares the hardcore gas to a perfect gas, allows to perform the thermodynamic limit. This Mayer expansion is the non-perturbative analog of the computation of the logarithm in the second part of a renormalization group step. In this way the renormalization group can be formulated correctly at the non-perturbative level, as a sequence of intertwinned cluster and Mayer expansions, and the flow of the critical parameters to renormalize, such as the mass, wave function and coupling constant can be computed in this framework.

Using this approach, models of non-trivial interacting field theories have been built over the past thirty years, which satisfy Osterwalder–Schrader's axioms, hence in turn have a continuation to Minkowski space that satisfies Wightman axioms [40, 41]. Such models are unfortunately yet restricted to space-time dimensions 2 or 3 but they include now both superrenormalizable models, such as $P(\phi)_2$ [43, 42, 44], ϕ^4_3 [45, 46, 47, 14] or the Yukawa model in 2 and 3 dimensions, as well as just renormalizable models such as the massive Gross–Neveu model in two dimensions [48, 49]. Most of these models have been built in the weak coupling regime, using expansions such as the cluster and Mayer expansions; the harder models require multiscale versions of these expansions, reshuffled according to the renormalization group philosophy.

In most cases the relationship of the non-perturbative construction to the perturbative one has been elucidated, the non-perturbative Green's functions being the Borel sum of the corresponding perturbative expansion [13, 14, 49]. In this sense one can say that in such cases constructive field theory has achieved the goal of resumming all Feynman graphs, although, as explained above, Borel resummation is not a naive ordinary summation but a clever reshuffling of the initial perturbative series.

Unfortunately constructing ϕ^4_4 itself, the initial goal of the constructive program has not been possible since it lacks ultraviolet asymptotic freedom. It has been possible to show numerically and through correlation inequalities that starting from a bare lattice action at short distance with some reasonable assumptions at short distance, the resulting theory is trivial i.e. not interacting [50, 51, 52].

But important partial results have been obtained for the construction of non-Abelian theories in 4 dimensions [53, 54]. New models not perturbatively renormalizable but asymptotically safe are also within reach of these techniques, such as the Gross–Neveu model in three dimensions [55]. In the infrared regime bosonic models of renormalizable power counting such as the critical (massless) ϕ^4_4 with an infrared cutoff [56, 57], or 4 dimensional weakly self-avoiding polymers have been controlled [58], and their asymptotics at large distance have been established. Nonperturbative mass generation has been established in the Gross–Neveu model in two dimensions and in the nonlinear σ model at large number of components with ultraviolet cutoff [59, 60]. Finally the RG when applied to condensed matter give rise to many rigorous results and programs, as sketched in the next section. Altogether this set of results strongly illustrate the power of functional integration in quantum field theory.

3.5 Extended singularities, the new RG frontier

During the last decade one of the main achievements in renormalization theory is the extension of the renormalization group of Wilson (which analyzes long-range behavior governed by simple scaling around the point singularity $p = 0$ in momentum space) to more general extended singularities [61, 62, 63]. This very natural and general idea is susceptible of many applications in various domains, including condensed matter and field theory in Minkowski space. In this section we will discuss the situation for interacting Fermions models such as those describing the conduction electrons in a metal.

The key features which differentiate electrons in condensed matter from Euclidean field theory, and makes the subject in a way mathematically richer, is that space-time rotation invariance is broken, and that particle density is finite. This finite density of particles creates the Fermi sea: particles fill states up to an energy level called the Fermi surface.

The field theory formalism is the best tool to isolate fundamental issues such as the existence of non-perturbative effects. In this formalism the usual Hamiltonian point of view with operators creating electrons or holes is replaced by anticommuting Fermion fields with two spin indices, and propagator

$$C_{ab}(k) = \delta_{ab} \frac{1}{ik_0 - [\epsilon(\vec{k}) - \mu]} \tag{3.18}$$

where $a, b \in \{1, 2\}$ are the spin indices. The momentum vector $\vec{k}$ has d spatial dimensions. and $\epsilon(\vec{k})$ is the energy for a single electron of momentum $\vec{k}$. The parameter μ corresponds to the chemical potential. The (spatial) Fermi surface is the manifold $\epsilon(\vec{k}) = \mu$ [5].

For a jellium isotropic model the energy function is invariant under spatial rotations

$$\epsilon(\vec{k}) = \frac{\vec{k}^2}{2m} \tag{3.19}$$

where m is some effective or "dressed" electron mass. In this case the Fermi surface is simply a sphere. This jellium isotropic model is realistic in the limit of weak electron densities, where the Fermi surface becomes approximately spherical. In general a propagator with a more complicated energy function $e(\vec{k})$ has to be considered. A very interesting case is the two dimensional Hubbard model corresponding to a square lattice. The momenta live on the dual "Brillouin zone" $[-\pi, \pi]^2$, and the energy function is

$$\epsilon(\vec{k}) = \cos k_1 + \cos k_2 \tag{3.20}$$

so that for $\mu = 0$ (the so-called half-filled model), the Fermi surface is a square.

Imaginary (Euclidean) time (in the form of a circle, with antiperiodic boundary conditions for Fermions) corresponds to finite temperature T. When T tends to 0, the imaginary time circle grows to $\mathbb{R}$. At finite temperature, since Fermionic fields have to satisfy antiperiodic boundary conditions, the component k_0 in (3.18) can take only discrete values (called the Matsubara frequencies) :

$$k_0 = \pm \frac{2n+1}{\beta\hbar} \pi \tag{3.21}$$

so the integral over k_0 is really a discrete sum over n. For any n we have $k_0 \neq 0$, so that the denominator in $C(k)$ can never be 0. This is why the temperature provides a natural infrared cut-off. But when $T \to 0$, k_0 becomes a continuous variable and the propagator diverges on the "space-time" Fermi surface, defined by $k_0 = 0$ and $\epsilon(\vec{k}) = \mu$.

[5] It may be convenient to add also an ultraviolet cut-off to this propagator to make its Fourier transformed kernel in position space well defined. Anyway, very high momenta should be suppressed in this non relativistic theory.

The interaction term is defined by:

$$S_\Lambda = \frac{g}{2}\int_\Lambda d^3x \, (\sum_a \bar\psi\psi)^2(x) \ . \tag{3.22}$$

Physically this interaction represents an effective interaction due to phonons or other effects. A more realistic interaction would not be completely local to include the short range nature of the phonon propagator, but we can consider the local action (3.22) as an idealization which captures all essential mathematical difficulties.

The basic new feature is that the singularity of the propagator is of codimension 2 in the $d+1$ dimensional space-time. This changes dramatically the power counting of the theory. Instead of changing with dimension, like in ordinary field theory, perturbative power counting is now independent of the dimension, and is the one of a just renormalizable theory. Indeed in a graph with 4 external legs, there are n vertices, $2n-2$ internal lines and $L = n-1$ independent loops. Each independent loop momentum gives rise to two transverse variables, for instance k_0 and $|\vec{k}|$ in the jellium case, and to $d-1$ inessential bounded angular variables. Hence the $2L = 2(n-1)$ dimensions of integration for the loop momenta exactly balance the $2n-2$ singularities of the internal propagators, as is the case in a just renormalizable theory.

In one spatial dimension, hence two space-time dimensions, the Fermi surface reduces to two points, and there is also no proper BCS theory since there is no continuous symmetry breaking in two dimensions (by the "Mermin–Wagner theorem"). Nevertheless the many Fermion system in one spatial dimension gives rise to an interesting non-trivial behavior, called the Luttinger liquid [61].

In two spatial dimensions or more, the key tool to correctly analyze the theory is a decomposition of the propagator analogous to (3.1), but both into discrete slices and in each slice into discrete angular sectors. The slices are defined by:

$$C = \sum_{j=1}^{\infty} C_j \quad ; \quad C_j(k) = \frac{f_j(k)}{ik_0 - e(\vec{k})} \tag{3.23}$$

where the slice function $f_j(k)$ effectively forces $|ik_0 - e(\vec{k})| \sim M^{-j}$, for some fixed parameter $M > 1$. These slices pinch more and more the Fermi surface as $j \to \infty$.

The slice propagator is further decomposed into sectors:

$$C_{(j)}(k) = \sum_{\sigma\in\Sigma_j} C_{j,\sigma}(k) \quad ; \quad C_{j,\sigma}(k) = \frac{f_{j,\sigma}(k)}{ik_0 - e(\vec{k})} \tag{3.24}$$

where Σ_j is a set of angular patches, called sectors, which cover the Fermi sphere. For instance if $d = 2$ we may simply cut the circle into M^j intervals of length $2\pi M^{-j}$, but a better idea is to make the patches as large as possible. What limits really the size of the patches is the curvature of the Fermi surface, so that the optimal number of such patches is really $M^{j/2}$ for the two dimensional jellium model [64], and only j^2 for the two dimensional Hubbard model at half-filling [65].

The RG applied to this problem means as before that higher slices give rise to local effects relatively to lower slices. Integrating the higher slices one obtains effective actions which govern larger distance physics. These effective actions are however more complicated than in the field theory context. In rotation invariant models, renormalization of the two point function can be absorbed in a change of normalization of the Fermi radius. It removes all infinities from perturbation theory at generic momenta [62]. But the flow for the four point function is a flow for an infinite set of coupling constants describing the momentum zero channel of the Cooper pairs [63]. In the case of an attractive interaction, when the temperature is lowered to zero, this flow diverges at the BCS scale. At this scale the symmetry linked to particle number conservation is spontaneously broken, giving rise to superconductivity, that is to the condensation of Cooper pairs.

This condensation is a nonperturbative phenomenon, like quark confinement. But in contrast with quark confinement, we know in principle how to investigate in a mathematically rigorous way this BCS condensation. Indeed sectors around the Fermi surface play a role analogous to components of a vector field, so that an expansion in $1/N$, where N is the number of such components, can control the BCS regime [66], in which ordinary perturbation is no longer valid. We may call this situation a "dynamical $1/N$" effect. Nevertheless the full mathematical construction of the BCS transition starting from weakly interacting fermions remains a long and difficult program which requires to combine together several ingredients.

The discussion of high temperature superconductivity lead also to some controversy about the nature of interacting fermions systems in the ordinary non-superconducting phase. In particular, validity of the standard Fermi liquid theory (which is essentially defined by the propagator (3.18) up to small corrections) has been questioned in two dimensions. According to a mathematical criterion designed by M. Salmhofer [69], it is now possible to distinguish rigorously between the so-called Fermi liquid behavior and Luttinger liquid behavior above the usual critical BCS temperature. Using renormalization group around the Fermi surface it should be possible to soon complete the proof of the following theorem:

Theorem 3.2 *In two dimensions an interacting fermion system above the condensation temperature can be either a Fermi or a Luttinger liquid, depending on the shape of the Fermi surface. The jellium model with round Fermi surface is a (slightly anomalous) Fermi liquid [68], but the half-filled Hubbard model with a square Fermi surface should be a (slightly anomalous) Luttinger liquid [65].*

The mathematically rigorous construction of a two-dimensional interacting Fermi liquid at zero temperature, corresponding to non-parity invariant Fermi surfaces like those obtained by switching on a generic "magnetic field cutoff", has also been completed recently [70].

Like in the previous section the key to these constructive theorems lies in the resummation of perturbation theory in a single slice, and then in the iteration of renormalization group steps. Curiously, although power counting does not depend on the dimension, momentum conservation in terms of sectors in a fixed slice depends on it. This has dramatic constructive consequences. In $d = 2$ we have the "rhombus rule": four momenta of equal length which add to zero at a given vertex must be roughly two by two parallel. This means that two dimensional condensed matter in a slice is again directly analogous to an N-vector model in which angles on the Fermi surface play the role of colors [67]. This remark is at the core of all rigorous constructions of interacting Fermi liquids [68, 70].

In three dimensions, we expect interacting fermions to behave as regular Fermi liquid above the BCS temperature, but this turns out to be surprisingly difficult to prove non-perturbatively. Indeed there is no longer any analog of the "rhombus rule". Two different momenta at a vertex in a given slice no longer determine the third and fourth: there is an additional torsion angle, since four momenta of same length adding to 0 are not necessarily coplanar. More sophisticated techniques have been designed to deal with this case [71] but until now it is not clear that these techniques allow a full constructive analysis of the model up to the scale where the BCS symmetry breaking takes place.

3.6 Conclusion

If we consider the universal character of the action principle both at the classical and quantum level, and observe that the relation between microscopic and macroscopic laws is perhaps the most central of all physical questions, it is probably not an exaggeration to conclude that the renormalization group is in some deep sense the "soul" of physics.

3.7 Acknowledgments

I thank B. Duplantier, C. Kopper and J. Magnen for their critical reading of the manuscript.

References

[1] M. Peskin and Daniel V. Schroeder (Contributor), *An Introduction to Quantum Field Theory*, Perseus Publishing, 1995.

[2] J. Glimm and A. Jaffe, Quantum Physics. A functional integral point of view. Mc Graw and Hill, New York, 1981.

[3] C. Itzykson and J.-B. Zuber, Quantum Field Theory, McGraw Hill, 1980.

[4] P. Ramond, Field Theory, Addison-Wesley, 1994.

[5] J. Zinn-Justin, Quantum Field Theory and Critical Phenomena, Oxford University Press, 2002.

[6] C. Itzykson and J.M Drouffe, Statistical Field Theory, Volumes 1 and 2, Cambridge University Press 1991.

[7] Giorgio Parisi, Statistical Field Theory, Perseus Publishing 1998.

[8] G. Benfatto, G. Gallavotti, Renormalization Group, Princeton University Press, 1995.

[9] V. Rivasseau, From Perturbative to Constructive Renormalization, Princeton University Press, 1991.

[10] Manfred Salmhofer, Renormalization: An Introduction, Texts and Monographs in Physics, Springer Verlag, 1999.

[11] R. Feynman and A. Hibbs, Quantum Mechanics and Path Integrals, Mc Graw and Hill, New York 1965.

[12] A. Sokal, An improvement of Watson's theorem on Borel summability, Journ. Math. Phys. **21**, 261 (1980).

[13] J.P. Eckmann, J. Magnen and R. Sénéor, Decay properties and Borel summability for the Schwinger functions in $P(\phi)_2$ theories, *Comm. Math. Phys.* **39**, 251 (1975).

[14] J. Magnen and R. Sénéor, Phase space cell expansion and Borel summability for the Euclidean ϕ_3^4 theory, *Comm Math. Phys.* **56**, 237 (1977).

[15] N. Nakanishi, Graph Theory and Feynman integrals, Gordon and Breach, New York 1971.

[16] K. Symanzik, in Local Quantum Theory, R. Jost, ed. (Varenna, 1968) Academic Press, New York (1969), 285.

[17] S. F. Edwards, *Proc. phys. Soc. Lond.* **85**, 613 (1965).

[18] P.-G. de Gennes, *Phys. Lett.* **38 A**, 339 (1972).

[19] J. des Cloizeaux, *J. Phys. France* **42**, 635 (1981).

[20] J. des Cloizeaux and G. Jannink, Polymers in Solution, their Modeling and Structure, Clarendon Press, Oxford (1990).

[21] N. Bogoliubov and Parasiuk, *Acta Math.* **97**, 227 (1957).

[22] K. Hepp, Théorie de la renormalisation, Berlin, Springer Verlag, 1969.

[23] W. Zimmermann, Convergence of Bogoliubov's method for renormalization in momentum space, *Comm. Math. Phys.* **15**, 208 (1969).

[24] M. Bergère and Y.M.P. Lam, Bogoliubov-Parasiuk theorem in the α-parametric representation, *Journ. Math. Phys.* **17**, 1546 (1976).

[25] M. Bergère and J.B. Zuber, Renormalization of Feynman amplitudes and parametric integral representation, *Comm. Math. Phys.* **35**, 113 (1974).

[26] C. de Calan and V. Rivasseau, Local existence of the Borel transform in Euclidean ϕ_4^4, *Comm. Math. Phys.* **82**, 69 (1981).

[27] G. 'tHooft and M. Veltman, *Nucl. Phys.* **B50**, 318 (1972).

[28] M. Benhamou and G. Mahoux, *J. Phys. France* **47** (1986) 559.

[29] F. David, B. Duplantier and E. Guitter, *Nucl. Phys.* **B394** (1993) 555-664.

[30] A. Connes and D. Kreimer, Renormalization in Quantum Field Theory and the Riemann-Hilbert problem, *J. High Energy Phys.* **09**, 024 (1999).

[31] A. Connes and D. Kreimer, Hopf algebras, renormalization and non commutative geometry, *Comm. Math. Phys.* **199**, 203 (1999).

[32] C. Callan, *Phys. Rev D.* **2**, 1541 (1970).

[33] K. Symanzik, Small distance behaviour in field theory and power counting, *Comm. Math. Phys.* **18**, 227 (1970).

[34] J. Polchinski, *Nucl. Phys. B* **231**, 269 (1984).

[35] V. Rivasseau, Construction and Borel summability of a planar 4 dimensional field theory, *Commun. Math.Phys.* **95**, 445 (1984).

[36] K. Wilson, Renormalization group and critical phenomena, II Phase space cell analysis of critical behavior, *Phys. Rev. B* **4**, 3184 (1974).

[37] G. Battle, Wavelets and Renormalization, World Scientific, 1999.

[38] Constructive Physics, Lecture Notes in Physics 446, Springer Verlag, 1995.

[39] V. Rivasseau, Constructive Field Theory and Applications: Perspectives and Open Problems, *Journ. Math. Phys.* **41**, 3764 (2000).

[40] A. Wightman, Quantum field theory in terms of vacuum expectation values, *Phys. Rev.* **101**, 860 (1956).

[41] K. Osterwalder and R. Schrader, Axioms for Euclidean Green's functions, *Comm. Math. Phys.* **31**, 83 (1973).

[42] J. Glimm and A. Jaffe, Quantum Physics. A functional integral point of view, Mc Graw and Hill, New York, 1981.

[43] J. Glimm, A. Jaffe and T. Spencer, The Wightman Axioms and Particle Structure in the $P(\phi)_2$ Quantum Field Model, *Ann. Math.* **100**, 585 (1974).

[44] B. Simon, The $P(\phi)_2$ (Euclidean) Quantum field theory, Princeton University Press, 1974.

[45] J. Glimm and A. Jaffe, Positivity of the ϕ_3^4 Hamiltonian, *Fortschr. Phys.* **21**, 327 (1973).

[46] J. Feldman, The $\lambda\phi_3^4$ field theory in a finite volume, *Comm. Math. Phys.* **37**, 93 (1974).

[47] J. Feldman and K. Osterwalder, The Wightman axioms and the mass gap for weakly coupled ϕ_3^4 quantum field theories, *Ann. Phys.* **97**, 80 (1976).

[48] K. Gawedzki and A. Kupiainen, Gross-Neveu model through convergent perturbation expansions, *Comm. Math. Phys.* **102**, 1 (1985).

[49] J. Feldman, J. Magnen, V. Rivasseau and R. Sénéor, A renormalizable field theory: the massive Gross-Neveu model in two dimensions, *Comm. Math. Phys.* **103**, 67 (1986).

[50] M. Aizenman, Geometric Analysis of ϕ^4 fields and Ising Models, *Comm. Math. Phys.* **86**, 1 (1982).

[51] J. Fröhlich. On the triviality of $\lambda\phi_d^4$ theories and the approach to the critical point in $d \geq 4$ dimensions, *Nucl. Phys. B* **200** FS4, 281 (1982).

[52] Roberto Fernandez, Jurg Fröhlich and Alan D. Sokal, Random Walks, Critical Phenomena, and Triviality in Quantum Field Theory, Springer Verlag, 1992.

[53] T. Balaban, *Comm. Math. Phys.* **109**, 249 (1987), **119**, 243 (1988), **122**, 175 (1989), **122**, 355 (1989) and references therein.

[54] J. Feldman, J. Magnen, V. Rivasseau and R. Sénéor, Construction of YM_4 with an infrared cutoff, *Commun. Math. Phys.* **155**, 325 (1993).

[55] P. Faria da Veiga, Construction de modèles non-perturbativement renormalisables en théorie quantique des champs, Thèse, Université de Paris XI, 1992.

[56] K. Gawedzki and A. Kupiainen, Massless ϕ_4^4 theory: Rigorous control of a renormalizable asymptotically free model, *Comm. Math. Phys.* **99**, 197 (1985).

[57] J. Feldman, J. Magnen, V. Rivasseau and R. Sénéor, Construction of infrared ϕ_4^4 by a phase space expansion, *Comm. Math. Phys.* **109**, 437 (1987).

[58] D. Iagolnitzer and J. Magnen, Polymers in a weak random potential in dimension 4: Rigorous Renormalization Group Analysis, *Commun. Math. Phys.* **162**, 85 (1994).

[59] C. Kopper, J. Magnen and V. Rivasseau, Mass Generation in the Large N Gross-Neveu Model, *Comm. Math. Phys.* **169** 121 (1995).

[60] C. Kopper, Mass Generation in the Large N Non-linear σ Model, to appear in *Comm. Math. Phys.* (1999).

[61] G. Benfatto and G. Gallavotti Perturbation theory of the Fermi surface in a quantum liquid. A general quasi-particle formalism and one dimensional systems, *Journ. Stat. Physics* **59**, 541, 1990.

[62] J. Feldman and E. Trubowitz, Perturbation Theory for Many Fermions Systems, *Helv. Phys. Acta* **63**, 156 (1990).

[63] J. Feldman and E. Trubowitz, The Flow of an Electron-Phonon System to the Superconducting State, *Helv. Phys. Acta* **64**, 213 (1991).

[64] J. Feldman, J. Magnen, V. Rivasseau and E. Trubowitz, An Infinite Volume Expansion for Many Fermion Green's functions, *Helv. Phys. Acta* **65**, 679 (1992).

[65] V. Rivasseau, The two dimensional Hubbard Model at half-filling: I.Convergent Contributions, *Journ. Stat. Phys.* **106**, 693 (2002).

[66] J. Feldman, J. Magnen, V. Rivasseau and E. Trubowitz, An Intrinsic 1/N Expansion for Many Fermion Systems, *Europhys. Letters* **24**, 437 (1993).

[67] J. Feldman, J. Magnen, V. Rivasseau and E. Trubowitz, Two dimensional Many Fermion Systems as Vector Models, *Europhys. Letters* **24**, 521 (1993).

[68] M. Disertori, V. Rivasseau, A Rigorous Proof of Fermi Liquid Behavior for Jellium Two-Dimensional Interacting Fermions, *Phys. Rev. Lett.* **85**, 361 (2000).

[69] M. Salmhofer, Continuous renormalization for fermions and Fermi liquid theory, *Comm. Math. Phys.* **194**, 249 (1998).

[70] J. Feldman, H. Knörrer and E. Trubowitz, A Two Dimensional Fermi Liquid, an impressive series of ten papers, all accessible at http://www.math.ubc.ca/~feldman/fl.html.

[71] M. Disertori, J. Magnen and V. Rivasseau, Interacting Fermi liquid in three dimensions at finite temperature: Part I: Convergent Contributions, *Annales Henri Poincaré* **2**, 733–806 (2001).

Vincent Rivasseau
Laboratoire de Physique Théorique
Université Paris XI
F-91405 Orsay Cedex

Poincaré Seminar 2002, 179 – 212

Poincaré Seminar 2002

Exact Renormalization Group

Giovanni Gallavotti

Abstract. Renormalization group methods are illustrated via examples.

Introduction

In theoretical physics are problems in which singular quantities appear in the basic equations, or problems in which the basic equations are apparently free of singularities but their solutions are nevertheless singular are among the most interesting. For instance

(a) The Coulomb potential is singular at contact (*i.e.* at zero distance: an "ultraviolet singularity")
(b) The electric potential is singular also at infinite distance because it decays too slowly to zero (an "infrared singularity").
(c) The gravitational potential is as singular.
(d) In an incompressible Euler fluid the singularity manifests itself because any nontrivial motion generates fluid velocity fields that involve phenomena observable on length and time scales as small as wished (again an ultraviolet singularity).
(e) In relativistic quantum field theory the relativistic covariance is implemented through a requirement of a "local interaction", which implies that physical quantities show interesting phenomena on all short length and time scales, again an ultraviolet singularity.
(f) In statistical mechanics even short range interactions generate phenomena that involve many long length scales (critical phenomena: an infrared singularity as it concerns large space scales).
(g) In mechanics systems able to oscillate with few frequencies generate, when interacting, motions in which all harmonics of the basic frequencies are present and in which some may become so important to change completely the behavior in comparison with the unperturbed one.

The list could continue for a while (to include Fermi liquids and superconductivity, Bose condensation and superfluids, for instance). Similar questions appear also in other fields of Science; an egregious example is the mathematical theory of Fourier transforms: a function can be reconstructed from its Fourier transform via a convolution with a singular kenel $K(x-y)$ (typically the Dirichlet kernel) which is singular at $x = y$. Therefore the convolution probes the behavior of the function on all scales and the question of the (pointwise) convergence of the Fourier series becomes an analysis of an ultraviolet singularity.

The reason for the success of the "renormalization group" is that it is a method that attempts to study problems of the above kinds from a *unified viewpoint*. It is remarkable that sometimes the attempts really solve the problems or, when the problem reamins open, at least provide new insights into it.

The method can be loosely defined as follows: one finds an explicit solution of the problem which, of course, involves quantities that one cannot really compute. A typical case is when the problem admits a perturbative solution, although this is not always the case: in any event the formal solution involves the analysis of singular integrals, *i.e.* of integrals involving functions with singularities.

The singular functions, we call them generically $C(x)$, are then expressed as sums of many very regular nonsingular functions each of which "lives on a fixed scale', which means that the regular "piece" of the singular function that lives on a scale l_0 is a smooth function of the form $l_0^\alpha \overline{C}(x\, l_0^{-1})$ where $\overline{C}$ is a smooth function rapidly decaying at infinity and α is a constant. The renormalization method is effective when the regular pieces into which $C(x)$ is decomposed differ only by the scale on which they live, possibly apart from a finite number of them. One says that in such cases the singlularity is "scale invariant" and the singularity is a power law behavior in $|x|$, and the exponent α is related to the value of the power. For instance this is for the Coulomb potential which can be written ("resolution of the *ultraviolet singularity*")

$$\frac{1}{|x|} = \frac{(1-e^{-|x|^2})}{|x|} + \sum_{k=1}^{\infty} \frac{(e^{-(\gamma^h|x|)^2} - e^{-(\gamma^{h+1}|x|)^2})}{|x|} = \frac{(1-e^{-|x|^2})}{|x|} + \sum_{k=1}^{\infty} \gamma^h \overline{C}(\gamma^h x)$$

where $\overline{C}(x) = \frac{e^{-|x|^2} - e^{-(\gamma|x|)^2}}{|x|}$; here $l_0 = 1$ and $\gamma > 1$ can be (arbitrarily) taken to be $\gamma = 2$.[1] Then one tries to show that the formal solution can be studied by breaking it into a sum of terms that have all the same structure apart form a change of scale: by summing infinitely many functions which are regular on infinitely many scales one can construct (hopefully in a controllable way) the singularities and the properties of the quantities that are formally expressed in terms of the original singular (often *a priori* even possibly meaningless) integrals.

Breaking a singular expression into many (infinitely many) parts allows to disentangle and to exhibit delicate cancellation phenomena which may allow us to give a meaning too expressions that seem meaningless at first. Sometimes the cancellations may be so effective that the apparent singularities are in fact not there.

[1]The reason for this (almost universal) choice seems to be, according to G. Parisi, that it is the only choice which does not generate the question "why is γ chosen $= 2$?": the choice $\gamma = \pi$ would be equally good but it would inevitaly raise uninteresting questions.

A typical example is the KAM theory where in the end no singularities are really there. Another archetypal example of success of the approach is the quantum theory of fields in dimensions $d = 2, 3$.

Here I select the latter two examples as their similarity is striking. Both are problems that arise in systems that are perturbations of simple systems (integrable systems in the first case and free fields in the second). However the singularities do not allow us to proceed straightforwardly.

The interest of the problems and their physical relevance is well known. For instance the first arose historically from the famous remark by Poincaré that the pertubation analysis, used in astronomy since Laplace and crowned by the well known successes of the theory of precessions, compilation of ephemeral tables, discovery of asteroids (Ceres) or major planets (Neptune) ..., could not be an approximation in the naive sense of the term because strictly speaking the series used could not possibly be convergent. It surfaced again in Fermi's juvenile work on the equipartition problem and again in his last work (the "incompiuto" and postumously published Fermi, Pasta Ulam experiment) on the same matter; and many still recall the frustration felt in trying to understand seemingly simple problems such as the computation of the error in the small oscillations thoery or in motions that are small perturbations of simple integrable ones: like two point masses on a circle interacting via a small mutual potential and subject to an external small potential, *i.e.* a system described by two angles $\underline{\alpha} = (\alpha_1, \alpha_2) \in T^2 \equiv [0, 2\pi]^2$ and a total potential function $\varepsilon f(\underline{\alpha})$ with equation of motion

$$\ddot{\underline{\alpha}} = -\varepsilon \underline{\partial}_{\underline{\alpha}} f(\underline{\alpha})$$

The motions of assigned angular velocities $\underline{\omega}_0 = (\omega_1, \omega_2)$ of such a system exist and remain similar to the corresponding free motions if the equation (3.1) below has a solution, as an elementary check would allows us to see. An existence proof is, however, rich of conceptual difficulties.

The second problem addresses the basic question of the very possibility of existence of a quantum theory of interacting particles which is, at the same time, relativistically covariant.

The approach allows us to solve completely the first problem (Sec. 3–6) at least for a large class of motions (the "non resonant" ones). For the second problem in the cases of space-time models of dimension $d = 2, 3, 4$ even the existence of a formal perturbative solution is not clear: but it can be established by the renormalization group method (Sec. 8,9). In the cases $d = 2, 3$ the renormalization approach becomes the basis for completing the solution of the problem (*i.e.* to go beyond the formal level) and thus it is a fundamental building block of the proof that, *at least in dimensions $d = 2, 3$, quantum fields and special relativity are compatible even in presence of nontrivial interactions.*[2]

[2]In dimension $d = 4$ the problem is still open, although via the renormalization group method one can show the existence of a well defined perturbative solution.

We choose the above two problems because of the elegance of their solution and of their pedagogical value: however they were not originally solved by the method discussed here. There are a number of other problems which have been first solved via a renormalization group method of the type we considered in the present review: the critical point of various classes of statistical mechanics models or the theory of the ground state of one dimensional spinless Fermi systems, the theory of the convergence of Fourier series, dipole gases, Anderson localization to mention a few.

Much larger is the set of problems that have been studied only heuristically in the physics literature: a permanent challenge is to understand them fully.

We shall introduce the KAM problem and the field theory renormalization in $d = 2, 3$ for scalar fields (typical multiscale problems) by first discussing their single scale counterparts (Sec. 1,2,7): this should induce appreciation of the power of a method to reduce a multiscale problem to a single scale one.

Sometimes the problems that are studied at a heuristic level involve drastic and uncontrolled approximations: therefore many physicists consider important to gain some control on what one would like to neglect. For this reason the applications of the renormalization group in which the results are obtained without concessions to uncontrolled approximations are called "*exact renormalization group*" results while the others do not receive the qualification of "exact" even though they are considered "better" than the results of perturbation theory (when possible) which in the Physics literature seems to be regarded with undeserved contempt. They are called "*non perturbative*": a name well deserved because they usually are (considered) reliable and are certainly remarkably different from predictions obtained by naively truncating perturbation series. The latter fact is in itself a really non trivial achievement as those working on the subject before the work of Wilson, Fisher, Kadanoff, Jona–Di Castro immediately realized.

Consistently I try here to keep the exposition essential but complete and self contained; certain really technical details are in Sec. 6 and in the appendices. Commented references to the (immediately relevant) literature can be found in the final pages.

1 Nonsingular perturbation theory

Examples of perturbation analysis abound: the simplest are the "single scale" problems. These are problems in which no "singularities" appear and, as a consequence, the perturbation expansions converge, or are at least asymptotic, for small pertubations. An example is the following implicit functions equations

$$\underline{h}(\underline{\psi}) = \varepsilon \frac{\partial f}{\partial \underline{\alpha}}(\underline{\psi} + \underline{h}(\underline{\psi})) \tag{1.1}$$

where $\underline{\psi} \in T^\ell$ is a point on the ℓ–dimensional torus $T^\ell = [0, 2\pi]^\ell$ and $f(\underline{\alpha})$ is a trigonometric polynomial $f(\underline{\alpha}) = \sum_{|\underline{\nu}| \le N} e^{i\underline{\nu}\cdot\underline{\alpha}} f_{\underline{\nu}}$ of degree N, $|\underline{\nu}| \stackrel{def}{=} \sum |\nu_j|$. The problem posed is to show the existence of a solution $\underline{h}$ analytic for $|\varepsilon|$ small and in $\underline{\psi}$. It is not the simplest of its kind but it is general enough to be useful also in the case of harder problems.[3]

A second example of the same type is the functional integral

$$E_\lambda(\varepsilon f) = \frac{\int P(d\varphi) e^{-\int_\Lambda d^d\underline{x}\,(\lambda\varphi^4_{\underline{x}} + \mu\varphi^2_{\underline{x}} + \varepsilon\varphi_{\underline{x}} f(\underline{x}))}}{\int P(d\varphi) e^{-\int_\Lambda d^d\underline{x}\,(\lambda\varphi^4_{\underline{x}} + \mu\varphi^2_{\underline{x}})}} \tag{1.2}$$

where $f(\underline{x})$ is a generic smooth test function, P is the Gaussian probability distribution on R^d, $d = 2, 3$, with covariance

$$C(\underline{x} - \underline{y}) \stackrel{def}{=} \langle \varphi_{\underline{x}} \varphi_{\underline{y}} \rangle_P = \frac{1}{(2\pi)^d} \int e^{i\underline{p}\cdot(\underline{x}-\underline{y})} \frac{1}{(\underline{p}^2+1)^2} d^d\underline{p} \tag{1.3}$$

and Λ is a finite cubic box.

The problem is to show that E_λ is a smooth function of $\lambda, \mu, \varepsilon$ for *all* test functions f and for $\lambda \ge 0, \mu, \varepsilon$ small enough. The functional derivatives with respect to $\varepsilon f(\underline{x})$ of $\log E_\lambda(\varepsilon f)$ are called the "*Schwinger functions*" of the functional integral in (1.2).

2 Tree expansions. Cancellations

Here we illustrate a technique to study the above problems. The technique is called "renormalization method": the appropriateness of the name is made manifest by its applications to the less trivial problems that will be discussed after Sec. 3 below.

Consider equation (1.1): we write the solution as $\underline{h}(\underline{\psi}) = \varepsilon \underline{h}^{(1)}(\underline{\psi}) + \varepsilon^2 \underline{h}^{(2)}(\underline{\psi}) + \varepsilon^3 \underline{h}^{(3)}(\underline{\psi}) + \ldots$ and note that the coefficients $\underline{h}^{(k)}$ satisfy an equation like $\underline{h}^{(k)} = \left[\partial_{\underline{\alpha}} f(\underline{\psi} + \underline{h}(\underline{\psi}))\right]^{(k-1)}$ where $[\cdot]^{(k)}$ denotes the k-th Taylor coefficient of an expansion in powers of ε of the function inside the square brackets. Hence

$$\underline{h}^{(k)}(\underline{\psi}) = \left[\sum_{\underline{s}\ge 0} \frac{1}{\underline{s}!} \partial_{\underline{\psi}} \partial^{\underline{s}}_{\underline{\psi}} f(\underline{\psi}) \underline{h}^{\underline{s}}\right]^{(k-1)} =$$

$$= \sum_{\underline{s}\ge 0} \frac{1}{\underline{s}!} \sum_{\sum k_{ij} = k-1} \partial_{\underline{\psi}} \partial^{\underline{s}}_{\underline{\psi}} f(\underline{\psi}) \prod_{i=1}^{\ell} \prod_{j=1}^{s_i} h^{(k_{ij})} \tag{2.1}$$

where $\underline{s} = (s_1, \ldots, s_\ell)$ is a multi–index with $s_i \ge 0$ integer, and we define $\underline{s}! \stackrel{def}{=} \prod_i s_i!$, $\partial^{\underline{s}}_{\underline{\psi}} \equiv (\partial^{s_1}_{\psi_1} \partial^{s_2}_{\psi_2} \ldots)$; the indices of the components of $\underline{h}$ are contracted

[3]The simplest equation of the kind would be $h = \varepsilon f(h)$ with $h \in R$: *i.e.* a "zero" dimensional version of (1.1), which could be studied by the same methods that we discuss below but which is too simple for our purposes.

with the corresponding indices of the components of $\underline{\partial}^{\,s}_{\underline{\psi}}$ as usual in a Taylor expansion.

Clearly the expression (2.1) is intricate: however we can find a quick graphical representation for it: the function $\underline{h}^{(k)}$ will be represented by

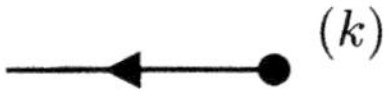

Figure 1: Representation of $\underline{h}^{(k)}$: adding a label $j = 1, \ldots, \ell$ on the line will indicate the j-th component of $\underline{h}^{(k)}$.

Therefore we shall represent the relation (2.1) as

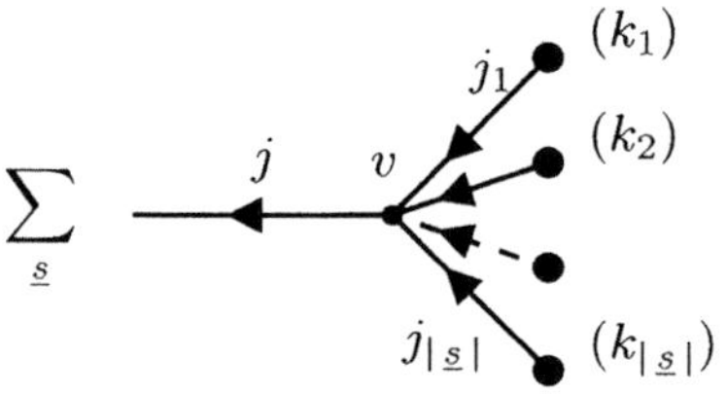

Figure 2: Representation of (2.1); here $k-1 = k_1 + \ldots + k_{|\underline{s}|}$. The "root" line carries the label j and the other lines carry the labels j_q corresponding to the components of $h^{(k_q)}_{j_q}$ for $q = 1, \ldots, |\underline{s}|$. The latter labels will not, in the following figures, continue to be marked (being dummy labels). The "node" v represents the derivatives $\partial_{\underline{\psi}} \partial^{\,\underline{s}}_{\underline{\psi}}$. Double counting is avoided by using the convention of assigning the label 1 to the first s_1 lines from top to bottom, the label 2 to the next s_2, and so on until all the $|\underline{s}|$ derivation label are considered

The recursive nature of (2.1) is clear and it is also quite clearly reflected in Fig. 2 above. The iteration of (2.1) will terminate in finitely many steps and the result can be naturally expressed graphically as in Fig. 3 below.

We imagine to draw the trees in Fig. 3 with lines of equal length and "coherently oriented" (*i.e.* the endpoint of an oriented line can only merge into the initial point of another oriented line) by assigning the line labels (not marked in the figures) as explained in the caption to Fig. 2: in this way one sees that the number of unlabeled distinct trees ϑ with k nodes does not exceed the number of closed $2k$–steps paths starting at the origin of a one dimensional lattice Z^1, *i.e.* it is $\le 2^{2k}$. A pair of consecutive (in the partial order fixed by the lines orientations) nodes $\lambda = (v'v)$ will also be called a "line" and it will also be denoted by λ_v if $v < v'$: the orientation of the lines allows us to say that a line follows another or that two lines are comparable, or that a node precedes another node or a line. On

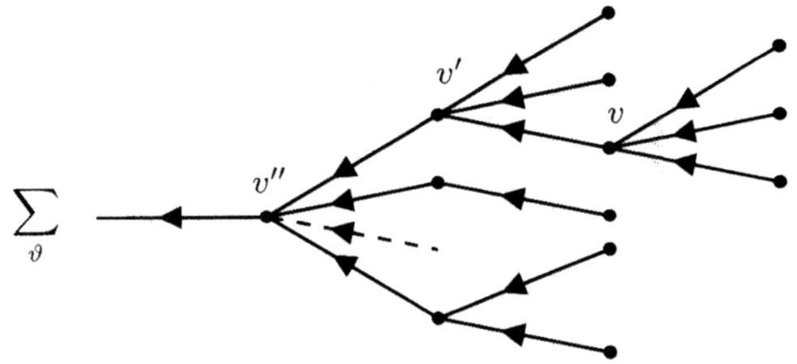

Figure 3: Representation of $\underline{h}^{(k)}$ as a sum of tree graphs ϑ with k nodes. Two pairs of consecutive nodes $v < v'$ and $v' < v''$ are also represented; the node v'' in the picture happens to be what we shall call the first or highest node of ϑ. With each tree graph ϑ a "value" $h_j(\vartheta)$ is assigned (see below) and the sum of the values of all trees with k nodes and root line bearing the label j yields $h_j^{(k)}$. A pair of consecutive nodes will be called a "line".

each line λ we attach a label $j_\lambda = 1, \ldots, \ell$ that we call a *component label.*

We call the highest line (*i.e.* the leftmost in the above figures) the "root" line but do not count the highest extreme of the highest line as a node and we may call it the "root": we shall call the number of nodes (hence of lines) k the "degree" $\deg(\vartheta)$ of ϑ.

Since the function $\underline{h}(\underline{\psi})$ is periodic it is convenient to look for its Fourier coefficients $\underline{h}_{\underline{\nu}}$ with $\underline{\nu} = (\nu_1, \ldots, \nu_\ell) \in Z^\ell$. A simple graphical representation can be given to the coefficients. It suffices to add to each node v of the trees in Fig. 3 a label $\underline{\nu}_v$, which we call a "node momentum", and to attribute to each line $\lambda = (v'v)$ a "line current" $\underline{\nu}(\lambda)$ defined as the sum of all the node momenta of the nodes $w < v$:

$$\underline{\nu}(\lambda) = \sum_{w<v} \underline{\nu}_w \tag{2.2}$$

Then if the root line start at the node denoted v_0 and carries a label j we define "*value*" of such labeled trees as

$$i(\underline{\nu}_{v_0})_j \prod_{\lambda\equiv(v'v)\in\vartheta} ((i\underline{\nu}_v)_{j_\lambda} i(\underline{\nu}_{v'})_{j_\lambda})) \prod_{v\in\vartheta} \frac{f_{\underline{\nu}_v}}{s_v!} \tag{2.3}$$

and we obtain $h_{j,\underline{\nu}}^{(k)}$ by summing over all trees with k nodes and with current $\underline{\nu}$ flowing on the highest line. In (2.3) we can imagine to have performed the summations over the labels j_λ assigned to the internal lines, thereby conveniently suppressing all of them except the label j on the root line, so that the summation of the values (2.3) is performed over the trees with k nodes v into which s_v lines merge and which carry a momentum label $\underline{\nu}_v$ over every node v.[4] Double counting is avoided by considering distinct any pair of trees that cannot be trivially superposed by pivoting the lines around the nodes into which they merge but *avoiding*, in the pivoting operations, line crossings.

However it will be convenient to remark that we may imagine that all the lines of such a tree of degree k bear a "*number label*" $1, \ldots, k$ (which will never be explicitly marked in the figures) that distinguishes them. Regard as different two trees that cannot be superposed (all labels included) by the operation of pivoting the lines around the nodes into which they merge while, this time, *allowing* crossing of the lines merging into the same node. Then we get many more trees and since we still require that the sum of the values of all trees is $h^{(k)}_{j\,\underline{\nu}}$ the definition of value has to be modified to avoid double countings into

$$\mathrm{Val}(\vartheta) = \frac{i(\underline{\nu}_{v_0})_j}{k!} \prod_{\lambda\equiv(v'v)\in\vartheta} (i\underline{\nu}_v \cdot i\underline{\nu}_{v'}) \prod_{v\in\vartheta} f_{\underline{\nu}_v} \tag{2.4}$$

With the new way of labeling the number of trees greatly increases, by a factor of order $k!$ being now bounded by $k!2^{2k}$, but the combinatorics becomes simpler for our purposes (even though the sum of the values of all trees is performed with great redundancy). Therefore we shall define the labeling of the trees by imagining that each line carries a number label that distinguishes it from the others.

Note that the node momenta can be supposed *bounded* by $|\underline{\nu}| \le N$ *i.e.* by the degree of f, *cfr* the factors $f_{\underline{\nu}_v}$ in (2.4). However the line currents $\underline{\nu}(\lambda)$ can only be bounded by $|\underline{\nu}(\lambda)| \le kN$ in a tree of degree k. This means that the number of trees with non zero value and degree k is finite and bounded by $2^{2k}(2N+1)^{k\ell}$; the momentum that flows in each line can be as large as kN.

We can say that the value of a tree is the product of node factors (the $f_{\underline{\nu}_v}$ in (2.3)) or "couplings" and of line factors (the $-\underline{\nu}_v \cdot \underline{\nu}_{v'}$ and $(i\underline{\nu}_{v_0})_j$) or "propagators". The perturbative series for $\underline{h}^{(k)}_{\underline{\nu}}$ acquires in this way the flavor of a Feynman graphs expansion, see Sec. 7.

Convergence for small $|\varepsilon|$ of the expansion for $\underline{h}_{\underline{\nu}}$ is immediately proved by bounding the sum of the values of the trees of degree k by

$$\sum_{\vartheta,\deg(\vartheta)=k} |\mathrm{Val}(\vartheta)| \le 2^{2k}(2N+1)^{\ell k} F^k (\ell N^2)^k \tag{2.5}$$

in fact there are at most $2^{2k}k!$ trees and the scalar products $\underline{\nu}_{v'} \cdot \underline{\nu}_v$ give at most ℓ^k terms of size N^2, while the $\underline{\nu}_v$ can be chosen in a number of ways bounded by $(2N+1)^{\ell k} < (3N)^{\ell k}$.

[4] Note that now s_v is just an integer ≥ 0 rather than a multiindex: this is due to the summation and to the consequent elimination of the component labels.

Taking into account that there are at most $(2Nk+1)^\ell$ harmonics (all of which $\le kN$) for which $\underline{h}^{(k)}_{\underline{\nu}}$ is not obviously zero we see that

$$\sum_{\underline{\nu}} e^{\kappa|\underline{\nu}|}|\underline{h}^{(k)}_{\underline{\nu}}| \le (2Nk+1)^\ell((2N+1)^\ell(\ell N^2)Fe^{\kappa N})^k \tag{2.6}$$

so that the function $\underline{h}$ is holomorphic in ε and in $\Im\psi_j$ for $|\Im\psi_j| < \kappa$ and for $|\varepsilon| < \varepsilon_0 = ((2N+1)^\ell \ell N^2 e^{\kappa N} F)^{-1}$. This concludes the "theory" of (1.1) by a "renormalization group approach".

Before passing to study less trivial problems it is worth remarking and stressing that there are important cancellations that occur in summing the trees values. A cancellation, noted in a different context by Lindstedt and Newcomb in particular cases and by Poincaré in general, shows that we can "just" consider trees *in which no line carries a zero current.* In fact if the current flowing through the root line of a tree ϑ is zero we can consider the collection of trees obtained from the given one by "detaching" the root line from the node v_0 ("*root node*" or "*first node*") from which it emerges and by attaching it successively to the other $k-1$ nodes: in this way we form a collection of k trees whose values differ only because the factor $i(\underline{\nu}_{v_0})_j$ changes as the root node v_0 varies among the tree nodes; therefore the sum of their values is proportional to $i\sum_v(\underline{\nu}_v)_j \equiv \underline{0}$.

The just exhibited cancellation implies that $\underline{h}_{\underline{0}} = \underline{0}$, *i.e.* $\underline{h}$ has zero average. In fact by a similar argument we see that $\underline{h}^{(k)}$ can be computed by considering only the sum of the values of trees in which no line is crossed by a zero current (in a different context this result was established by Poincaré, see below).

Other cancellations are possible: for instance consider a tree ϑ which contains two comparable lines $\lambda_+ < \lambda_-$ (*i.e.* two lines on the same path to the root) on which the same current flows. This means that if $v_1, \ldots, v_s$ are the nodes that precede λ_- but which do not precede λ_+ it is $\sum_{p=1}^{s} \underline{\nu}_{v_p} = \underline{0}$. Then we form the collection of trees obtained by detaching the entering line λ_+ and attaching it successively to the nodes $v_1, \ldots, v_s$: we see that the values of all the trees differ only because they contain the factor associated with the propagator of the line λ_+, *i.e.* $-\underline{\nu}_{v_p} \cdot \underline{\nu}_+$ (if v_p is the node to which the line λ_+ is attached and $\underline{\nu}_+$ is the node momentum of the other node of λ_+). Therefore the sum of the values of the collection of trees is proportional to $\sum_j \underline{\nu}_{v_p} \cdot \underline{\nu}_- \equiv \underline{0}$. Of course the same argument applies if we use the line λ_-.

The latter cancellations do not imply that we can compute $\underline{h}$ by summing only the values of trees in which no pair of comparable lines carry zero current: the reason is that the same tree may be necessary to achieve the cancellation relative to "overlapping pairs" of comparable lines, *i.e.* pairs of lines such that the paths joining them along the tree lines overlap. In the case of more difficult problems this is an important obstacle whose proper understanding has been one of the central problems of renormalization theory: in the above case it is not necessary to understand how to disentangle and turn into a useful tool the "overlapping cancellations".

3 Infrared singularities: the problem of KAM theory

Using the terminology of field theory the above is a "one scale problem" because the propagators $-\underline{\nu}_v \cdot \underline{\nu}_{v'}$ are bounded.

The matter becomes much more interesting if one studies what we shall call "Lindstedt equation"

$$(\underline{\omega}_0 \cdot \underline{\partial}_\psi)^2 \underline{h}(\underline{\psi}) = -\varepsilon(\partial_{\underline{\alpha}} f)(\underline{\psi} + \underline{h}(\underline{\psi})) \tag{3.1}$$

where $\underline{h}, f, \varepsilon$ are as in Sec. 1,2 above and $\underline{\omega}_0 = (\omega_1, \ldots, \omega_\ell) \in R^\ell$ is a *Diophantine vector*, *i.e.* a vector with the property that there exist two constants $C, \tau > 0$ such that for all non zero integer components vectors $\underline{\nu} \in Z^\ell$ it is

$$|\underline{\omega}_0 \cdot \underline{\nu}| > \frac{1}{C|\underline{\nu}|^\tau}, \qquad \underline{0} \neq \underline{\nu} \in Z^\ell \tag{3.2}$$

Equation (3.1) is substantially more difficult than its "naive" version (1.1). It admits, however, a very similar formal solution: namely $\underline{h}^{(k)}_{\underline{\nu}}$ is given by a "tree expansion" in terms of all the trees (with the same labels and the same counting) considered in the previous case *discarding the trees which contain a line with zero current*; the difference just consists in a different definition of the propagators which change, if $\lambda = (v'v)$ is a line carrying a current $\underline{\nu}(\lambda)$, so that

$$-\underline{\nu}_v \cdot \underline{\nu}_{v'} \;\longrightarrow\; -\frac{\underline{\nu}_v \cdot \underline{\nu}_{v'}}{(\underline{\omega}_0 \cdot \underline{\nu}(\lambda))^2}, \qquad i(\underline{\nu}_{v_0})_j \;\longrightarrow\; \frac{i(\underline{\nu}_{v_0})_j}{(\underline{\omega}_0 \cdot \underline{\nu})^2} \tag{3.3}$$

where $\underline{\nu}$ is the the root line current. The new "value" of the labeled trees is, therefore,

$$\mathrm{Val}(\vartheta) = \frac{1}{k!} \frac{i(\underline{\nu}_{v_0})_j}{(\underline{\omega}_0 \cdot \underline{\nu})^2} \prod_{\lambda \equiv (v'v) \in \vartheta} \frac{-\underline{\nu}_v \cdot \underline{\nu}'_v}{(\underline{\omega}_0 \cdot \underline{\nu}(\lambda))^2} \prod_{v \in \vartheta} f_{\underline{\nu}_v} \tag{3.4}$$

and the new difficulty is easily seen. The Diophantine inequality could be "saturated" for large values of the currents $\underline{\nu}(\lambda)$ without *any* of the couplings $f_{\underline{\nu}_v}$ vanishing (*i.e.* $|\underline{\nu}(\lambda)|$ can be very large, of order kN in trees of degree k): therefore if many, say bk for some $b > 0$, lines "resonate" in the sense that $\underline{\omega}_0 \cdot \underline{\nu}(\lambda)$ is of the order of a power $k^{-\tau}$ the bound on the tree value can become greater, by a factor $k^{\tau b' k} \simeq k!^{b' \tau}$ for some $b' > 0$, than the estimate in (2.5) and the bound is no longer sufficient to achieve a convergence proof no matter how small ε is.

The just described difficulty is called an "*infrared*" problem as it arises from propagators with denominators being *too close to zero at small frequencies*: the quantities $\underline{\omega} \cdot \underline{\nu}(\lambda)/2\pi$ have the interpretation of frequencies that can arise in the Fourier transform of solutions of the mechanical problem that is behind the equation (3.1), *cfr* introduction. It contrasts with *ultraviolet* problems which arise, on the contrary, from propagators with denominators which are away from zero but *not large enough at large frequencies*: an example of such problems is the theory of the boundedness of the functional integral in (1.2) and it will be discussed later.

The key idea for showing that the sum of the tree values in (3.4) over all trees of degree k can still be bounded by B^k for some $B > 0$ is to show that whenever a tree graph appears which contains too many small divisors then cancellations similar to the ones exhibited at the conclusion of Sec. 2 take place almost exactly and make the value of the graph small enough for convergence to follow.

It is interesting to check first that the problem really exists. For the purpose it is sufficient to exhibit a single graph of degree k whose value has actually size of order of a power of a factorial of k. The graph is drawn in Fig. 4.

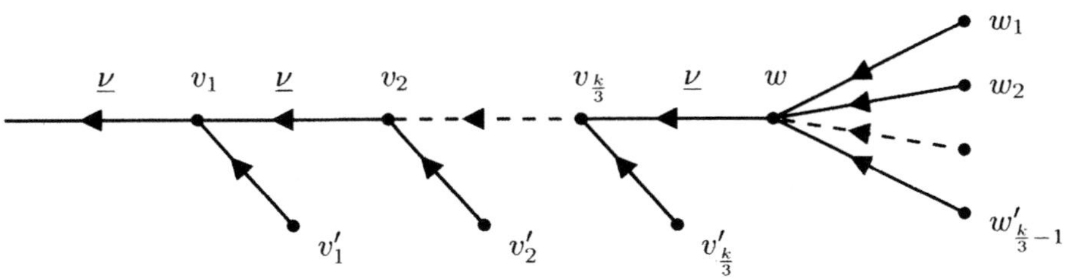

Figure 4: An example of a resonant graph. The graph consists of $\frac{k}{3}$ nodes $v_1, \ldots, v_{\frac{k}{3}}$, drawn on a horizontal line, each attached by a line to a side node, $v'_1, \ldots, v'_{\frac{k}{3}}$ respectively, and the last $\frac{k}{3}$ nodes are the initial nodes $w_1, w_2, \ldots$ of a bunch of $\frac{k}{3} - 1$ lines merging into the node w where the horizontal lines begin. The last $\frac{k}{3} - 1$ nodes carry momenta $\underline{\nu}_{w_1}, \ldots, \underline{\nu}_{w_{\frac{k}{3}-1}}$ (not marked in the figure) which together with the momentum $\underline{\nu}_w$ generate at the beginning of the horizontal strectch a current $\underline{\nu}$ suitably constructed to resonate maximally in the sense that $\underline{\omega}_0 \cdot \underline{\nu} \simeq ak^{-\tau/3}$ for some $a > 0$. The nodes $v_1, \ldots, v_{\frac{k}{3}}$ carry a small momentum $\underline{\nu}_0$ while the corresponding nodes $v'_1, \ldots, v'_{\frac{k}{3}}$ carry momentum $-\underline{\nu}_0$ so that the current flowing in the $\frac{k}{3} + 1$ horizontal lines is steadily $\underline{\nu}$, *i.e.* steadily resonant.

The value of the tree in Fig. 4 can be immediately written down from (3.4) (a useful exercise) and one readily sees that there are so many small divisors that the value has size of the order of a power of $k!$.

Formula (3.4) would provide immediately a proof of convergence for small ε if certain trees, among which the one in Fig. 4, were not present. The idea, realized in the later sections, is that the "unwanted" trees cancel each other to the extent that their sums behave well enough for not spoiling the bounds.

4 Exhibiting cancellations. The overlapping problem

To clarify the last paragraph of Sec. 3, *for the purpose of illustration*, we shall first restrict the sum of the contributions (3.4) to $\underline{h}^{(k)}_{\underline{\nu}}$ to a sum over trees which, besides the property that for all lines one has $\underline{\nu}(\lambda) \neq \underline{0}$ (as discussed above), satisfy the property

(P) $\underline{\nu}(\lambda_v) \neq \underline{\nu}(\lambda_{v'})$ *for all pairs of comparable nodes* v, v' *(not necessarily next to each other in the tree order), with* $v' > v$.

There are at most $2^{2k}k!$ trees (as in the simple case of Sec. 2) and the scalar products $\underline{\nu}_{v'} \cdot \underline{\nu}_v$ give at most ℓ^k terms of size N^2, while the $\underline{\nu}_v$ can be chosen in a number of ways bounded by $(2N+1)^{\ell k} < (3N)^{\ell k}$. Therefore, if $F = \max_{\underline{\nu}} |f_{\underline{\nu}}|$,

$$\begin{aligned} \left| \underline{h}^{(k)}_{\underline{\nu}} \right| &\leq (3N)^{\ell k} 2^{2k} \ell^k F^k C^{2k} N^{2k-1} \max_{\vartheta \in \Theta_{k,\underline{\nu}}} \prod_{\lambda \in V(\vartheta)} (C\underline{\omega}_0 \cdot \underline{\nu}(\lambda))^{-2} \\ &\leq (FC^2)^k N^{(\ell+2)k-1} (4\ell 3^\ell)^k M, \end{aligned} \tag{4.1}$$

where M is an estimate of the indicated maximum which is over the k-th degree trees ϑ verifying property (P) above. Hence the whole problem is reduced to find an estimate for M.

Let q be large. By the Diophantine condition in (3.2) one has $C|\underline{\omega}_0 \cdot \underline{\nu}| \geq q^{-1}$ if $0 < |\underline{\nu}| \leq q^{1/\tau}$: we say that the harmonic with Fourier label $\underline{\nu} \in Z^\ell$ is "q–*singular*" if $C|\underline{\omega}_0 \cdot \underline{\nu}| < q^{-1}$ and the following (extension) of a lemma by Bryuno holds for trees of degree k verifying property (P) above:

Fixed $q \geq 1$ *let* $N(k,q)$ *be the number of "q–singular lines" (i.e. of lines corresponding to q–singular harmonics) in a tree* ϑ *with k nodes. Then*

$$N(k,q) \leq \text{const}\, \frac{k}{q^{1/\tau}}, \tag{4.2}$$

and the constant could be taken $2N2^{3/\tau}$.

Remark: The intuition behind (4.2) is very simple. In order to achieve a current $\underline{\nu} = \underline{\nu}(\lambda_v)$ with $C\underline{\omega}_0 \cdot \underline{\nu}$ of size q^{-1} one needs at least $|\underline{\nu}| \geq q^{1/\tau}$, *i.e.* by (3.2) the node v must be preceded by *at least* $N^{-1}q^{1/\tau}$ nodes. Once a q–singular line λ has been generated the following lines λ' will have non-q–singular momentum until the number of lines not preceding λ and preceding λ' has grown large of the order of $q^{-1/\tau}$, *i.e.* we must collect *about as many new nodes* (*i.e.* $O(q^{1/\tau})$) to generate a second q–singular line and so on, *at least if* the new singular line λ' does not have the *same* momentum (a case excluded by hypothesis). The latter event would mean that the nodes that precede λ' but do not precede λ have node momenta adding exactly to $\underline{0}$; their presence would invalidate the argument as this is a situation which can be realized *already* with *just* 2 intermediate nodes as in the case of Fig. 4. Since the total number of nodes is k it follows that the number of q–singular lines is bounded proportionally to $k/q^{1/\tau}$. The actual estimate of the constant in (4.2) is irrelevant for our immediate purposes.

To proceed we shall assume, for simplicity, that $f_{\underline{\nu}} = f_{-\underline{\nu}}$, *i.e.* that f is an even function. Fix an exponentially decreasing sequence γ^n, $n = 1, 0, -1, -2, \ldots$; we shall make the choice $\gamma = 2$, which recommends itself. The number of 2^{-n}–singular harmonics which are not also $2^{-(n-1)}$-singular is bounded by $2N2^{3/\tau}k2^{-n}$, (by (4.2), being trivially bounded by the number of 2^{-n}-singular harmonics!). Hence

$$\prod_{\lambda\in\vartheta} \frac{1}{(C\underline{\omega}_0 \cdot \underline{\nu}(\lambda))^2} \le \prod_{n=-\infty}^{1} 2^{-(n-1)4N2^{3/\tau}2^{n/\tau}k} = e^{cN\tau k} \equiv M, \tag{4.3}$$

where $c > 0$ is a suitable constant (τ–independent); therefore the series for the approximation to $\underline{h}_{\underline{\nu}}^{(k)}$, that we are considering because of the extra restriction (P) on the sum, has radius of convergence in ε bounded below by ε_0' given by

$$(\varepsilon_0')^{-1} = (FC^2J^{-1})^k N^{(\ell+2)}(4\ell 3^\ell)e^{cN\tau}. \tag{4.4}$$

The key remark in order to take into account the trees that we have excluded by imposing the *unphysical property* (P) above is that they cancel *almost exactly*. The reason is very simple. Let $\underline{\nu}(\lambda_v) = \underline{\nu}(\lambda')$ with λ' coming out of a node following v and ending in the node v' then we can imagine to detach from the tree ϑ the subtree ϑ_2 with last node v. Then attach it, successively, to all the remaining nodes w which precede λ' but do not precede v. The simplest case is illustrated in Fig. 5 with $\lambda' = (v', w_1)$.

We obtain a family of trees whose contributions to $\underline{h}_{\underline{\nu}}^{(k)}$ differ because
(1) some of the lines below λ' changed the current by the amount $\underline{\nu} \equiv \underline{\nu}(\lambda_v)$: this means that some of the denominators $(\underline{\omega}_0 \cdot \underline{\nu}(\lambda_w))^{-2}$ have become $(\underline{\omega}_0 \cdot \underline{\nu}(\lambda_w) + \delta)^{-2}$ if $\delta \equiv \underline{\omega}_0 \cdot \underline{\nu}$ (see the line $\lambda_{w_2} \equiv w_1 w_2$ in Fig. 5) and:
(2) the scalar product $\underline{\nu}_v \cdot \underline{\nu}_w$ changes because of the successive changes of the factor $\underline{\nu}_w$, where $w \in \vartheta/\vartheta_2$ is the node to which the line λ_v is reattached.

Hence the sum of the values of all the trees considered plus those obtained by a simultaneous change of the signs of the node momenta of the nodes w preceding λ' but not preceding v would build a quantity which is even in δ. Factoring the common δ^{-4} due to the propagators of the lines λ_v and λ' the remaining sum is a function of δ which for $\delta = 0$ would be proportional to: $\sum \underline{\nu}_w = \underline{\nu}(\lambda') - \underline{\nu}(\lambda_v)$ which is zero (note that the simplifying parity in $\underline{\nu}$ assumed on $f_{\underline{\nu}}$ has to be used here). Since $\delta \neq 0$ we can expect to see that it has order δ^2 which would "cancel" one of the divisors of the lines λ', λ_v.

This is indeed true in the case of Fig. 4: by performing the operation depicted in Fig. 5 *for each* of the $\frac{k}{3}$ pairs of nodes following the initial bunch of $k/3$ nodes one checks that the result of the sum of the values of the $2\,2^{\frac{k}{3}}$ trees thus

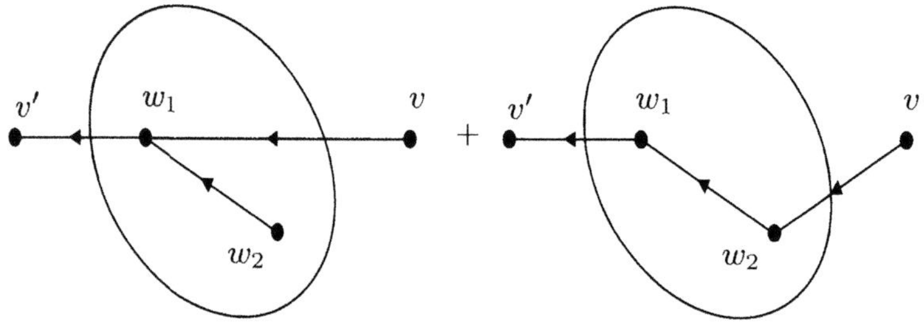

Figure 5: The simplest cancellation: the circle represents a violation of property (P) (which we shall call later a *self-energy graph*), provided $\underline{\nu}_{w_1} + \underline{\nu}_{w_2} = \underline{0}$. The parts of the tree ϑ above v' and below v are not drawn. Imagine that the line momentum $\underline{\nu}$ of the line coming out of v is very large so that $\delta \equiv \underline{\omega}_0 \cdot \underline{\nu}$ is very small and note that in the two trees one has $(\underline{\omega}_0 \cdot \underline{\nu}(\lambda_{w_2}))^2 = (\underline{\omega}_0 \cdot \underline{\nu}_{w_2})^2$ and $(\underline{\omega}_0 \cdot \underline{\nu}(\lambda_{w_2}))^2 = (\underline{\omega}_0 \cdot \underline{\nu}_{w_2} + \delta)^2$, respectively. If the signs of the node momenta of w_1, w_2 are *simultaneously* changed and the values of the four trees obtained in this way are summed we obtain an even function of δ.

obtained is bounded proportionally to $\delta^{2\frac{k}{3}}$ (*i.e.* we get a δ^2 from the cancellation in Fig. 5 for each pair of nodes) which compensates the division by $\delta^{2\frac{k}{3}+1}$ due to the small divisors at the cost of adding a factor exponential in k (harmless for the purposes of convergence as it affects only the size of the convergence radius estimate). Therefore although there are too many small divisors *things go as if the whole chain in Fig. 4 had only one!*

In general this can be true only if $|\delta| \ll |\underline{\omega}_0 \cdot \underline{\nu}(\lambda)|$ for all lines λ between λ_v and λ'. If the latter property is not true then δ must be small of order $\underline{\omega}_0 \cdot \underline{\nu}(\lambda)$ at least and, hence, this means that there are many nodes w with $v' < w$ but not $\le v$: indeed in a number of the order needed to create a momentum with small divisors of order $\underline{\omega}_0 \cdot \underline{\nu}(\lambda)$.

The intuitive argument about Bryuno's lemma following (4.2) shows that such an extreme case would be also treatable: after all also in this case the repetition of the small divisor in the lines λ', λ_v is accompanied by a great number of nodes w between v and v' so that the argument given in the remark for estimate (4.2) remains valid. Therefore the problem is to show that the two regimes just envisaged (and their "combinations") do exhaust all possibilities.

5 Multiscale decomposition. Clusters and self energy graphs. Hierarchical organization of cancellations and overlapping control

Such problems are very common in renormalization theory where they are called *overlapping divergences* problems. Their systematic analysis is made through the "renormalization group methods".

We fix a *scaling* parameter γ, and we take $\gamma = 2$ for consistency with (4.3) (see also the footnote [1] in the introduction); we also define $\underline{\omega} \equiv C \underline{\omega}_0$: it is a dimensionless frequency. Then we say that a propagator $-\underline{\nu}_{v'} \cdot \underline{\nu}_v / (\underline{\omega}_0 \cdot \underline{\nu}(\lambda))^{-2}$ is *on scale* n if $2^{n-1} \le |\underline{\omega} \cdot \underline{\nu}(\lambda)| < 2^n$, for $n \le 0$, and we set $n = 1$ if $1 \le |\underline{\omega} \cdot \underline{\nu}(\lambda)|$.

We make at this point a second simplifying assumption (which can be removed easily, as discussed in the literature quoted below). Namely we want to suppose more than the Diophantine condition (3.2): the condition will be the existence of constants $C, \tau, \gamma > 1$ such that

$$\begin{aligned} &(1) \qquad C\,|\underline{\omega}_0 \cdot \underline{\nu}| \ge |\underline{\nu}|^{-\tau}, && \underline{0} \ne \underline{\nu} \in Z^l, \\ &(2) \qquad \min_{0 \ge p \ge n} \big|C\,|\underline{\omega}_0 \cdot \underline{\nu}| - \gamma^p\big| > \gamma^{n+1} && \text{if } n \le 0,\ 0 < |\underline{\nu}| \le (\gamma^{n+3})^{-\tau^{-1}}, \end{aligned} \tag{5.1}$$

where we shall again take $\gamma = 2$. The property in (5.1) will called the *strong Diophantine condition.* One can check that the set of strongly Diophantine vectors contained in any ball Σ_r of radius r in R^ℓ has measure which tends to $volume(\Sigma_r)$ for $C \to \infty$, *i.e.* the set of strongly Diophantine vectors has full volume.

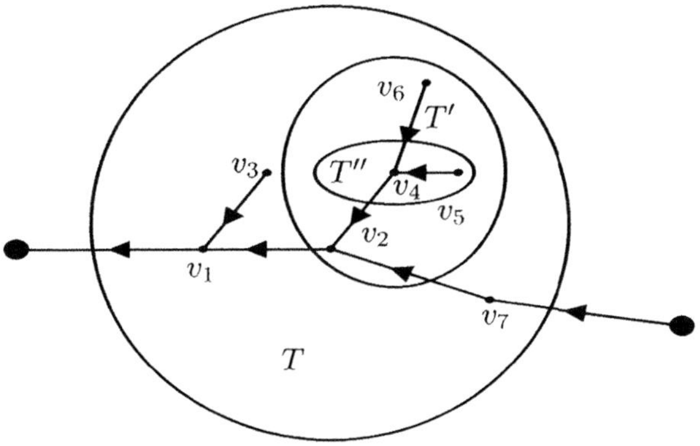

Figure 6: An example of three clusters symbolically delimited by circles, as visual aids, inside a tree (whose remaining lines and clusters are not drawn and are indicated by the bullets); not all labels are explicitly shown. The scales (not marked) of the lines increase as one crosses inward the circles boundaries: recall, however, that the scale labels are ≤ 0. If the mode labels of (v_4, v_5) add up to $\underline{0}$ the cluster T'' is a self-energy graph. If the mode labels of (v_4, v_5, v_2, v_6) add up to $\underline{0}$ the cluster T' is a self-energy graph and such is T if the mode labels of $(v_1, v_2, v_7, v_4, v_5, v_2, v_6)$ add up to $\underline{0}$. The cluster T' is maximal in T.

Given a tree ϑ and calling $\Lambda(\vartheta)$ the set of its lines including the line ending in the root, we can attach a *scale label* to each line $\lambda \in \Lambda(\vartheta)$: it is equal to n if n is the scale of the line propagator. Note that the scale labels thus attached to a tree are *uniquely determined* by the other tree labels: they will have only the function of help in visualizing the orders of magnitude of the divisors associated with the various tree lines.

Looking at such labels we identify the connected *clusters T of scale n_T* formed by a set of lines
(i) connected by a continuous path in the tree consisting of lines with scale labels $\ge n_T$,
(ii) which contain at least one line of scale n_T
(iii) and which are maximal with the latter two properties.
We shall say that the "*cluster T has scale n_T*".

We shall denote by $V(T)$ the set of nodes in T, and by $\Lambda(T)$ the set of lines connecting them. We also denote by $\Lambda_1(T)$ the set of lines in $\Lambda(T)$ plus the entering and exiting lines of T. Finally call $\mathcal{T}(\vartheta)$ the set of all clusters in ϑ.

Among the clusters we consider the ones with the property that there is only one tree line entering them and only one exiting and both carry the same momentum. Here we recall that the tree lines carry an arrow pointing to the root to give a meaning to the words "entering" and "exiting".

If T is one such cluster and λ_T is the line entering it we say that T is a *self-energy cluster* if the number $M(T)$ of lines contained in T is "not too large"

$$M(T) \stackrel{def}{=} \text{number of lines contained in } T \le E\,2^{-n\eta}, \tag{5.2}$$

where $n = n_{\lambda_T}$, and E, η are defined by:[5] $E \equiv 2^{-3\eta} N^{-1}$, $\eta = \tau^{-1}$. We call n_{λ_T} the *self-energy-scale* of T, and λ_T a *self-energy line.*

To refer to self-energy clusters T we need some terminology
(a) we denote $\lambda_T \equiv (w_{-T} v_{-T})$ the entering line: its scale $n = n_{\lambda_T}$ is smaller than the smallest scale n_T of the lines inside T; likewise we denote $\lambda_T^+ \equiv (v_{+T} w_{+T})$ the exiting line. Hence w_{-T} is the node inside T into which the entering line λ_T ends.
(b) Let $\widetilde{T}$ be the set of nodes of $V(T)$ *outside* the self-energy clusters *contained* in T (if any).
(c) Denote by $\Lambda(\widetilde{T})$ the set of lines λ contained in T and with at least one point in $\widetilde{T}$, and by $\Lambda_1(\widetilde{T})$ the set of lines in $\Lambda(\widetilde{T})$ *plus* the lines entering and exiting T; note that all lines $\lambda \in \Lambda(\widetilde{T})$ have a scale $n_\lambda \ge n_T$.

Remarks. (1) The self-energy clusters are called *resonances* in Eliasson's terminology, see references.
(2) Note that the self-energy-scale n of a self-energy cluster T (*i.e.* the scale of the entering line) is different from the scale n_T of T as a cluster (*i.e.* the lowest scale of the lines inside the cluster): one has $n < n_T$.

Let us consider a tree ϑ and its clusters. We wish to estimate the number $N_n(\vartheta)$ of lines in $\Lambda(\vartheta)$ with scale $n \le 0$.

---

[5]This is just a convenient definite choice.

Denoting by T a cluster of scale n let q_T be the number of self-energy clusters of self-energy-scale n contained in T (hence with entering lines of scale n), we have the following inequality.

For all trees $\vartheta \in \Theta_{k,\underline{\nu}}$ *one has*

$$N_n(\vartheta) \le \frac{4k}{E\,2^{-\eta n}} + \sum_{\substack{T\in\mathcal{T}(\vartheta)\\ n_T=n}} (-1+q_T), \tag{5.3}$$

with $E = N^{-1}2^{-3\eta}, \eta = \tau^{-1}$.

Remark. This is a version of Bryuno's lemma; a proof is given for completeness in the Appendix below. Intuitively the above inequality has the same content as (4.2): if there are self energy clusters one simply adds to the bound (4.2) (first term in the r.h.s. of (5.3)) the number of such graphs (*i.e.* the sum in (5.3)). For the apparently "extra -1" see appendix A1.

Consider a tree ϑ^1; we define the family $\mathcal{F}(\vartheta^1)$ generated by ϑ^1 as follows. Given a self-energy cluster T of ϑ^1 we detach the part of ϑ^1 which has λ_T as root line and attach it successively to the points $w \in \widetilde{T}$ (note that the endpoint $w_1 \in V(T)$ of λ_T is necessarily among them).

The above procedure is then repeated for all self-energy clusters in ϑ. For each self-energy cluster T of ϑ^1 we shall call V_T the number of nodes in $\widetilde{T}$, *i.e.* $V_T = |V(\widetilde{T})|$. To the just defined set of trees we add the trees obtained by reversing simultaneously the signs of the node modes $\underline{\nu}_w$, for $w \in \widetilde{T}$: the change of sign is performed independently on the various self-energy clusters. This defines a family of $\prod 2V_T$ trees that we call $\mathcal{F}(\vartheta_1)$ (the product is over all self-energy clusters in ϑ). The number $\prod 2V_T$ will be bounded by $\exp\sum 2V_T \le e^{2k}$.

It is important to note that the definition of self-energy graph is such that the above operation (of shift of the node to which the line entering the self-energy cluster is attached) cannot change too much the sizes of the propagators of the lines inside the self-energy clusters.

This is called the "non overlapping lemma" and the reason behind its validity is simply that inside a self-energy cluster of self-energy-scale n the number of lines is not very large, being $\le \overline{N}_n \equiv E\,2^{-n\eta}$.

Indeed let λ be a line contained inside the self-energy clusters $T = T_1 \subset T_2 \subset \ldots$ of self-energy-scales $n = n_1 > n_2 > \ldots$; then the shifting of the lines λ_{T_i} can cause a change in the size of the propagator of λ by at most

$$2^{n_1} + 2^{n_2} + \ldots < 2^{n+1}. \tag{5.4}$$

For any line λ in $\Lambda(T)$ the quantity $\underline{\omega}_0 \cdot \underline{\nu}_\lambda$ has the form $\underline{\omega}_0 \cdot \underline{\nu}^0_\lambda + \sigma_\lambda \underline{\omega}_0 \cdot \underline{\nu}(\lambda_T)$ if $\underline{\nu}^0_\lambda$ is the momentum of the line λ "inside the self-energy cluster T", *i.e.*

it is the sum of all the node momenta of the nodes preceding λ in the sense of the line arrows, but contained in T; and $\sigma_\lambda = 0, 1$.

Therefore not only $|\underline{\omega} \cdot \underline{\nu}^0(\lambda)| \geq 2^{n+3}$ (because $\underline{\nu}^0(\lambda)$ is a sum of $\leq \overline{N}_n$ node momenta, so that $|\underline{\nu}^0(\lambda)| \leq N\overline{N}_n$) but $\underline{\omega} \cdot \underline{\nu}^0(\lambda)$ is "in the middle" of the diadic interval containing it and, by the strong Diophantine property(5.1), does not get out of it if we add a quantity bounded by 2^{n+1} (like $\sigma_\lambda \underline{\omega}_0 \cdot \underline{\nu}(\lambda_T)$). Hence no line changes scale as ϑ varies in $\mathcal{F}(\vartheta^1)$, if $\underline{\omega}_0$ verifies (5.1).

By the strong Diophantine condition (5.1) *on* $\underline{\omega}_0$ *the self-energy clusters of the trees in* $\mathcal{F}(\vartheta^1)$ *all contain the same sets of lines, and the same lines enter or exit each self-energy cluster (although they are attached to generally distinct nodes inside the self-energy clusters: the identity of the lines is here defined by the number label that each of them carries in* ϑ^1*). Furthermore the scales of the self-energy clusters, and in fact of all the lines, do not change.*

Let ϑ^2 be a tree not in $\mathcal{F}(\vartheta^1)$ and construct $\mathcal{F}(\vartheta^2)$, *etc*, obtaining in this way a collection $\{\mathcal{F}(\vartheta^i)\}_{i=1,2,\ldots}$ of pairwise disjoint families of trees. We shall sum all the contributions to $\underline{h}^{(k)}_{\underline{\nu}}$ coming from the individual members of each family and then sum over the families. This is a realization of *Eliasson's resummation*: it is more detailed than his original one, where no subdivision of the trees in classes was considered and the cancellation implied by the one that we exhibit in Sec. 6 was derived from an argument involving all graphs at the same time. Thus the Eliasson cancellation can be regarded as a cancellation due to a special symmetry of the problem (analogous to the Ward identities of field theory) and the above analysis shows that more symmetry is present as the cancellation takes place already at a lower level in which less trees are added together.

This completes the organization of the tree values which makes evident the cancellations necessary to show, see Sec. 6, that not only the problem associated with the tree in Fig. 4 but also the analogous problem in the most general graph can be solved by the above considerations.

6 Cancellations and dimensional bounds

The above hierarchical organization of the sum of the terms giving rise to the k–th order contribution $h^{(k)}_j$ is sufficient for our purposes. One can proceed to bound the sum of the contributions from each collection of terms straightforwardly by using repeatedly the maximum principle (namely the bound of the value of an analytic function at a point by its maximum modulus in a (complex) region around the point divided by the distance to the boundary of the region).

Referring to the notions associated with the self energy clusters, see items a,b,c following (5.2), we call η_T the quantity $\underline{\omega}_0 \cdot \underline{\nu}(\lambda_T)$ associated with the self-energy cluster T. If λ is a line in $\Lambda(\widetilde{T})$, defined after (5.2), we can imagine to write the quantity $\underline{\omega}_0 \cdot \underline{\nu}(\lambda)$ as $\underline{\omega}_0 \cdot \underline{\nu}^0(\lambda) + \sigma_\lambda \eta_T$, with $\sigma_\lambda = 0, 1$: the product of the propagators of the lines *inside* $\widetilde{T}$ is

$$\prod_{\lambda \equiv (v'v) \in \Lambda(\widetilde{T})} \frac{-\underline{\nu}_v \cdot \underline{\nu}_{v'}}{(\underline{\omega}_0 \cdot \underline{\nu}^0(\lambda) + \sigma_\lambda \eta_T)^2}. \tag{6.1}$$

For simplicity we do not explicitly distinguish the possibility that λ is the root line: in that case the corresponding factor in (6.1) has the slightly different form $-i(\underline{\nu}_{v_0})_j / (\underline{\omega}_0 \cdot \underline{\nu}(\lambda))^2$.

If the tree does not contain any self-energy clusters, we say that it has *height* 0; if the only self-energy clusters do not contain other self-energy clusters, we say that the tree has height 1, see Fig. 4 for an example; more generally if the maximum number of self-energy clusters that contain a given self-energy cluster is p, we say that the *tree has height* p. Similarly we say that a *self-energy cluster has height* p if it contains at least one self-energy cluster that is contained in exactly p self-energy clusters and none which is contained in more ($p = 0$ corresponds to a self-energy cluster which does not contain any other self-energy clusters). Given a tree ϑ, call $V(\vartheta)$ the set of all self-energy clusters in ϑ, and set

$$\Lambda(V(\vartheta)) = \cup_{T \in V(\vartheta)} \Lambda(T), \qquad \Lambda_1(V(\vartheta)) = \cup_{T \in V(\vartheta)} \Lambda_1(T). \tag{6.2}$$

Of course in (6.2) the union could be restricted only to the maximal self-energy clusters.

First consider the simple case of a tree ϑ of height 1 and let us denote by T any of its self-energy clusters: if we regard the quantities η_T as independent variables we see that (6.1) is holomorphic in η_T for $|\eta_T| < 2^{n_T - 3}$. While η_T varies in such complex disk the quantity $|\underline{\omega}_0 \cdot \underline{\nu}^0(\lambda) + \sigma_\lambda \eta_T|$ does not become smaller than 2^{n_T-3}.[6] The main point here is that the quantity 2^{n_T-3} will usually be $\gg 2^{n_{\lambda_T}}$ which is the value that η_T actually can reach in every tree in $\mathcal{F}(\vartheta)$; this is what happens in the special case of Fig. 4 and it can be exploited in applying the maximum principle, as done below.

Note that the quantities η_T do not depend on the element of the family $\mathcal{F}(\vartheta)$ so that we could factor out of the sum of the values of the graphs in $\mathcal{F}(\vartheta)$ the factors η_T^{-2}; we can write the product of the propagators of any tree as

$$\Bigg(\prod_{\substack{\lambda \in \Lambda(\vartheta) \setminus \Lambda_1(V(\vartheta)) \\ \lambda \equiv (v'v)}} \frac{-\underline{\nu}_{v'} \cdot \underline{\nu}_v}{(\underline{\omega}_0 \cdot \underline{\nu}(\lambda))^2} \Bigg) \cdot \Bigg(\prod_{T \in V(\vartheta)} \prod_{\lambda \in \Lambda(T)} \frac{-\underline{\nu}_{v'} \cdot \underline{\nu}_v}{(\underline{\omega}_0 \cdot \underline{\nu}^0(\lambda) + \sigma_\lambda \eta_T)^2} \Bigg) \cdot$$
$$\cdot \Bigg(\prod\nolimits^{+}_{T \in V(\vartheta)} \frac{-\underline{\nu}_{v_T^+} \cdot \underline{\nu}_{w_T^+}}{\eta_T^2} \Bigg) \cdot \Bigg(\prod\nolimits^{-}_{T \in V(\vartheta)} \frac{-\underline{\nu}_{w_T^-} \cdot \underline{\nu}_{v_T^-}}{\eta_T^2} \Bigg) \tag{6.3}$$

[6] In fact $|\underline{\omega}_0 \cdot \underline{\nu}^0(\lambda)| \geq 2^{n+3}$ because T is a self-energy cluster; therefore $|\underline{\omega}_0 \cdot \underline{\nu}(\lambda)| \geq 2^{n+3} - 2^{n+1} > 2^{n+2}$ so that $n_T \geq n+3$. On the other hand we note that $|\underline{\omega}_0 \cdot \underline{\nu}^0(\lambda)| > 2^{n_T-1} - 2^{n+1}$, so that it follows that $|\underline{\omega}_0 \cdot \underline{\nu}^0(\lambda) + \sigma_\lambda \eta_T| \geq 2^{n_T-1} - 2^{n+1} - 2^{n_T-3} \geq 2^{n_T-3}$, for $|\eta_T| < 2^{n_T-3}$.

where the first product is over the lines λ which neither enter nor exit nor are inside a self-energy cluster of ϑ (so that their momentum is the same in all trees of the family $\mathcal{F}(\vartheta)$), the second product is over the lines λ contained in $V(\vartheta)$, the third product is over the self-energy clusters $T \in V(\vartheta)$ and takes into account the lines exiting T *but not entering another self energy cluster* and the last product is over the lines that *enter* the self energy clusters.

As said above the denominators η_T^2 factor out of the sum of the values of the trees in $\mathcal{F}(\vartheta)$ at fixed ϑ. We can therefore consider the sum of the $\prod 2V_T \le e^{2k}$ values of the graphs members of the family $\mathcal{F}(\vartheta)$ *divided by the product of the factors* η_T^{-2} *associated with the lines entering or exiting the self energy clusters, i.e.* we consider the sum of the values in (6.3) computed without the denominators in the last two products.

Each such sum is holomorphic in the region $|\eta_T| < 2^{n_T-3}$ and in the latter region it is bounded by $\prod 2^{-2(n_\lambda-3)} \le 2^{6k} \prod_\lambda 2^{-2n_\lambda}$, if n_λ the scale of the line λ in ϑ and if the product is over the lines neither entering nor exiting a self-energy cluster. This even holds if the η_T are regarded as independent complex parameters.

By construction the just considered sum of the $\prod 2V_T \le e^{2k}$ terms from the trees in $\mathcal{F}(\vartheta)$, vanishes to second order in *each* of the η_T parameters (by the approximate cancellation discussed above due to the fact that the sum $\sum_{w\in T} \underline{\nu}_w = \underline{0}$ and to the parity property supposed for $f_{\underline{\nu}}$). *By the maximum principle* this means that if we bound the sum by the number of terms times the maximum among them (which is easy to estimate because the propagators have all well defined seizes fixed by their scales) we can multiply the result by a further factor of the order of $2^{2n_{\lambda_T}}/2^{2(n_T-3)}$ and still obtain a valid bound.

Hence by the maximum principle and, recalling that each $\underline{\nu}_v$ can be bounded by N, we can bound the contribution to $\underline{h}_{\underline{\nu}}^{(k)}$ from the family $\mathcal{F}(\vartheta^1)$ by

$$\Big[\frac{1}{k!}\Big(FC^2N^2\Big)^k 2^{6k}e^{2k}\prod_{n\le 0} 2^{-2nN_n}\Big]\Big[\prod_{n\le 0}\prod_{\substack{T\in\mathcal{T}(\vartheta)\\ n_T=n}}\prod_{i=1}^{q_T} 2^{2(n-n_i+3)}\Big], \tag{6.4}$$

where
(1) $N_n = N_n(\vartheta)$ is the number of propagators of scale n in ϑ^1 ($n = 1$ does not appear because $|\underline{\omega}\cdot\underline{\nu}| \ge 1$ in such cases);
(2) the first square bracket is the bound on the product of individual elements in the family $\mathcal{F}(\vartheta^1)$ times the bound e^{2k} on their number: this takes into account *also* the last product in (6.3);
(3) the second square bracket is the part coming from the maximum principle, applied to bound the resummations, and is explained as follows.
(4) The dependence on the variables $\eta_{T_i}{\equiv}\eta_i$ relative to self-energy clusters $T_i \subset T$ with self-energy-scale $n_{\lambda_{T_i}} = n$ is holomorphic for $|\eta_i| < 2^{n_i-3}$, if $n_i{\equiv}n_{T_i}$, provided $n_i > n+3$ (see above).
(5) The resummation says that the dependence on the η_i's has a second order zero in each. Hence the maximum principle tells us that we can improve the bound given

by the third factor in (6.3) by the product of factors $(|\eta_i|\,2^{-n_i+3})^2$ as $n_i \ge n+3$ which yield the product in the second square bracket.

The above would be sufficient if there were no trees of height higher than 1. In fact substituting (5.3) into (6.4) we see that the q_T is taken away by the first factor in $2^{2n}2^{-2n_i}$, while the remaining 2^{-2n_i} are compensated by the -1 before the $+q_T$ in (5.3), taken from the factors with $T=T_i$, (note that there are always enough -1's). It follows that the product (6.4) is bounded by

$$\frac{1}{k!}\,(C^2FN^2)^k e^{2k} 2^{12k} \prod_{n\le 0} 2^{-8nkE^{-1}2^{\eta n}} \le \frac{1}{k!}\,B_0^k, \tag{6.5}$$

with B_0 suitably chosen.

To sum over the trees we note that fixed ϑ the collection of clusters is fixed. Therefore we only have to multiply (6.5) by the number of tree shapes for ϑ, $(\le 2^{2k}k!)$, by the number of ways of attaching momentum labels, $(\le (3N)^{\ell k})$, by the number of ways of contracting the tensor labels, $(\le \ell^k)$, so that we can bound $|\,\underline{h}_{\underline{\nu}}^{(k)}|$ by

$$\varepsilon_0^{-k} \equiv (b_\ell C^2 F N^{2+\ell} e^{cN})^k, \tag{6.6}$$

with b_ℓ suitably chosen.

To treat the general case we can proceed inductively and suppose that the bound in (6.4) holds for trees of height $1,2,\ldots,p-1$ and for values of the $\eta_T = \underline{\omega}_0 \cdot \underline{\nu}(\lambda_T)$ of the lines that enter the maximal self-energy clusters T which are in the complex disk $|\eta_T| < 2^{n_T-3}$, see Appendix.

7 Feynman graphs for the integral in (1.2)

The analysis of the problem (1.2),(1.3) is started by checking the existence of a formal series expansion in λ: which, of course, has to be followed by the study of the convergence (in fact of the asymptoticity) properties of the series.

The first problem is an easy one and its solution is classical: it is most easily described in terms of graphs: consider the following *graphical elements*

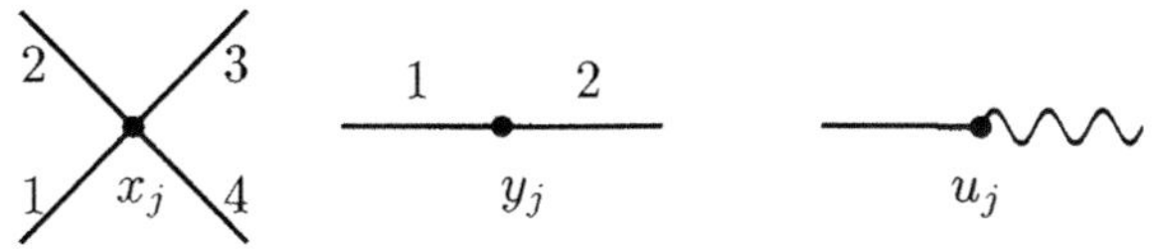

Figure 7: The three graphical elements for the Feynman graphs expansion of $\log E_{\lambda,\mu}(\varepsilon f)$. The labels $1,2\ldots$ signify that the lines of a single graph element must be considered as distinct.

The coefficient of $\lambda^{k_4}\mu^{k_2}\varepsilon^{k_1}$ in the expansion of $\log E_{\lambda,\mu}(\varepsilon f)$ is obtained by considering the *connected* graphs Γ that can be formed with k_4 graph elements of the first type in Fig. 7, *i.e.* with 4 lines, and with k_2 graph elements of the second type and k_1 elements of the third type and then merging pairwise the solid lines to form a connected graph with k_1 wiggly lines left unpaired. One often refers to lines obtained by merging a pair of lines by calling it a "*contraction*" so that the various graphs are obtained by contracting pairwise the non wavy lines of the graph elements in Fig. 7. With each such Feynman graph we associate an "*amplitude*" which is simply

$$\int (-\lambda)^{k_4}(-\mu)^{k_2}(-\varepsilon)^{k_1} \prod_{(\xi_i\xi_j)\in\Gamma} C(\xi_i-\xi_j)\frac{\prod d\xi_j}{k_4!k_2!k_1!} \tag{7.1}$$

where $(\xi_1,\ldots,\xi_{k_4+k_2+k_1}) = (x_1,\ldots,x_{k_4},y_1,\ldots,y_{k_2},u_1,\ldots,u_{k_1})$.

For instance for $k_2 = 1, k_1 = 2, k_4 = 0$ and $k_2 = 0, k_1 = 2, k_4 = 2$ we get, respectively, graphs like

Figure 8: Two graphs contributing to the orders $\mu\varepsilon^2$ and $\lambda^2\varepsilon^2$, respectively, to $\log E_{\lambda,\mu}(\varepsilon f)$.

Many graphs which differ only by the identity of the lines that are contracted yield the same value. If we do not write the identity labels on the lines then each graph Γ has to be multiplied by a suitable combinatorial factor $n(\Gamma)$ for an appropriate count. The total number $\sum_{\Gamma\ labeled} 1 = \sum_{\Gamma\ unlabeled} n(\Gamma)$ is of the order of $(2(k_4+k_2+k_1))!$.

Since the propagator $C(\underline{x}-\underline{y})$ in (1.2) is a continuous function which decays exponentially as $|\underline{x}-\underline{y}|\to\infty$ it is clear that the integrals in (7.1) are finite for all graphs: hence the formal power series for the generating function $\log E_\lambda(\varepsilon f)$ of the Schwinger functions is well defined to all orders.

In the present case we certainly cannot have convergence of the latter well defined expansion for the obvious reason that the integral in (1.2) is (almost) obviously divergent for $\lambda < 0, \mu, \varepsilon = 0$ (divergence can be established because in this case all integrals are non negative and admit a lower bound B^k that grows exponentially in k so that the order k coefficient grows as $B^k(2k)!/k!$, *i.e.* too fast).

Nevertheless convergence of the integral can be proved as well as the asymptoticity of the formal power series to which it is formally equal: this is a consequence of an important inequality due to Nelson. However here we shall not discuss this point further.

8 An ultraviolet problem: φ_d^4 for $d = 2, 3$. Multiscale decomposition and dimensional estimates

One of the most studied problems in renormalization theory is the analysis of the integral (1.2) in which the probability distribution $P(d\varphi)$ is Gaussian with a covariance different from (1.3) and given by

$$C(\underline{x} - \underline{y}) \stackrel{def}{=} \langle \varphi_{\underline{x}} \varphi_{\underline{y}} \rangle_P = \frac{1}{(2\pi)^d} \int e^{i\underline{p}\cdot(\underline{x}-\underline{y})} \frac{1}{\underline{p}^2+1} d^d \underline{p} \tag{8.1}$$

We see that $C(\underline{x})$ decays exponentially as $|\underline{x}| \to \infty$ *but* $C(0) = \infty$ which means that with P-probability 1 the functions $\varphi_{\underline{x}}$ are in fact rather singular and, more precisely, are distributions. The rate of divergence of $C(\underline{x})$ as $\underline{x} \to 0$ is $\simeq -\log|\underline{x}|$ if $d = 2$ and $\simeq |\underline{x}|^{-1}$ if $d = 3$.

Therefore not only the problem is harder, but it is not even clear whether it makes sense at all since the function in the exponent in (1.2) is no longer meaningful.

The logarithmic divergence in $d = 2$ is "very weak" that one checks that all graphs without self contractions, *i.e.* without any pairing of lines emerging from the same graph element, are finite. However in general such contractions occur and therefore yield factors $C(\underline{0}) = +\infty$. Hence, clearly, the problem is not well posed.

The physical interpretation of the integral (1.2) with P with covariance (8.1) does not require that the integral be well defined for all λ, μ small but "just" that there is a function $\mu(\lambda)$ (possibly depending also on $\underline{x}$) such that the integral is meaningful. In other words one asks whether one can find $\mu(\lambda)$ such that the $\log E_\lambda(\varepsilon f)$ is well defined and smooth in λ, ε for $\lambda > 0$ small and $\mu = \mu(\lambda)$. It is not surprising that the $\mu(\lambda)$ if at all existent should be *infinite*! Of course the fault can be attributed to the fact that the integrand itself is not well defined.

One tries to attach a meaning to the integral by replacing the probability distribution $P(d\varphi)$ by a *regularized* distribution $P_N(d\varphi)$ where N is a "*cut-off*" parameter and P_N is a Gaussian functional integral with covariance

$$C^{(N)}(\underline{x} - \underline{y}) = \frac{1}{(2\pi)^d} \int \frac{e^{i\underline{p}\cdot(\underline{x}-\underline{y})}}{1+\underline{p}^2} \chi_N(\underline{p})\, d^d \underline{p} \tag{8.2}$$

where $\chi_N(\underline{p}) \xrightarrow[N\to\infty]{} 1$. Possible choices are

$$\chi_N(\underline{p}) = \begin{cases} 1 & if |\underline{p}| \le 2^N \\ 0 & if |\underline{p}| > 2^N \end{cases} \qquad \text{or} \qquad \chi_N(\underline{p}) = \frac{2^{2N}-1}{2^{2N}+\underline{p}^2} \tag{8.3}$$

where the first choice is is perhaps the most natural while the second might be the easiest technically. The matter is debated: the two choices however lead to the same result in the limit as $N \to \infty$ (*i.e.* to the same $N = \infty$ limit value for the integral (1.2)). Here I shall follow the traditional approach that uses a smooth cut–off, *e.g.* the second choice in (8.3), for expository reasons (brevity): however the first choice is gaining grounds in the recent research works on "exact renormalization group".

If C is replaced by $C^{(N)}$ as defined by the second of (8.3) the integral becomes well defined essentially because it becomes like the one studied in Sec. 4: the Fourier transform of the propagator goes to 0 as $\underline{p}^{-4}$ and $C^{(N)}(\underline{x})$ is a continuous exponentially decreasing function so that all integrals in the perturbative expansion for $\log Z$ are finite and Z itself is well defined thanks to the Nelson inequality.

The possibility of taking advantage of the freedom of the choice of $\mu(\lambda)$ then leads to consider the integral

$$E_{\lambda,N}(\varepsilon f) = \frac{\int P_N(d\varphi)e^{-\int (V_N(\varphi_x)+\varepsilon\varphi_{\underline{x}} f(\underline{x}))d^d\underline{x}}}{\int P_N(d\varphi)e^{-\int V_N(\varphi_x)d^d\underline{x}}}$$
$$V_N(\varphi_x) = \lambda\varphi_{\underline{x}}^4 + \mu_N(\underline{x})\varphi_{\underline{x}}^2 \tag{8.4}$$

and the problem is to find μ_N so that the limit as $N \to \infty$ of (8.4) exists and is smooth in λ, ε, f. The quantity μ is allowed to depend on $\lambda, N, \underline{x}$ (but not $\varphi_{\underline{x}}$) with the only condition that it should be bounded uniformly in $\underline{x}$ *at fixed* N so that at fixed N the formal perturbation expansion is well defined.

If $d = 2$ the above remark that the only divergent (as $N \to \infty$) graphs are the ones with self contractions leads immediately to try to determine μ_N in such a way that all such graphs cancel each other.

From the theory of Gaussian integrals it is well known that elimination of the graphs with self contractions is possible simply by requiring that V_N be a linear combination of "Wick monomials" defined as $:\varphi_{\underline{x}}^n: = \sqrt{2C^{(N)}(0)}^n H_n(\frac{\varphi_{\underline{x}}}{\sqrt{2C^{(N)}(0)}})$ where $H_n(x)$ is the n–th Hermite polynomial ($H_4(z) = z^4 - 3z^2 + \frac{3}{2}$, $H_2(z) = z^2 - \frac{1}{2}, \ldots$) and $C^{(N)}(0) = \langle\varphi_{\underline{x}}^2\rangle_{P_N}$.

It is therefore very convenient to start with a V_N of the form

$$\begin{aligned} V_N(\varphi) &= \lambda : \varphi_{\underline{x}}^4 : + \overline{\mu}_N(\underline{x}) : \varphi_{\underline{x}}^2 : \equiv \\ &\equiv \lambda\,(\varphi_{\underline{x}}^4 - 6\,C^{(N)}(0)\,\varphi_{\underline{x}}^2) + \overline{\mu}_N(\underline{x})\,\varphi_{\underline{x}}^2 + const \end{aligned} \tag{8.5}$$

and in the case $d = 2$ one can simply take $\overline{\mu}_N \equiv 0$ which in terms of the notation in (8.4) means $\mu_N(\underline{x}) = -6\,\lambda\,C^{(N)}(0)$ (which diverges as $N \to \infty$ as expected and depends on λ). The perturbative analysis is complete and gives a finite result (uniformly in N) order by order: it remains the hard part of the job which is to prove the existence of the limit as $N \to \infty$ of $E_{\lambda,N}(\varepsilon f)$, (8.4). This is not discussed here because we only want to show the analogy between the KAM and the field theory problems. Therefore we shall eventually concentrate attention on the much more interesting problem of the perturbation analysis of (8.4) in the $d = 3$ case.

Of course, even with the choice in (8.5) if $d = 3$ all integrals depend on N with divergences occurring as $N \to \infty$ at least if one does not attempt to use the freedom in the choice of $\overline{\mu}(\lambda)$. Nevertheless at fixed N we can write the perturbation expansion, or better its formal coefficients, just as in the case discussed in Sec. 4 and we therefore imagine to have written the expansion of $\log E_{\lambda,\mu}(\varepsilon f)$ in powers of $\lambda, \mu, \varepsilon$ via exactly the Feynman graphs of Sec. 4 with the new propagator $C^{(N)}$ and with the constant $\overline{\mu}$ allowed to depend on N, $\underline{x}$, λ and *no Feynman graphs with self contractions.*

Since $C^{(N)}$ is the Fourier transform of the function $\frac{2^{2N}-1}{(2^{2N}+\underline{p}^2)(1+\underline{p}^2)} \equiv \frac{1}{1+\underline{p}^2} - \frac{1}{2^{2N}+\underline{p}^2}$ we can write

$$\begin{aligned} C^{(N)}(\underline{x} - \underline{y}) &= \frac{1}{(2\pi)^d}\int d^d\underline{p}\, e^{i\underline{p}\cdot(\underline{x}-\underline{y})}\sum_{h=0}^{N-1}\Big(\frac{1}{2^{2h}+\underline{p}^2} - \frac{1}{2^{2(h+1)}+\underline{p}^2}\Big) = \\ &= \frac{1}{(2\pi)^d}\int d^d\underline{p}\, e^{i\underline{p}\cdot(\underline{x}-\underline{y})}\sum_{h=0}^{N-1} 2^{2h}\frac{3}{(2^{2h}+\underline{p}^2)(2^{2(h+1)}+\underline{p}^2)} = \\ &= \sum_{h=0}^{N-1} 2^{(d-2)h}\overline{C}(2^h(\underline{x} - \underline{y})) \stackrel{def}{=} \sum_{h=0}^{N-1} C_h(\underline{x} - \underline{y}) \end{aligned} \tag{8.6}$$

where $\overline{C}$ has a Fourier transform $3/(1+\underline{p}^2)(4+\underline{p}^2)$, which gives a "*multiscale decomposition*" of the "ultraviolet" singularity of the propagator in the limit as $N \to \infty$.

The "scale covariant" representation of the propagator achieved by the decomposition in (8.6) can be used to decompose finely the integrals corresponding to the Feynman graphs and defining the perturbation expansion coefficients and to rearrange the sums of the terms thus obtained in a way that exhibits the remarkable cancellations that will allow us to show that *if* $\mu_N(\lambda)$ *is suitably defined* then the perturbation expansion of $E_\lambda(\varepsilon f)$ (defined as the limit of $E_{\lambda,N}(\varepsilon f)$) is well defined order by order in λ, ε thus proving renormalizability (and performing renormalization).

The latter result will not be sufficient yet to show that the generating function of the Schwinger functions is actually well defined and defines a new non trivial (*i.e.* non Gaussian) probability distribution over the fields $\varphi_{\underline{x}}$: again the reason is that we cannot expect that the perturbative series be convergent since the expressions that they should define are not defined for $\lambda < 0$ already for $N < \infty$. Nevertheless renormalizability is a key ingredient and a first step in the proof of existence of the no cut–off limit of the generating function for the Schwinger functions.

We imagine to associate with each line $\ell = (\underline{x}, \underline{y})$ of the Feynman graphs a "*scale label*" $h_\ell = N, N-1, \ldots, 0$ and we shall define its value by replacing the propagator $C^{(N)}(\underline{x} - \underline{y})$ with $C_{h_\ell}(\underline{x} - \underline{y}) = 2^{(d-2)h_\ell}\overline{C}(2^{h_\ell}(\underline{x} - \underline{y}))$, *i.e.* with the propagator on scale h_ℓ.

Fixed a Feynman graph Γ of the latter type we can associate with it a tree graph ϑ: we define first the "*clusters*" of lines in Γ: a "cluster of scale h" will be a connected subset of lines whose scales are $\geq h$ which, furthermore, contains at least one line of scale h and which is a maximal set with the latter two properties.

It will be convenient and natural to regard each vertex of the Feynman graph as a cluster, in fact as a cluster of scale $h+1$ if h is the highest scale of the graph lines that merge into the vertex, even though they contain no lines. In this way the number of vertices of a Feynman graph and the number of nodes of the corresponding tree are equal.

The clusters are by definition arranged hierarchically and are naturally partially ordered by inclusion: therefore they can be represented by the nodes of a tree ϑ. The set of all graphs which give rise to the same tree will be called the set of graphs Γ compatible with ϑ and we shall write, somewhat improperly, $\Gamma \subset \vartheta$.

Note that the lines of the tree have nothing to do with the lines of the Feynman graph. An illustration of the above definitions is provided by the following Fig. 9 (with a Feynman graph with 7 vertices).

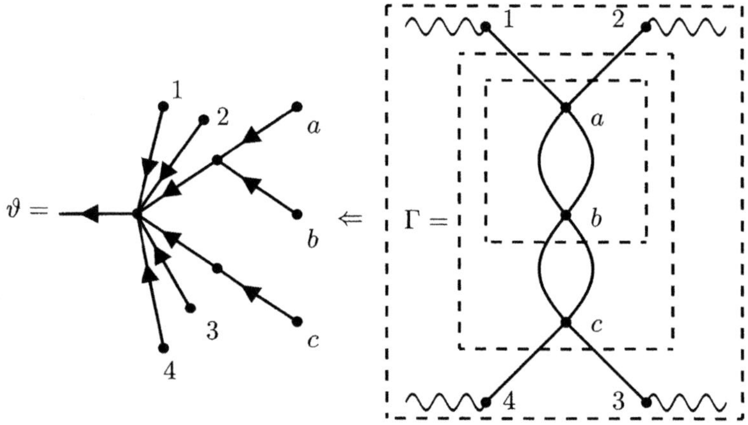

Figure 9: An example of a Feynman graph Γ with its clusters. The cluster structure uniquely identifies a tree ϑ. The nodes a, b, c are supposed to represent the first graph element of Fig. 7 while the nodes $1,2,3,4$ represent the third graph element: altogheter such seven vertices correspond to the seven lowest nodes of the tree, *i.e.* they are the innermost clusters. The clusters are here represented by dashed rectangles (rather than by ellipses as in Fig. 5,6, to avoid confusion with the graph loops). The three unlabeled nodes of ϑ correspond to the three dashed rectangles. It is instructive to draw the corresponding, quite different, picture in the case the line $1a$ (for instance) has a scale larger than all the others.

Instead of thinking the coefficient of orders k_4, k_2, k_1 in $\lambda, \mu, \varepsilon$ as a sum of values of Feynman graphs we can think of it as a double sum over all trees with $k_4 + k_2 + k_1$ nodes and over all graphs compatible with each tree.

Given a tree ϑ and a graph Γ compatible with it

(1) we say that a graph vertex v belongs to a cluster if at least one of the lines that end in it is part of the cluster;

(2) we say that a node $v \in \vartheta$ corresponding to a cluster of scale h_v has degree p_v if there are p_v lines "emerging" from it, *i.e.* p_v lines of the graph have an extreme vertex in the cluster v but have scale $h < h_v$. It is possible that both ends of the line are vertices in a cluster but the line has scale lower than that of the cluster: in this case the line counts twice in the definition of p_v (because we imagine that two lines emerge from the cluster and are then contracted on a lower scale).
(3) the number p_v of lines external to the cluster v should not be confused with the number s_v defined as the number of tree nodes that precede v or, equivalently, as the number of clusters contained in the cluster v but not in smaller clusters.

The cluster structure sets a natural order in the integrations necessary to evaluate the graph value. Taking into account the expression of the graph values in (7.1) and the form $2^{(d-2)h}C_1(2^h(\underline{x} - \underline{y}))$ of the propagator on scale h we can bound the graph value simply by bounding the propagators on scale h by $B_0 2^{(d-2)h} e^{-\kappa 2^h |\underline{x} - \underline{y}|}$, for some $B_0, \kappa > 0$, and use the exponential decay to bound the results of the integrals.

Let h_0 be the scale of the root line. Let $M_{4,v}, M_{2,v}, M_{1,v}$ be the numbers of vertices of the graph elements in the cluster v which correspond, respectively, to the first, second and third graph element in Fig. 7 and let $m_{4,v}, m_{2,v}, m_{1,v}$ be the numbers of vertices in the cluster v of scale h_v *which are not contained inside inner clusters* and which correspond to the mentioned graph elements. One remarks that

$$\sum_v (h_v - h_0)(s_v - 1) = \sum_v (h_v - h_{v'})(M_{4,v} + M_{2,v} + M_{1,v} - 1) \tag{8.7}$$

and it follows that the bound on the value of a graph Γ compatible with the given tree ϑ is $|\Lambda| B^{k-1} |\lambda^{M_4} \overline{\mu}_{max}^{M_2} \varepsilon^{M_1}|$, with $k = M_4 + M_2 + M_1$, $\mu_{max} = \max_{\underline{x}} |\mu_N(\underline{x})|$ and B depending on B_0 and $\int e^{-\kappa |\underline{x}|} d^d \underline{x}$, times

$$\begin{aligned} \prod\nolimits_v 2^{h_v (d-2)\frac{1}{2}(4m_{4,v}+2m_{2,v}+m_{1,v}-n_v^e+\sum_{j=1}^{s_v} n_{v_j}^e)} 2^{-h_v d(s_v-1)} = \\ = 2^{h_0((d-4)M_4-2M_2-(d-\frac{d-2}{2})M_1+d)} \,. \\ \cdot \prod\nolimits_v 2^{(h_v-h_{v'})((d-4)M_{4,v}-2M_{2,v}-(d-\frac{d-2}{2})M_{1,v}-n_v^e\frac{d-2}{2}+d)} \end{aligned} \tag{8.8}$$

where M_4, M_2, M_1 are the total number of graph elements of the first,second and third type in Fig. 7, $M_{4,v}, M_{2,v}, M_{1,v}$ are the total number of graph elements of the first,second and third type in Fig. 7 contained in the cluster v, and n_v^e are the total number of lines emerging from the cluster v.

Having obtained the above *dimensional estimates* we must sum over the graphs and trees, which essentially means summing over the scales h_v, and show that the sum converges as $N \to \infty$: this will be true but only for certain choices of $\overline{\mu}_N$, as dicussed in Sec. 9 below. The case $d = 2$ is very simple (and in any event we have already shown the existence of the perturbation analysis so that discussing the bound (8.8) is not necessary): therefore we concentrate attention on the more interesting $d = 3$ case.

9 Determination of the counterterms and renormalizability ($d = 3$)

Since we consider contributions with $M_1 > 0$ ($M_1 = 0$ corresponds to "vacuum graphs" which do not contribute to the Schwinger functions) the exponent in the first factor is always < 0 so that the sum over h_0 (which has to be performed together with the sum over the other scales to take into account all graphs) converges; noting that $M_{1,v}$ has to be even (otherwise not all lines can be paired) the exponents in the product are also < 0 unless

(1) $n_v^e = 0$ which also corresponds to a vacuum graph

(2) $n_v^e = 2$ if $M_{2,v} = 1, M_{1,v} = M_{4,v} = 0$ which would give an exponent 0 but which cannot arise in a cluster

(3) $n_v^e = 4$ if $M_{4,v} = 1, M_{1,v} = M_{2,v} = 0$ which would give an exponent 0 but which cannot arise in a cluster

(4) $n_v^e = 2$ if $M_{4,v} = 1, M_{1,v} = M_{2,v} = 0$ which gives an exponent 1

(5) $n_v^e = 2$ if $M_{4,v} = 2, M_{1,v} = M_{2,v} = 0$ which gives an exponent 0.

Therefore we only have to study the cases (4) and (5) which correspond to graphs containing a subgraph like the ones in Fig. 10 below.

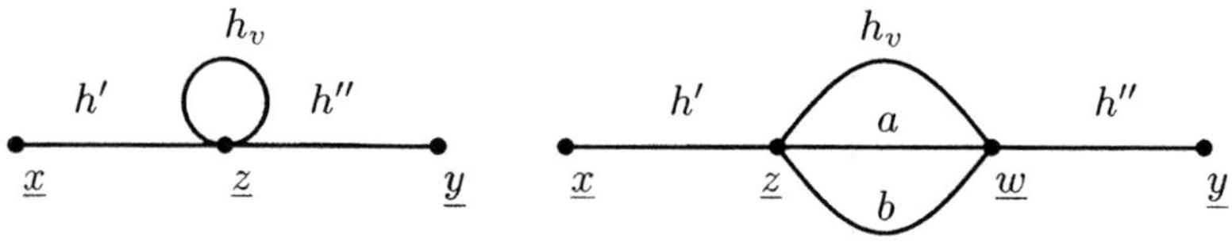

Figure 10: The two subgraphs for which the estimate above diverges, as $N \to \infty$, when summed over the scales: we suppose $h' \le h'' < h_v \le a, b$.

By our choice (8.5) of the interaction V_N the first graph in Fig. 10 contains a self contraction and therefore does not arise (as commented above). Furthermore we can associate the second graphs with the subgraph

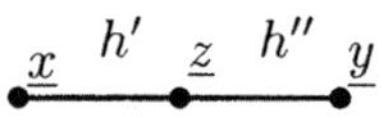

Figure 11: The subgraph whose contribution will be summed to the one of the second subgraph in Fig. 10.

and their contribution to a graph value will be respectively

$$\overline{\mu}\, C_{h'}(\underline{x} - \underline{z}) C_{h''}(\underline{z} - \underline{y}) \tag{9.1}$$
$$6\,\lambda^2\, C_{h_v}(\underline{z} - \underline{w}) C_a(\underline{z} - \underline{w}) C_b(\underline{z} - \underline{w}) C_{h'}(\underline{x} - \underline{z}) C_{h''}(\underline{w} - \underline{y})$$

which, if summed together, exonerate us from considering Feynman graphs containing clusters like the second one in Fig. 10.

It will be useful to rewrite the sum of the two terms in (9.1) as the sum of

$$\Big(\overline{\mu}+6\,\lambda^2\sum_{h_v>h'',a,b\ge h_v}\int d^d\,\underline{w}\;C_{h_v}(\,\underline{z}\,-\,\underline{w}\,)C_a(\,\underline{z}\,-\,\underline{w}\,)C_b(\,\underline{z}\,-\,\underline{w}\,)\Big)\cdot$$
$$\cdot\,C_{h'}(\,\underline{x}\,-\,\underline{z}\,)\,C_{h''}(\,\underline{z}\,-\,\underline{y}\,) \tag{9.2}$$

and of

$$6\,\lambda^2\textstyle\sum_{h_v>h'',a,b\ge h_v}\int d^d\,w\,C_{h_v}(\,\underline{z}\,-\,\underline{w}\,)C_a(\,\underline{z}\,-\,\underline{w}\,)C_b(\,\underline{z}\,-\,\underline{w}\,)\cdot$$
$$\cdot C_{h'}(\,\underline{x}\,-\,\underline{z}\,)\Big(C_{h''}(\,\underline{w}\,-\,\underline{y}\,)-C_{h''}(\,\underline{z}\,-\,\underline{y}\,)\Big) \tag{9.3}$$

Therefore we can imagine to consider graphs without subgraphs like the ones in Fig. 10 but which can instead contain subgraphs of the two forms

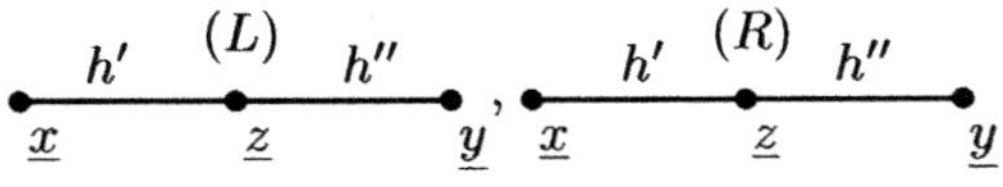

Figure 12: The two subgraphs into which the sum of the contributions from the second subgraph in Fig. 9 and from the one in Fig. 11 is decomposed.

which contribute to the value of a graph in which they appear, respectively, a factor given by (9.2) or by (9.3).

Note that in (9.3) the difference $\Delta \stackrel{def}{=} \big(C_{h''}(\,\underline{w}\,-\,\underline{y}\,)-C_{h''}(\,\underline{z}\,-\,\underline{y}\,)\big)$ appears multiplied by the factor $C_{h_v}(\,\underline{z}\,-\,\underline{w}\,)$: therefore the points $\underline{z}$, $\underline{w}$ can be considered to be at distance $O(2^{-h_v})$ typical of the scale of the lines linking $\underline{z}$, $\underline{w}$. However the covariance $C_{h''}$ "lives" on scale h'' so that the difference Δ can be bounded proportionally to $2^{-(h_v-h'')}$.

The factor (9.3) will yield in the bound (8.8) *an extra factor* $2^{-(h''-h_v)}$ while if $\overline{\mu}_N(\lambda,\,\underline{x}\,)$ is *defined* as

$$\overline{\mu}_N(\lambda,\,\underline{x}\,)=-\Big(6\,\lambda^2\sum_{h=0,a,b\ge h}^{N}\int d^d\,\underline{w}\;C_h(\,\underline{z}\,-\,\underline{w}\,)C_a(\,\underline{z}\,-\,\underline{w}\,)C_b(\,\underline{z}\,-\,\underline{w}\,)\Big) \tag{9.4}$$

then the first graph in Fig. 12 will contribute to the bound (8.8) an extra factor

$$-\Big(6\,\lambda^2\sum_{h=0,a,b\ge h}^{h''}\int d^d\,\underline{w}\;C_h(\,\underline{z}\,-\,\underline{w}\,)C_a(\,\underline{z}\,-\,\underline{w}\,)C_b(\,\underline{z}\,-\,\underline{w}\,)\Big) \tag{9.5}$$

which is bounded proportionally to h'': *i.e.* in the bound corresponding to (8.8) there will be an extra power of $h_v^{m_{2,v}}$ which does not affect convergence of the sums over the scales.

Thus we see that formal perturbation theory can be well defined at each order so that the theory is "*renormalizable*", *i.e.* it admits a formal power series in λ, ε for the Schwinger functions generator $E_\lambda(\varepsilon f)$ with coefficients uniformly bounded as $N \to \infty$: the *harder problem* of showing that, with the choice (9.4) for $\overline{\mu}(\underline{x}, \lambda)$, the limit as $N \to \infty$ exists and the perturbation series is asymptotic to it is of course far more interesting: the above analysis is an essential tool for obtaining the result but new ideas need to be introduced, see references.

Note that the dependence of $\overline{\mu}$ on $\underline{x}$ is essential (it can be avoided only if periodic boundary conditions on Λ are adopted): a point that is often not mentioned in the literature.

The case $d = 4$ can only be studied in perturbation theory and often it is conjectured that there is no way to find a probability distribution over the fields $\varphi_{\underline{x}}$ which can be associated with the formal perturbation series (however the problem is wide open).

It becomes necessary, however, to allow also the coefficient of $\varphi^4_{\underline{x}}$ to depend on the cut–off N and to add another "counterterm" $\alpha_N : (\underline{\partial}_{\underline{x}} \varphi_{\underline{x}})^2$:

$$V_N(\varphi_{\underline{x}}) = \lambda_N(\underline{x}) : \varphi^4_{\underline{x}} : + \mu_N(\underline{x}) : \varphi^2_{\underline{x}} : + \alpha_N(\underline{x}) : (\underline{\partial}_{\underline{x}} \varphi_{\underline{x}})^2 : \qquad (9.6)$$

and the question is whether one can find functions $\lambda_N, \mu_N, \alpha_N$ of a parameter λ such that the limit of $E_{\lambda,N}(\varepsilon f)$ as $N \to \infty$ exists and is smooth in λ, ε as well as *non trivial*, *i.e.* not quadratic in f. The problem remains physically interesting because the physical interpretation of the theory as a quantum field model would allow such an extension of the problem.

The method followed in dimension $d = 2, 3$ can also be applied to show renormalizability in the case $d = 4$ with, however, some rather major modifications that we cannot discuss here: see the references.

The reader should not be surprised that the analysis of the quantum fields models is apparently simpler than the one met in studying the KAM theory. The reason is simply that while in the KAM case we have presented a complete discussion in the case of quantum fields we only presented a complete solution to problem of the existence of a *formal perturbation series*. As already stressed another very important (and not easy) part of the work remains to be done and it is to show that the functions $E_{\lambda,N}(\varepsilon f)$ really have a limit as $N \to \infty$ and that the limit admits the formal pertirbation series as an asymptotic series: this problem is often called the *large fields problem* and it can be solved in dimension $d = 2, 3$ by a deeper use of the multiscale analysis, see the literature below.

Appendix A: Siegel–Bryuno bound on the number of self-energy clusters

Call $N^*_n(\vartheta)$ the number of non-self-energy lines carrying a scale label $\le n$ in a tree ϑ with k nodes. We shall prove first that $N^*_n(\vartheta) \le 2k(E2^{-\eta n})^{-1} - 1$ if $N_n(\vartheta) > 0$

(recall that $E = N^{-1}2^{-3\eta}$ and $\eta = 1/\tau$). We fix n and denote $N_n^*(\vartheta)$ as $N^*(\vartheta)$.

If ϑ has the root line λ_0 with scale $> n$ then calling $\vartheta_1, \vartheta_2, \ldots, \vartheta_m$ the subtrees of ϑ emerging from the last node of ϑ and with $k_j > E\,2^{-\eta n}$ lines, one has $N^*(\vartheta) = N^*(\vartheta_1) + \ldots + N^*(\vartheta_m)$ and the statement is inductively implied from its validity for $k' < k$ provided it is true that $N^*(\vartheta) = 0$ if $k < E2^{-\eta n}$, which is is certainly the case if E is chosen as in equation (5.3).[7]

In the other case, call $\lambda_1, \ldots, \lambda_m$ the $m \geq 0$ lines on scale $\leq n$ which are the nearest to λ_0:[8] such lines are the entering lines of a cluster T on scale $n_T > n$. If ϑ_i is the tree with λ_i as root line one has $N^*(\vartheta) \leq 1 + \sum_{i=1}^m N^*(\vartheta_i)$, and if $m = 0$ the statement is trivial, while if $m \geq 2$ the statement is again inductively implied by its validity for $k' < k$.

If $m = 1$ we once more have a trivial case unless the order k_1 of ϑ_1 is $k_1 > k - E\,2^{-n\eta}/2$. Finally, and this is the real problem as the analysis of a few examples shows, we claim that in the latter case either the root line of ϑ_1 is a self-energy line or it cannot have scale $\leq n$.

To see this, note that $|\underline{\omega}_0 \cdot \underline{\nu}(\lambda_0)| \leq 2^n$ and $|\underline{\omega}_0 \cdot \underline{\nu}(\lambda_1)| \leq 2^n$, hence $\delta \equiv |(\underline{\omega}_0 \cdot (\underline{\nu}(\lambda_0) - \underline{\nu}(\lambda_1))| \leq 2^{n+1}$, and the Diophantine condition implies that either $|\underline{\nu}(\lambda_0) - \underline{\nu}(\lambda_1)| > 2^{-(n+1)\eta}$ or $\underline{\nu}(\lambda_0) = \underline{\nu}(\lambda_1)$. The latter case being discarded as $k - k_1 < E\,2^{-n\eta}/2$ (and we are not considering the self-energy clusters), it follows that $k - k_1 < E\,2^{-n\eta}/2$ is inconsistent: it would in fact imply that $\underline{\nu}(\lambda_0) - \underline{\nu}(\lambda_1))$ is a sum of $k - k_1$ node momenta and therefore $|\underline{\nu}(\lambda_0) - \underline{\nu}(\lambda_1)| < NE\,2^{-n\eta}/2$, hence $\delta > 2^3\,2^n$ which contradicts the above opposite inequality.

A similar, far easier, induction can be used to prove that if $N_n^*(\vartheta) > 0$ then the number $p_n(\vartheta)$ of clusters of scale n verifies the bound $p_n(\vartheta) \leq 2k\,(E2^{-\eta n})^{-1} - 1$. Thus equation (5.3) is proved.

Remark. The above argument is a minor adaptation of Bryuno's proof of Siegel's theorem, as remarkably exposed by Pöschel.

Appendix B: The KAM bound for graphs containing overlapping self energy graphs.

Let ϑ be a tree with height p: then each of its maximal self-energy clusters V contains a tree of height $< p$. We imagine that all the resummations relative to the lines that enter the self-energy clusters that are not maximal have been performed

[7]Note that if $k \leq E\,2^{-n\eta}$ one has, for all momenta $\underline{\nu}$ of the lines, $|\underline{\nu}| \leq NE\,2^{-n\eta}$, *i.e.* $|\underline{\omega}_0 \cdot \underline{\nu}| \geq (NE\,2^{-n\eta})^{-\tau} = 2^3\,2^n$ so that there are *no* clusters T with $n_T = n$ and $N^*(\vartheta) = 0$. The choice $E = N^{-1}2^{-3\eta}$ is convenient: but this, as well as the whole lemma, remains true if 3 is replaced by any number larger than 1. The choice of 3 is made only to simplify some of the arguments based on the self-energy cluster concept.

[8]*i.e.* such that no other lin along the paths connecting the lines $\ell_1, \ldots, \ell_m$ to the root is on scale $\leq n$.

so that we only have to consider the trees that are obtained by attaching the lines that enter the maximal self-energy clusters T to the nodes in $\widetilde{T}$.

Suppose for simplicity that there is only one maximal self-energy cluster T of height p. Then the sum of the values of the trees of the family $\mathcal{F}(\vartheta)$ obtained by shifting the entrance node into the self-energy clusters of lower height will have the form

$$\Big(\prod_{\lambda\in\Lambda(\vartheta)\backslash\Lambda_1(T)}\frac{\underline{\nu}_{v'}\cdot\underline{\nu}_{v}}{(\underline{\omega}\cdot\underline{\nu}(\lambda))^2}\Big)\cdot$$
$$\cdot\Big(\prod_{\lambda\in\Lambda(\widetilde{T})}\frac{\underline{\nu}_{v'}\cdot\underline{\nu}_{v}}{(\underline{\omega}\cdot\underline{\nu}^0(\lambda)+\sigma_\lambda\eta_T)^2}\Big)\cdot\frac{1}{\eta_T^4}\cdot\Big(\prod_{T_i}F(T_i,\underline{\nu}_{v_i},\underline{\nu}_{v_i'})\Big),\qquad\text{(B.1)}$$

where α is 1 if the lines entering and exiting the cluster the last product is over all the maximal self-energy clusters T_i contained in T, v_i' and v_i are the nodes in $\widetilde{T}_i$ from which exits (or enters, respectively) the line that enters (or exits) the self-energy cluster T_i, and $F(T_i,\underline{\nu}_{v_i},\underline{\nu}_{v_i'})$ is the sum of the values of all the trees that we have to sum in shifting the entrance node of the lines that enter the self-energy clusters of lower order inside T_i. Note that the lines λ in the first product are the lines external to T (but neither entering or exiting T), while the ones in the second product are the lines internal to T (*i.e.* lines connecting to nodes in $\widetilde{T}$).

We can then remark that when η_T varies in the complex disk $|\eta_T|\le 2^{n_T}$ the divisors of the lines that enter the inner self-energy clusters T' (of any height) do not exceed in modulus $2^{n_{\lambda_{T'}}}$. Therefore we can bound the quantities $F(T_i,\underline{\nu}_{v_i},\underline{\nu}_{v_i'})$ via the inductive bound and obtain that (6.4) is valid also for trees of height p which contain only one maximal self-energy cluster.

The case in which there are many self-energy clusters of height p is reducible to the case in which there is only one such cluster, see references, the conclusion is the validity of the inequality (6.4) in general. It would also possible to give a proof of the inequality that is not based on an inductive argument but we leave it as a problem for the reader.

Acknowledgments

I am grateful to G.Gentile, A. Giuliani and V. Mastropietro for their critical comments. Support from IHES and INFN is also acknowledged.

Bibliographical notes

The KAM analysis is taken from

[Ga94] G. Gallavotti, Twistless KAM tori, *Communications in Mathematical Physics* **164**, 145–156 (1994).

and it is a reinterpretation of the original work of Eliasson

[El96] H. Eliasson, Absolutely convergent series expansions for quasi-periodic motions, *Mathematical Physics Electronic Journal* **2**, 1996, (preprint published in 1988).

In [Ga94] the cancellations are shown to occur by collecting suitable families of terms (each containing a rather small number of terms) appearing in the Lindstedt series for $\underline{h}$ while in [El96] the necessary cancellations are shown to occur if one sums all the Lindstedt terms as a consequence of a general symmetry argument. The extension of Brjuno's lemma used here is taken from

[Pö86] J. Pöschel, J.: Invariant manifolds of complex analytic mappings, Les Houches, XLIII (1984), Vol. II, p. 949–964, eds. K. Osterwalder and R. Stora, North Holland, 1986.

Closely related developments (and generalizations) can be found in

[GG95] G. Gallavotti, G. Gentile, Majorant series convergence for twistless KAM tori, *Ergodic theory and dynamical systems* **15**, 857–869, (1995). See also G. Gallavotti, Invariant tori: a field theoretic point of view on Eliasson's work, in *Advances in Dynamical Systems and Quantum Physics*, 117–132, ed. R. Figari, World Scientific, 1995; and Renormalization group in Statistical Mechanics and Mechanics: gauge symmetries and vanishing beta functions, *Physics Reports* **352**, 251–272 (2001); and G. Gentile, V. Mastropietro, KAM theorem revisited, *Physica D* **90**, 225–234 (1996). And Methods of analysis of the Lindstedt series for KAM tori and renormalizability in classical mechanics. A review with some applications, *Reviews in Mathematical Physics* **8**, 393–444 (1996).

The field theory analysis is based on Wilson's approach to renormalization and it is taken from

[BCGNPOS78] G. Benfatto, M. Cassandro, G. Gallavotti, F. Nicolò, E. Presutti, E. Olivieri and E. Scacciatelli, Some probabilistic techniques in field theory, *Communications in Mathematical Physics* **59**, 143–166 (1978) and **71**, 95–130 (1980).

For the "exact renormalization group" analysis in the case $d = 4$ see

[Pl84] J. Polchinski, Renormalization group and effective lagrangians, *Nuclear Physics* **B231**, 269–295 (1984).
[GN85] G. Gallavotti and F. Nicolò, Renormalization theory for four dimensional scalar fields, *Communications in Mathematical Physics* **100**, 545–590 (1985) and **101**, 1–36 (1985).

The two (independent) approaches differ only because in [Pl84] a continuous variation of the cut–off is considered while in [GN85] the variation takes place discretely: one obtains a continuous flow of the effective potential in the first case described by differential equations and a discrete evolution descibed by a map, respectively.

For reviews on the approach followed here see

[Ga85] G. Gallavotti, Renormalization theory and ultraviolet stability via renormalization group methods, *Reviews of Modern Physics* **57**, 471–569 (1985). And G. Gallavotti, The structure of renormalization theory: renormalization, form factors, and resummations in scalar field theory, ed. K. Osterwalder, R. Stora, Les Houches, XLIII, 1984, Phénomènes critiques, Systèmes aléatoires, Théories de jauge, Elseviers Science, 1986, p. 467–492. The beta function method for resummations in field theory, Constructive quantum field theory, II, Scuola di Erice, 1988, ed. G. Velo, A. Wightman, NATO ASI series B, vol. **234**, 69–88 (1990).

For a general review on the exact renormalization group see

[BG95] G. Benfatto, G. Gallavotti, Renormalization group, p. 1–144, Princeton University Press, 1995.

Giovanni Gallavotti
IHES
35, route de Chartres
F-91440 Bures sur Yvette
and
Fisica
Università di Roma "La Sapienza
P.le Moro 2
00185 Roma, Italie

Poincaré Seminar 2002, 213 – 239

Poincaré Seminar 2002

Phase Transitions and Renormalization Group: from Theory to Numbers

Jean Zinn-Justin

Abstract. During the last century, in two apparently distinct domains of physics, the theory of fundamental interactions and the theory of phase transitions in condensed matter physics, one of the most basic ideas in physics, the decoupling of physics on different length scales, has been challenged. To deal with such a new situation, a new strategy was invented, known under the name of renormalization group. It has allowed not only explaining the survival of universal long distance properties in a situation of coupling between microscopic and macroscopic scales, but also calculating precisely universal quantities.

We here briefly recall the origin of renormalization group ideas; we describe the general renormalization group framework and its implementation in quantum field theory. It has been then possible to employ quantum field theory methods to determine many universal properties concerning the singular behaviour of thermodynamical quantities near a continuous phase transition. Results take the form of divergent perturbative series, to which summation methods have to be applied. The large order behaviour analysis and the Borel transformation have been especially useful in this respect.

As an illustration, we review here the calculation of the simplest quantities, critical exponents.

More details can be found in the work

J. Zinn-Justin, *Quantum Field Theory and Critical Phenomena*, Clarendon Press 1989, (Oxford 4th ed. 2002).

1 Renormalization Group: Motivation and Basic Ideas

During the last century, in two apparently distinct domains of physics, the theory of fundamental interactions and the theory of phase transitions in condensed matter physics, one of the most basic ideas in physics has been challenged:

We have all been taught that physical phenomena should be described in terms of degrees of freedom adapted to their typical scale. For instance, we conclude from dimensional considerations that the period of the pendulum scales like the square root of its length. This result implicitly assumes that other lengths in the problem, like the size of constituent atoms or the radius of the earth, are not relevant because they are much too small or much too large. In the same way, in newtonian mechanics the motion of planets around the sun can be determined, to a very good approximation, by considering planets and sun as point-like, because their sizes are very small compared with the size of the orbits.

It is clear that if this property also called *the decoupling of different scales of physics*, would not generally hold, progress in physics would have been very slow, maybe even impossible.

However, starting from about 1930, it was discovered that the quantum extension of Electrodynamics was plagued with infinities due to the point-like nature of the electron. The basic reason for this disease, the non-decoupling of scales, was understood only much later, but in the mean time physicists had discovered empirically a recipe to do finite calculations, called *renormalization*. Superficially, the renormalization idea is conventional: to describe physics, use parameters adapted to the scale of observation, like the observed strength of the electromagnetic interaction and the observed mass of the electron, rather than the initial parameters of the quantum lagrangian. However, there remained two peculiarities, the relation between initial parameters and so-called renormalized parameters involved infinities and the values of the renormalized parameters varied with the length or energy scale at which they were defined. This effect was eventually observed very directly in experiments; for example, the fine structure constant $\alpha = e^2/4\pi\hbar c$ is about $1/137$ at the scale given by the electron mass, but increases to $1/128$ at the scale of Z vector boson mass (one of the particles mediating weak interactions). The relation between the strength of interactions at different scales was called renormalization group (RG).

Later, similar difficulties were discovered in another branch of physics, in the study of continuous phase transitions (liquid–vapour, ferromagnetic, superfluid helium). Near a continuous phase transition a length, called the *correlation length*, becomes very large. This means that dynamically a length scale is generated, which is much larger than the scale characterizing the microscopic interactions. In such a situation, some non-trivial macroscopic physics is generated and it could have been expected that phenomena at the scale of the correlation length could be described by a small number of degrees of freedom adapted to this scale. Such an assumption leads to universal quasi-gaussian or mean field critical behaviour, but it became slowly apparent that critical phenomena could not be described by mean field theory. Again the deep reason for this failure is the non-decoupling of scales, that is the initial microscopic scale is never completely forgotten.

Both in the theory of fundamental interactions and in statistical physics, this coupling of very different scales is the sign that an *infinite* number of "stochastic" (i.e. subject to quantum or statistical fluctuations) degrees of freedom are involved.

One could then have feared that even at large scales physics remained completely dependent on the initial microscopic interactions, rendering a predictive theory impossible. However, this is not what empirically was discovered. Instead, phenomena could be gathered in *universality classes* that shared a number of universal properties, a situation that indicated that only a limited number of qualitative properties of the initial microscopic interactions were important.

Remark. We have already referred to the correlation length without defining it. In statistical systems, the correlation length ξ describes the exponential decay of

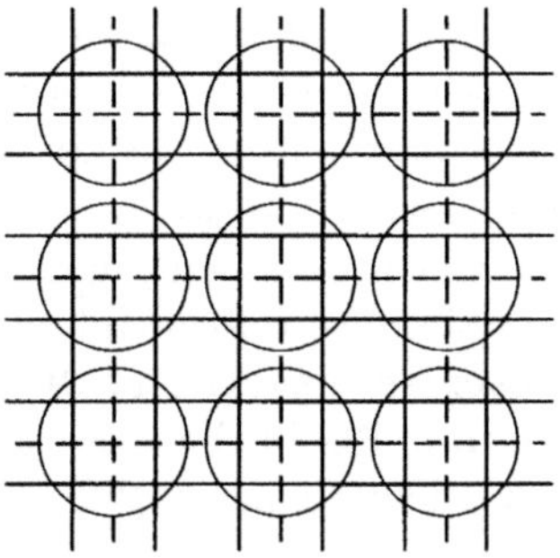

Figure 1: Initial lattice and lattice with double spacing.

correlation functions in the disordered phase. For instance, for a system where the degrees of freedom are spins $S(x)$ at space position x, the two-point correlation function $\langle S(x)S(y)\rangle$ decays exponentially at large distance like

$$\frac{\ln \langle S(x)S(y)\rangle}{|x-y|} \underset{|x-y|\to\infty}{\sim} -\frac{1}{\xi}.$$

The renormalization group idea. To explain this puzzling situation a new concept had to be invented, which was given again the name of RG. The idea that we will shortly describe, involved determining inductively the effective interactions at a given scale. The relation between effective interactions at neighbouring scales is called a RG transformation. A way to construct such a RG was proposed initially by Kadanoff. One considers a statistical model initially defined in terms of classical spin variables on some lattice of spacing a and configuration energy $\mathcal{H}_a(S)$. The partition function is obtained by summing over all spin configurations with a Boltzmann weight $\mathrm{e}^{-\mathcal{H}_a(S)/T}$. The idea then is to sum over the initial spins, keeping their average on the coarser lattice of spacing $2a$ fixed (figure 1). After this summation, the partition function is given by summing over the average spins on a lattice of spacing $2a$ with an *effective* configuration energy $\mathcal{H}_{2a}(S)$. It is clear that this transformation can be iterated as long as the lattice spacing remains much smaller than the correlation length ξ that describes the decay of correlation functions. This defines effective hamiltonians $\mathcal{H}_{2^n a}(S)$ on lattices of spacing $2^n a$. The recursion relation

$$\mathcal{H}_{2^n a}(S) = \mathcal{T}\left[\mathcal{H}_{2^{n-1}a}(S)\right],$$

is a renormalization group transformation. If the transformation $\mathcal{T}$ has fixed points:

$$\mathcal{H}_{2^n a}(S) \underset{n\to\infty}{\to} \mathcal{H}^*(S),$$

or fixed surfaces, then both the non-gaussian behaviour and universality can be understood. Wilson transformed this idea based on an iterative summation of short distance degree of freedom, whose initial formulation was somewhat vague, into a more precise framework, replacing, in particular, RG in space by integration over large momenta in the Fourier representation. Wegner, Wilson and others then discovered exact functional RG equations in the continuum with fixed points.

However, these general equations do not provide a very efficient framework for finding fixed points and calculating explicitly universal quantities. On the other hand, it can be argued that the simplest universality classes contain some standard quantum field theories. Moreover, the field theory RG that had been identified previously, appeared as an asymptotic RG in the more general framework. Therefore, previously developed quantum field theory (QFT) techniques could be used to prove universality and devise efficient methods of calculation, a domain in which the Saclay group has been especially active.

A strong limitation of this strategy is that the construction is possible only when fixed points are gaussian or, in the sense of some external parameter, close to a gaussian fixed point. This explains the role of Wilson–Fisher's ε-expansion, where ε is the deviation from the dimension 4: in dimension 4, non-trivial IR fixed points relevant for many simple phase transitions merge with the gaussian fixed point.

Note, however, that a combination of clever tricks has allowed doing calculations also at fixed dimensions, like the physical dimension 3.

Finally, let me notice that the understanding of non-decoupling of scales and universality resulting from RG fixed points, has also led to an understanding of the renormalization procedure in the theory of fundamental interactions. The quantum field theory that describes almost all known phenomena in particle physics except gravitation (the Standard Model) is now viewed as an *effective* low energy theory in the RG sense, and the cut-off as the remnant of some initial still unknown microscopic physics.

2 Renormalization Group: The General Idea

Even, if initially a statistical model is defined in terms of lattice variables taking a discrete set of values, asymptotically after RG transformations, the averaged variables will have a continuous distribution, and space will also be continuous. Therefore, RG fixed points belong to the class of statistical field theories in the continuum.

We thus consider a general statistical model defined in terms of some, translation invariant, hamiltonian $\mathcal{H}(\phi)$, function of a field $\phi(x)$ ($x \in \mathbb{R}^d$), which is assumed to be expandable in powers of ϕ:

$$\mathcal{H}(\phi) = \sum_{n=0} \frac{1}{n!} \int \mathrm{d}^d x_1 \mathrm{d}^d x_2 \ldots \mathrm{d}^d x_n \, \mathcal{H}_n(x_1, x_2, \ldots, x_n) \phi(x_1) \ldots \phi(x_n), \qquad (2.1)$$

and has all the properties of the thermodynamic potential of Landau's theory. For example, the Fourier transforms of the functions $\mathcal{H}_n$, after factorization of a δ function of momentum conservation, are regular at low momenta (assumption of short-range forces or locality). In this framework, the space of all possible hamiltonians is infinite dimensional.

To a hamiltonian $\mathcal{H}(\phi)$ (really a configuration energy), corresponds a set of connected correlation functions $W^{(n)}(x_1, \ldots, x_n)$:

$$W^{(n)}(x_1, x_2, \ldots, x_n) = \left[\int [\mathrm{d}\phi]\, \phi(x_1) \ldots \phi(x_n)\, \mathrm{e}^{-\beta\mathcal{H}(\phi)}\right]_{\text{connect.}} . \tag{2.2}$$

Connected correlation functions decay at large distance. One of the central problems is the determination of the long distance behaviour of correlation functions, that is the behaviour of $W^{(n)}(\lambda x_1, \ldots, \lambda x_n)$ when the dilatation parameter λ becomes large, near a continuous phase transition. In what follows we will only discuss critical correlation functions, that is correlation functions at the critical temperature where the correlation length is infinite ($T = T_c$, $\xi = \infty$), although universal behaviour extends to the neighbourhood of the critical temperature where the correlation length is large.

2.1 The renormalization group idea. Fixed points

The RG idea is to trade the initial problem, studying the behaviour of correlation functions as a function of dilatation parameter λ acting on space variables, for the study of the flow of a scale-dependent hamiltonian $\mathcal{H}_\lambda(\phi)$ which has essentially the same correlation functions at fixed space positions. More precisely, one wants to construct a hamiltonian $\mathcal{H}_\lambda(\phi)$ which has correlation functions $W_\lambda^{(n)}(x_i)$ satisfying

$$W_\lambda^{(n)}(x_1, \ldots, x_n) = Z^{-n/2}(\lambda) W^{(n)}(\lambda x_1, \ldots, \lambda x_n). \tag{2.3}$$

The mapping $\mathcal{H}(\phi) \mapsto \mathcal{H}_\lambda(\phi)$ is called a RG transformation. We define the transformation such that $\mathcal{H}_{\lambda=1}(\phi) \equiv \mathcal{H}(\phi)$. The choice of the function $Z(\lambda)$ depends on RG transformations.

In the case of models invariant under space translations, equation (2.3) after a Fourier transformation reads

$$\widetilde{W}_\lambda^{(n)}(p_1, \ldots, p_n) = Z^{-n/2}(\lambda) \lambda^{(1-n)d} \widetilde{W}^{(n)}(p_1/\lambda, \ldots, p_n/\lambda) . \tag{2.4}$$

The simplest such RG transformation corresponds to rescalings of space and field. However, this transformation has a fixed point only in exceptional cases (gaussian models) and thus more general transformations have to be considered.

The fixed point hamiltonian. Let us assume that a RG transformation has been found such that, when λ becomes large, the hamiltonian $\mathcal{H}_\lambda(\phi)$ has a limit $\mathcal{H}^*(\phi)$, the fixed point hamiltonian. If such a fixed point exists in hamiltonian space, then the correlation functions $W_\lambda^{(n)}$ have corresponding limits $W_*^{(n)}$ and equation (2.3) becomes

$$W^{(n)}(\lambda x_1, \ldots, \lambda x_n) \underset{\lambda \to \infty}{\sim} Z^{n/2}(\lambda) W_*^{(n)}(x_1, \ldots, x_n). \tag{2.5}$$

We now introduce a second scale parameter μ and calculating $W^{(n)}(\lambda\mu x_i)$ from equation (2.5) in two different ways, we obtain a relation involving only $W_*^{(n)}$:

$$W_*^{(n)}(\mu x_1,\ldots,\mu x_n) = Z_*^{n/2}(\mu) W_*^{(n)}(x_1,\ldots,x_n) \tag{2.6}$$

with

$$Z_*(\mu) = \lim_{\lambda\to\infty} Z(\lambda\mu)/Z(\lambda). \tag{2.7}$$

Equation (2.6) being valid for arbitrary μ immediately implies that Z_* forms a representation of the dilatation semi-group. Thus, under reasonable assumptions,

$$Z_*(\lambda) = \lambda^{-2d_\phi}. \tag{2.8}$$

The fixed point correlation functions have a power law behaviour characterized by a positive number d_ϕ which is called the dimension of the field or order parameter $\phi(x)$.

Returning now to equation (2.7), we conclude that $Z(\lambda)$ also has asymptotically a power law behaviour. Equation (2.5) then shows that the correlation functions $W^{(n)}$ have a scaling behaviour at large distances:

$$W^{(n)}(\lambda x_1,\ldots,\lambda x_n) \underset{\lambda\to\infty}{\sim} \lambda^{-nd_\phi} W_*^{(n)}(x_1,\ldots,x_n) \tag{2.9}$$

with a power d_ϕ which is a property of the fixed point. The r.h.s. of the equation, which determines the critical behaviour of correlation functions, therefore, depends only on the fixed point hamiltonian. In other words, the correlation functions corresponding to all hamiltonians which flow after RG transformations into the same fixed point, have the same critical behaviour. This property is an example of *universality*. The space of hamiltonians is thus divided into *universality classes*. Universality, beyond the gaussian theory, relies upon the existence of IR fixed points in the space of hamiltonians.

2.2 Hamiltonian flows. Scaling operators

Let us consider an infinitesimal dilatation which leads from the scale λ to the scale $\lambda(1+\mathrm{d}\lambda/\lambda)$. The variation of the hamiltonian $\mathcal{H}_\lambda$, consistent with equation (2.5), takes the form of a differential equation which involves a mapping $\mathcal{T}$ of the space of hamiltonians into itself and a real function η defined on the space of hamiltonians:

$$\lambda\frac{\mathrm{d}}{\mathrm{d}\lambda}\mathcal{H}_\lambda = \mathcal{T}[\mathcal{H}_\lambda], \tag{2.10}$$

$$\lambda\frac{\mathrm{d}}{\mathrm{d}\lambda}\ln Z(\lambda) = -2d_\phi[\mathcal{H}_\lambda]. \tag{2.11}$$

Equation (2.10) is a RG transformation in differential form. Moreover, we look only for markovian flows as a function of the "time" $\ln\lambda$, that is such that $\mathcal{T}$ does not depend on λ.

A fixed point hamiltonian $\mathcal{H}^*$ is then a solution of the fixed point equation

$$\mathcal{T}[\mathcal{H}^*] = 0. \tag{2.12}$$

The dimension d_ϕ of the field ϕ follows

$$d_\phi = d_\phi[\mathcal{H}^*]. \tag{2.13}$$

Linearized flow equations. To study the local stability of fixed points, we apply the RG transformation (2.10) to a hamiltonian $\mathcal{H}_\lambda = \mathcal{H}^* + \Delta\mathcal{H}_\lambda$ close to the fixed point $\mathcal{H}^*$. The linearized RG equation takes the form

$$\lambda \frac{\mathrm{d}}{\mathrm{d}\lambda} \Delta\mathcal{H}_\lambda = L^*(\Delta\mathcal{H}_\lambda), \tag{2.14}$$

where L^* is a linear operator, also independent of λ, acting on hamiltonian space. Let us assume that L^* has a discrete set of eigenvalues l_i corresponding to a set of eigenoperators $\mathcal{O}_i$. Then, $\Delta\mathcal{H}_\lambda$ can be expanded on the $\mathcal{O}_i$'s:

$$\Delta\mathcal{H}_\lambda = \sum h_i(\lambda)\mathcal{O}_i, \tag{2.15}$$

and the transformation (2.14) becomes

$$\lambda \frac{\mathrm{d}}{\mathrm{d}\lambda} h_i(\lambda) = l_i h_i(\lambda) \;\Rightarrow\; h_i(\lambda) = \lambda^{l_i} h_i(1). \tag{2.16}$$

Classification of eigenvectors or scaling fields. The eigenvectors $\mathcal{O}_i$ can be classified into four families depending on the corresponding eigenvalues l_i:

(i) Eigenvalues with a positive real part. The corresponding eigenoperators are called *relevant*. If $\mathcal{H}_\lambda$ has a component on one of these operators, this component will grow with λ, and $\mathcal{H}_\lambda$ will move away from the neighbourhood of $\mathcal{H}^*$. Operators associated with a deviation from criticality are clearly relevant since a dilatation decreases the effective correlation length.

(ii) Eigenvalues with $\mathrm{Re}(l_i) = 0$. Then, two situations can arise: either $\mathrm{Im}(l_i)$ does not vanish, and the corresponding component has a periodic behaviour, or $l_i = 0$. Eigenoperators corresponding to a vanishing eigenvalue are called *marginal*. To determine the behaviour of the corresponding component h_i, it is necessary to expand beyond the linear approximation. Generically, one finds

$$\lambda \frac{\mathrm{d}}{\mathrm{d}\lambda} h_i(\lambda) \sim B h_i^2. \tag{2.17}$$

Depending on the sign of the constant B and the initial sign of h_i, the fixed point then is marginally unstable or stable. In the latter case, the solution takes for λ large the form

$$h_i(\lambda) \sim -1/(B \ln \lambda). \tag{2.18}$$

A marginal operator generally leads to a logarithmic approach to a fixed point. In section 3.2, we show that in the ϕ^4 field theory, the operator $\phi^4(x)$ is marginally irrelevant in four dimensions.

An exceptional example is provided by the XY model in two dimensions ($O(2)$ symmetric non-linear σ-model) which instead of an isolated fixed point, has a line of fixed points. The operator which corresponds to a motion along the line is obviously marginal.

(iii) Eigenvalues with a negative real part. The corresponding operators are called *irrelevant.* The effective components on these operators go to zero for large dilatations.

(iv) Finally, some operators do not affect the physics. An example is provided by the operator realizing a constant multiplicative renormalization of the dynamical variables $\phi(x)$. These operators are called *redundant.* In QFT, quantum equation of motions correspond to redundant operators with vanishing eigenvalue.

Classification of fixed points. Fixed points can be classified according to their local stability properties, that is, to the number of relevant operators. This number is also the number of conditions a general hamiltonian must satisfy to belong to the surface which flows into the fixed point.

The critical domain. Universality is not limited to the critical theory. For temperatures close to T_c, and more generally for theories in which the hamiltonian is the sum of a critical hamiltonian and a linear combination of relevant operators with very small amplitudes, universal properties can be derived. Indeed, for small dilatations, the RG flow is hardly affected. After some large dilatation, the flow starts deviating substantially from the flow of the critical hamiltonian. But at this point the components of the hamiltonian on all irrelevant operators are already small.

This argument indicates that the behaviour of correlation functions as a function of amplitudes of relevant operators is universal in the limit of asymptotically small amplitudes. One calls *critical domain* the domain of parameters in which universality can be expected.

3 Critical Behaviour: The Effective ϕ^4 Field Theory

In the discussion, we restrict ourselves to Ising-like systems, the field ϕ having only one component. A generalization to the N-vector model with $O(N)$ symmetry is straightforward.

The main difficulty with the general RG approach is that it requires an explicit construction of RG transformations for hamiltonians, which have a chance to possess fixed points. The general idea is to integrate over the large momentum modes of the dynamical variables, but its practical implementation is far from being straightforward. In the continuum, RG equations, known under the name of Exact or Functional RG, have been discovered, which in simple examples have indeed fixed points. They can be written

$$\begin{aligned}\lambda\frac{\mathrm{d}}{\mathrm{d}\lambda}\mathcal{H}(\phi,\lambda) &= -\int \mathrm{d}^d x\,\frac{\delta\mathcal{H}(\phi,\lambda)}{\delta\phi(x)}\left[d_\phi(\mathcal{H})+\sum_\mu x^\mu\frac{\partial}{\partial x^\mu}\right]\phi(x)\\ &\quad -\frac{1}{2}\int \mathrm{d}^d x\mathrm{d}^d y\,D(x-y)\left[\frac{\delta^2\mathcal{H}}{\delta\phi(x)\delta\phi(y)}-\frac{\delta\mathcal{H}}{\delta\phi(x)}\frac{\delta\mathcal{H}}{\delta\phi(y)}\right]\\ &\quad -\int \mathrm{d}^d x\,\mathrm{d}^d y\,L(x-y)\frac{\delta\mathcal{H}}{\delta\phi(x)}\phi(y),\end{aligned} \tag{3.1}$$

where the functions D and L are defined in terms of a propagator Δ, whose Fourier transform $\tilde{\Delta}(k)$ can be written

$$\tilde{\Delta}(k)=C(k^2)/k^2\,,\quad C(0)=1\,,$$

the regular function $C(k^2)$ decreasing faster than any power for $|k|\to\infty$. Then, the Fourier transform of the function D is

$$\tilde{D}(k^2)=2C'(k^2),$$

and

$$L(x)=\frac{1}{(2\pi)^d}\int \mathrm{d}^d k\;\mathrm{e}^{ikx}\,\tilde{D}(k)\tilde{\Delta}^{-1}(k),$$

Various approximation schemes like derivative expansions reduce these equations to partial differential equations, which can be studied numerically. They are quite useful for exploring general properties, beyond perturbative expansions, but are somewhat complicated for precise calculations of universal quantities.

Another strategy is to start from the only fixed point that can be analyzed completely, the gaussian fixed point, the statistical analogue of free QFT. It corresponds to the hamiltonian

$$\mathcal{H}_{\mathrm{G}}(\phi)=\frac{1}{2}\int \mathrm{d}^d x\sum_\mu\left(\partial_\mu\phi(x)\right)^2.$$

An analysis of local perturbations, even functions of ϕ, shows that $\phi^2(x)$, which affects the correlation length, is always relevant. For $d>4$, all other perturbations are irrelevant. For $d=4$, $\phi^4(x)$ becomes marginal and relevant for $d<4$. For lower dimensions eventually other terms become relevant too. The idea then is to work in dimension $d=4$ or in the neighbourhood of dimension 4 (the famous $\varepsilon=4-d$ expansion) and try to write an asymptotic RG (in a sense that will be explained in next section) for the simplified effective local hamiltonian

$$\mathcal{H}(\phi)=\int \mathrm{d}^d x\left\{\frac{1}{2}\nabla\phi(x)K(-\nabla^2/\Lambda^2)\nabla\phi(x)+\frac{1}{2}r\phi^2(x)+\frac{1}{4!}g\Lambda^{4-d}\phi^4(x)\right\}, \tag{3.2}$$

where K is a positive differential operator, $K(z)=1+O(z)$, r and g are *regular* functions of the temperature for T close to T_c and Λ is a large momentum analogous to the cut-off used to regularize QFT, that is $1/\Lambda$ represents the scale of distance at which this effective hamiltonian is no longer generally valid. The parameter g is chosen here dimensionless.

The hamiltonian (3.2) generates a perturbative expansion of field theory type, which can be described in terms of Feynman, diagrams. The quadratic term in (3.2) contains additional higher order derivatives, corresponding to irrelevant operators, reflexion of the initial microscopic structure. They are needed to render perturbation theory finite and this is another manifestation of the non-decoupling of scales.

At g fixed, the correlation length ξ diverges at a value $r = r_c$, which thus corresponds to the critical temperature T_c. In terms of the scale Λ, the critical domain, where universality is expected, is then defined by $|r - r_c| \ll \Lambda^2$, distances large compared to $1/\Lambda$ or momenta much smaller than Λ, and magnetization $M \equiv \langle \phi(x) \rangle \ll \Lambda^{(d/2)-1}$. These conditions are met if Λ is identified with the cut-off of a usual QFT. However, an inspection of the action (3.2) also shows that, in contrast with conventional QFT, the ϕ^4 coupling constant has a dependence in Λ given *a priori*. This follows from the assumption that the effective hamiltonian is derived from some initial microscopic model, and, thus, all operators have coefficients proportional to powers of the cut-off given by their dimension at the gaussian fixed point. For $d < 4$, the ϕ^4 coupling is thus very large in terms of the scale relevant for the critical domain. In the usual formulation of QFT, by contrast, the coupling constant is also an adjustable parameter and the resulting QFT thus is less generic.

3.1 Renormalization group equations near dimension 4

The hamiltonian (3.2) can be studied by QFT methods. Rather than writing RG equations for the hamiltonian, it appears that it is simpler to first derive RG equations for correlation or vertex functions directly, as we now explain. Using a power counting argument, one verifies that the critical theory does not exist in perturbation theory for any dimension smaller than 4. If one defines, by dimensional continuation, a critical theory in dimension $d = 4 - \varepsilon$, even for arbitrarily small ε there always exists an order in perturbation ($\sim 2/\varepsilon$) at which infrared (IR, i.e. zero momentum) divergences appear. Therefore, the idea, originally due to Wilson and Fisher, is to perform a double series expansion in powers of the coupling constant g and ε. Order by order in this expansion, the critical behaviour differs from the gaussian behaviour only by powers of logarithm, and one can construct a perturbative critical theory by adjusting r to its critical value $r_c(T = T_c)$.

In the critical theory, correlation functions have the form

$$W^{(n)}(x_i, g, \Lambda) = \Lambda^{n(d-2)/2} W^{(n)}(\Lambda x_i, g, 1).$$

Therefore, studying the large distance behaviour is equivalent to studying the large cut-off behaviour. One then can use methods developed for the construction of the renormalized massless ϕ^4 field theory. One considers correlation functions of the

Fourier components of the field, after factorization of the δ function of momentum conservation due to translation invariance. Furthermore, it is more convenient to work with algebraic combinations of correlation functions called vertex functions, denoted below by $\Gamma^{(n)}$, and derived from the Legendre transform of the generating functional of connected correlation functions. For example,

$$\widetilde{W}^{(2)}(p; g, \Lambda)\Gamma^{(2)}(p; g, \Lambda) = 1 .$$

One then introduces rescaled (or renormalized) vertex functions characterized by a new scale $\mu \ll \Lambda$ at which universal behaviour is expected,

$$\Gamma_{\mathrm{r}}^{(n)}(p_i; g_{\mathrm{r}}, \mu, \Lambda) = Z^{n/2}(g, \Lambda/\mu)\Gamma^{(n)}(p_i; g, \Lambda), \tag{3.3}$$

where $Z(g, \Lambda/\mu)$ is a field renormalization constant and g_{r} a renormalized coupling constant, which characterizes the strength of the ϕ^4 interaction at scale μ. At criticality

$$\Gamma^{(2)}(p = 0; g, \Lambda) = \Gamma_{\mathrm{r}}^{(2)}(0; g_{\mathrm{r}}, \mu, \Lambda) = 0 .$$

The renormalization factor $Z(g, \Lambda/\mu)$ and the renormalized coupling constant g_{r} are then determined by additional conditions, for example, by renormalization conditions of the form

$$\begin{aligned} \frac{\partial}{\partial p^2}\Gamma_{\mathrm{r}}^{(2)}(p; g_{\mathrm{r}}, \mu, \Lambda)\big|_{p^2=\mu^2} &= 1 , \\ \Gamma_{\mathrm{r}}^{(4)}(p_i = \mu\theta_i; g_{\mathrm{r}}, \mu, \Lambda) &= \mu^{\varepsilon} g_{\mathrm{r}} , \end{aligned} \tag{3.4}$$

in which θ_i is a numerical vector ($\theta_i \neq 0$).

From renormalization theory (more precisely a slightly extended version adapted to the ε-expansion), one then infers that the functions $\Gamma_{\mathrm{r}}^{(n)}(p_i; g_{\mathrm{r}}, \mu, \Lambda)$ of equation (3.3) have at p_i, g_{r} and μ fixed, large cut-off limits which are the renormalized vertex functions $\Gamma_{\mathrm{r}}^{(n)}(p_i; g_{\mathrm{r}}, \mu)$. Moreover, renormalized functions $\Gamma_{\mathrm{r}}^{(n)}$ do not depend on the specific cut-off procedure and, given the normalization conditions (3.4), are universal. Since the renormalized functions $\Gamma_{\mathrm{r}}^{(n)}$ and the initial ones $\Gamma^{(n)}$ are asymptotically proportional, both functions have the same small momentum or large distance behaviour. The renormalized functions thus contain the whole information about the asymptotic universal critical behaviour. One could, therefore, study only renormalized correlation functions, which indeed are the ones useful for many explicit calculations of universal quantities. However, universality is not limited to the asymptotic critical behaviour; leading corrections have also some interesting universal properties. Moreover, renormalized quantities are not directly obtained in non-perturbative calculations. For these various reasons, it is also useful to study the implications of equation (3.3) directly for the initial correlation functions.

RG equations. Differentiating equation (3.3) with respect to Λ at g_{r} and μ fixed, one obtains

$$\Lambda\frac{\partial}{\partial\Lambda}\bigg|_{g_{\mathrm{r}},\mu \text{ fixed}} Z^{n/2}(g, \Lambda/\mu)\Gamma^{(n)}(p_i; g, \Lambda) = O(\Lambda^{-2}(\ln\Lambda)^L). \tag{3.5}$$

We now neglect corrections subleading by powers of Λ order by order in the double series expansion of g and ε. We assume that these corrections, generated by operators irrelevant from the point of view of the gaussian fixed point, remain, after summation, corrections, that is that irrelevant operators are continuously deformed into irrelevant operators for the non-trivial fixed points.

Then, using chain rule, one infers from equation (3.5):

$$\left[\Lambda\frac{\partial}{\partial\Lambda}+\beta(g,\Lambda/\mu)\frac{\partial}{\partial g}-\frac{n}{2}\eta(g,\Lambda/\mu)\right]\Gamma^{(n)}(p_i;g,\Lambda)=0\,. \tag{3.6}$$

The functions β and η, which are dimensionless and may thus depend only on the dimensionless quantities g and Λ/μ, are defined by

$$\beta(g,\Lambda/\mu) = \Lambda\frac{\partial}{\partial\Lambda}\bigg|_{g_r,\mu} g\,, \tag{3.7}$$

$$\eta(g,\Lambda/\mu) = -\Lambda\frac{\partial}{\partial\Lambda}\bigg|_{g_r,\mu} \ln Z(g,\Lambda/\mu). \tag{3.8}$$

However, the functions β and η can also be directly calculated from equation (3.6) in terms of functions $\Gamma^{(n)}$ which do not depend on μ. Therefore, the functions β and η cannot depend on the ratio Λ/μ and equation (3.6) simplifies as

$$\left[\Lambda\frac{\partial}{\partial\Lambda}+\beta(g)\frac{\partial}{\partial g}-\frac{n}{2}\eta(g)\right]\Gamma^{(n)}(p_i;g,\Lambda)=0\,. \tag{3.9}$$

Equation (3.9), consequence of the existence of a renormalized theory, is satisfied, when the cut-off is large, by the physical vertex functions of statistical mechanics which are also the *bare* vertex functions of QFT. It follows implicitly from the solution of equation (3.9) (see section 3.2) that, conversely, the equation implies the existence of a renormalized theory.

This RG is only asymptotic because the r.s.h. of equation (3.5) and thus (3.6) have been neglected.

3.2 Solution of the RG equations: The ε-expansion

Equation (3.9) can be solved by the method of characteristics. One introduces a dilatation parameter λ and looks for functions $g(\lambda)$ and $Z(\lambda)$ such that

$$\lambda\frac{\mathrm{d}}{\mathrm{d}\lambda}\left[Z^{-n/2}(\lambda)\Gamma^{(n)}(p_i;g(\lambda),\lambda\Lambda)\right]=0\,. \tag{3.10}$$

Consistency with equation '3.9) implies

$$\lambda\frac{\mathrm{d}}{\mathrm{d}\lambda}g(\lambda) = \beta\big(g(\lambda)\big),\quad g(1)=g\,, \tag{3.11}$$

$$\lambda\frac{\mathrm{d}}{\mathrm{d}\lambda}\ln Z(\lambda) = \eta\big(g(\lambda)\big)),\quad Z(1)=1\,. \tag{3.12}$$

The function $g(\lambda)$ is the effective coupling at the scale λ, and is governed by the flow equation (3.11). Equation (3.10) implies

$$\Gamma^{(n)}(p_i; g, \Lambda) = Z^{-n/2}(\lambda)\Gamma^{(n)}(p_i; g(\lambda), \lambda\Lambda).$$

It is actually convenient to rescale Λ by a factor $1/\lambda$ and write the equation

$$\Gamma^{(n)}(p_i; g, \Lambda/\lambda) = Z^{-n/2}(\lambda)\Gamma^{(n)}(p_i; g(\lambda), \Lambda). \tag{3.13}$$

Equations (3.11)-(3.12) and (3.13) implement approximately (because terms subleading by powers of Λ have been neglected) the RG ideas as presented in section 2: since the coupling constant $g(\lambda)$ characterizes the hamiltonian $\mathcal{H}_\lambda$, equation (3.11) is the equivalent of equation (2.10) (up to the change $\lambda \mapsto 1/\lambda$); equations (2.11) and (3.12) differ only by the definition of $Z(\lambda)$.

The solutions of equations (3.11)-(3.12) can be written as

$$\int_g^{g(\lambda)} \frac{\mathrm{d}g'}{\beta(g')} = \ln\lambda\,, \tag{3.14}$$

$$\int_1^\lambda \frac{\mathrm{d}\sigma}{\sigma}\eta(g(\sigma)) = \ln Z(\lambda). \tag{3.15}$$

Equation (3.9) is the RG equation in differential form. Equations (3.13) and (3.14)-(3.15) are the integrated RG equations. In equation (3.13), we see that it is equivalent to increase Λ or to decrease λ. To investigate the large Λ limit we, therefore, study the behaviour of the effective coupling constant $g(\lambda)$ when λ goes to zero. Equation (3.14) shows that $g(\lambda)$ increases if the function β is negative, or decreases in the opposite case. Fixed points correspond to zeros of the β-function which, therefore, play an essential role in the analysis of critical behaviour. Those where the β-function has a negative slope are IR repulsive: the effective coupling moves away from such zeros, except if the initial coupling has exactly a fixed point value. Conversely, those where the slope is positive are IR attractive.

The RG functions have been calculated in perturbation theory and one finds

$$\beta(g, \varepsilon) = -\varepsilon g + \frac{3g^2}{16\pi^2} + O(g^3, g^2\varepsilon). \tag{3.16}$$

The explicit expression (3.16) of the β-function shows that above dimension 4, that is, $\varepsilon < 0$, if initially g is small, $g(\lambda)$ decreases approaching the origin $g = 0$. We recover that the gaussian fixed point is IR stable.

By contrast, below four dimensions, if initially g is very small, $g(\lambda)$ first increases, a behaviour reflecting the instability of the gaussian fixed point.

However, the explicit expression (3.16) shows that, in the sense of an expansion in powers of ε, $\beta(g)$ has another zero

$$g^* = 16\pi^2\varepsilon/3 + O(\varepsilon^2) \;\Rightarrow\; \beta(g^*) = 0\,, \tag{3.17}$$

with a positive slope for ε infinitesimal:

$$\omega \equiv \beta'(g^*) = \varepsilon + O(\varepsilon^2) > 0 . \tag{3.18}$$

Then, equation (3.14) shows that $g(\lambda)$ has g^* as an asymptotic limit. Below dimension 4, at least for ε infinitesimal, this non-gaussian fixed point is IR stable. In dimension 4, it merges with the gaussian fixed point and the eigenvalue ω vanishes indicating the appearance of the marginal operator.

From equation (3.15), we then derive the behaviour of $Z(\lambda)$ for λ small. The integral in the l.h.s. is dominated by small values of σ. It follows that

$$\ln Z(\lambda) \underset{\lambda \to 0}{\sim} \eta \ln \lambda , \tag{3.19}$$

where we have set

$$\eta = \eta(g^*).$$

Equation (3.13) then determines the behaviour of $\Gamma^{(n)}(p_i; g, \Lambda)$ for Λ large:

$$\Gamma^{(n)}(p_i; g, \Lambda/\lambda) \sim \lambda^{-n\eta/2} \Gamma^{(n)}(p_i; g^*, \Lambda). \tag{3.20}$$

On the other hand, from simple dimensional considerations, we know that

$$\Gamma^{(n)}(p_i; g, \Lambda/\lambda) = \lambda^{-d+(n/2)(d-2)} \Gamma^{(n)}(\lambda p_i; g, \Lambda). \tag{3.21}$$

Combining this equation with equation (3.20), we obtain

$$\Gamma^{(n)}(\lambda p_i; g, \Lambda) \underset{\lambda \to 0}{\sim} \lambda^{d-(n/2)(d-2+\eta)} \Gamma^{(n)}(p_i; g^*, \Lambda). \tag{3.22}$$

This equation shows that critical vertex functions have a power law behaviour for small momenta, independent of the initial value of the ϕ^4 coupling constant g, at least if g initially is small enough for perturbation theory to be meaningful, or if the β-function has no other zero.

Equation (3.22) yields for $n = 2$ the small momentum behaviour of the vertex two-point function, and thus of the two-point correlation function

$$\widetilde{W}^{(2)}(p) = \left[\Gamma^{(2)}(p)\right]^{-1} \underset{|p| \to 0}{\sim} 1/p^{2-\eta}. \tag{3.23}$$

The spectral representation of the two-point function implies $\eta > 0$. A short calculation yields

$$\eta = \frac{\varepsilon^2}{54} + O(\varepsilon^3). \tag{3.24}$$

Finally, we note that equation (3.22) can be interpreted by saying that the field $\phi(x)$, which had at the gaussian fixed point a canonical dimension $(d-2)/2$, has now acquired an anomalous dimension (equation (2.13))

$$d_\phi = \tfrac{1}{2}(d - 2 + \eta).$$

Universality. Within the framework of the ε-expansion, all correlation functions have, for $d < 4$, a long distance behaviour different from the one predicted by a quasi-gaussian or mean field theory. In addition, the critical behaviour does not depend on the initial value of the ϕ^4 coupling constant g. Therefore, the critical behaviour is *universal*, although less universal than in the quasi-gaussian theory, in the sense that it depends only on some small number of qualitative properties of the physical system under study.

4 Calculation of Universal Quantities

We present here some explicit results obtained within the framework of the $O(N)$ symmetric $(\phi^2)^2$ statistical field theory. The results of the $(\phi^2)^2$ field theory, or N-vector model do not apply only to ferromagnetic systems. The superfluid helium transition corresponds to $N = 2$, the $N = 0$ limit is related to the statistical properties of polymers and the Ising-like $N = 1$ model also describes the physics of the liquid–vapour transition.

We discuss only critical exponents, although a number of other universal quantities have been calculated like the scaling equation of state or ratios of critical amplitudes.

4.1 The ε-expansion

Critical exponents in the N-vector model are known up to order ε^5. The higher order calculations have been done using dimensional regularization and a minimal subtraction scheme. The equation of state is known up to order ε^2 for arbitrary N and to order ε^3 for $N = 1$. A number of results have also been obtained for the two-point correlation function.

4.2 Critical exponents

Although the RG functions of the $(\phi^2)^2$ theory and, therefore, the critical exponents are known up to five-loop order, we give here only two successive terms in the expansion for illustration purpose, referring to the literature for higher order results. In terms of the variable

$$\tilde{g} = N_d \, g \,, \qquad N_d = \frac{2}{(4\pi)^{d/2}\Gamma(d/2)} \,, \tag{4.1}$$

the RG functions $\beta(\tilde{g})$ and $\eta_2(\tilde{g})$ at two-loop order, $\eta(\tilde{g})$ at three-loop order are

$$\beta(\tilde{g}) = -\varepsilon\tilde{g} + \frac{(N+8)}{6}\tilde{g}^2 - \frac{(3N+14)}{12}\tilde{g}^3 + O\left(\tilde{g}^4\right), \tag{4.2}$$

$$\eta(\tilde{g}) = \frac{(N+2)}{72}\tilde{g}^2\left[1 - \frac{(N+8)}{24}\tilde{g}\right] + O\left(\tilde{g}^4\right), \tag{4.3}$$

$$\eta_2(\tilde{g}) = -\frac{(N+2)}{6}\tilde{g}\left[1 - \frac{5}{12}\tilde{g}\right] + O\left(\tilde{g}^3\right), \tag{4.4}$$

k	0	1	2	3	4	5
γ	1.000	1.1667	1.2438	1.1948	1.3384	0.8918
η	0.0...	0.0...	0.0185	0.0372	0.0289	0.0545

Table 1: Sum of the successive terms of the ε-expansion of γ and η for $\varepsilon = 1$ and $N = 1$

The zero $\tilde{g}^*(\varepsilon)$ of the β-function then is $\tilde{g}^*(\varepsilon) = 6\varepsilon/(N+8) + O(\varepsilon^2)$. The values of the critical exponents η, γ and the correction exponent ω,

$$\eta = \eta(\tilde{g}^*), \quad \gamma = \frac{2-\eta}{2+\eta_2(\tilde{g}^*)}, \quad \omega = \beta'(\tilde{g}^*),$$

follow

$$\eta = \frac{\varepsilon^2(N+2)}{2(N+8)^2}\left[1 + \frac{(-N^2+56N+272)}{4(N+8)^2}\varepsilon\right] + O\left(\varepsilon^4\right), \tag{4.5}$$

$$\gamma = 1 + \frac{(N+2)}{2(N+8)}\varepsilon + \frac{(N+2)}{4(N+8)^3}\left(N^2+22N+52\right)\varepsilon^2 + O\left(\varepsilon^3\right), \tag{4.6}$$

$$\omega = \varepsilon - \frac{3(3N+14)}{(N+8)^2}\varepsilon^2 + O\left(\varepsilon^3\right). \tag{4.7}$$

Other exponents can be obtained from scaling relations. Note that the results at next order involve $\zeta(3)$. At higher orders $\zeta(5)$ and $\zeta(7)$ successively appear. In table 1, we give the values of the critical exponents γ and η obtained by simply adding the successive terms of the ε-expansion for $\varepsilon = 1$ and $N = 1$.

One immediately observes a striking phenomenon: the sums first seem to settle near some reasonable value and then begin to diverge with increasing oscillations. This is an indication that the ε-expansion is divergent for all values of ε. Divergent series can be used for small values of the argument. However, only a limited number of terms of the series can then be taken into account. The last term added gives an indication of the size of the irreducible error. For the exponents γ and η we roughly conclude from the series

$$\gamma = 1.244 \pm 0.050\,, \qquad \eta = 0.037 \pm 0.008\,,$$

where the errors are only indicative.

4.3 The perturbative expansion at fixed dimension

Critical exponents and various universal quantities have also been calculated within the framework of the massive $(\phi^2)^2$ field theory, as perturbative series at fixed dimension 3. The basic reason is that in dimension 3 one-loop diagrams have simple analytic expressions that can be used to simplify most higher order diagrams. It has, therefore, been possible to calculate the RG functions of the N-vector model up to six- and partially seven-loop order.

Note that this massive ϕ^4 field theory is a somewhat artificial construction: when the correlation length increases, simultaneously the coefficient of the relevant ϕ^4 operator is tuned to decrease like $g \propto (\Lambda\xi)^{d-4}$. Then, all correlation functions have a limit for $\Lambda \to \infty$, order by order in an expansion in powers of g at fixed dimension $d < 4$. However, the usual critical theory corresponds in this framework to an infinite coupling constant. Therefore, correlation functions renormalized at zero momentum are introduced, and correspondingly a renormalized coupling constant g_r, which is a universal function of g. Within the framework of the ε-expansion, one proves that g_r has a finite limit g_r^* when $g \to \infty$. To the mapping $g \mapsto g_r$ is associated a function $\beta(g_r)$. For example, the RG β-function in three dimensions, for $N = 1$, has the expansion

$$\begin{aligned}\beta(\tilde{g}) &= -\tilde{g} + \tilde{g}^2 - \tfrac{308}{729}\tilde{g}^3 + 0.3510695978\tilde{g}^4 - 0.3765268283\tilde{g}^5 \\ &\quad +0.49554751\tilde{g}^6 - 0.749689\tilde{g}^7 + O\left(\tilde{g}^8\right) \qquad (4.8)\end{aligned}$$

with the normalization

$$\tilde{g} = 3g_r/(16\pi). \qquad (4.9)$$

To calculate exponents or other universal quantities, one has first to find the IR stable zero g_r^* of the function $\beta(g_r)$, which is given by a few terms of a divergent expansion. An obvious problem is the absence of any small parameter: g^* is a number of order 1. Already at this stage a summation method is required. Estimates of critical exponents are displayed in table 3. In recent years universal ratios of critical amplitudes as well as the equation of state for Ising-like systems ($N = 1$) have also been calculated. Note, however, that in this framework, the calculation of physical quantities in the ordered phase leads to additional technical problems because the theory is parameterized in terms of the disordered phase correlation length m^{-1} which is singular at T_c. Also, the normalization of correlation functions is singular at T_c. This required developing a combination of techniques based on series summation, parametric representation and a method of order-dependent mapping.

4.4 Series summation

Because all series, ε-expansion or perturbative expansions at fixed dimension, are divergent, summation methods had to be developed. We describe here methods based on generalized Borel transformations. The necessary analytic continuation of the Borel transform outside its circle of convergence is then achieved by a conformal mapping.

The method. Several different variants based on the Borel–Leroy transformation have been implemented and tested. Let $R(z)$ be the function whose expansion has to be summed (z here stands for ε or the coupling constant $\tilde{g}$):

$$R(z) = \sum_{k=0} R_k z^k. \tag{4.10}$$

A plausible assumption is that the Borel transform is analytic in a cut-plane. One thus transforms the series into

$$R(z) = \sum_{k=0}^{\infty} B_k(\rho) \int_0^{\infty} t^{\rho}\, \mathrm{e}^{-t}\, [u(zt)]^k \, \mathrm{d}t\,, \tag{4.11}$$

$$u(z) = \frac{\sqrt{1+az}-1}{\sqrt{1+za}+1}. \tag{4.12}$$

The coefficients B_k are calculated by identifying the expansion of the r.h.s. of equation (4.11) in powers of z with the expansion (4.10). The constant a is known from the large order behaviour analysis in QFT based on instantons,

$$a(d=3) = 0.147774232 \times \bigl(9/(N+8)\bigr), \tag{4.13}$$

and ρ is a free parameter, adjusted empirically to improve the convergence of the transformed series by weakening the singularities of the Borel transform near $z = -a$. Eventually, the method has been refined, which involved also introducing two additional free parameters.

Needless to say, with three parameters and short initial series it becomes possible to find occasionally some transformed series whose apparent convergence is deceptively good. It is, therefore, essential to vary the parameters in some range around the optimal values to examine the sensitivity of the results upon their variations. Finally, it is useful to sum independently series for exponents related by scaling relations. An underestimation of the apparent errors leads to inconsistent results. It is clear from these remarks that the errors quoted in the final results are educated guesses based on a large number of consistency checks.

A few examples of transformed series are displayed in table 2 to illustrate the convergence. *The $(\phi^2)^2$ field theory at fixed dimensions.* The RG β-function

has been determined up to six-loop order in three dimensions, while the series for the dimensions of the fields ϕ and ϕ^2 have recently been extended to seven loops. The series of the RG β-function has been first summed and its zero $\tilde{g}^*$ determined ($\tilde{g} = g_{\mathrm{r}}(N+8)/(48\pi)$ for $d=3$. The series of the other RG functions have then been summed for $\tilde{g} = \tilde{g}^*$. Examples of convergence are given in table 2.

The ε-expansion. The ε-expansion has one advantage: it allows connecting the results in three and two dimensions. In particular, in the cases $N=1$ and $N=0$, it is possible to compare the ϕ^4 results with exact results coming from lattice models and to test both universality and the reliability of the summation procedure. Moreover, it is possible to improve the three-dimensional results by imposing the exact two-dimensional values or the behaviour near two dimensions for $N > 1$.

k	2	3	4	5	6	7
$\tilde{g}^*$	1.8774	1.5135	1.4149	1.4107	1.4103	1.4105
ν	0.6338	0.6328	0.62966	0.6302	0.6302	0.6302
γ	1.2257	1.2370	1.2386	1.2398	1.2398	1.2398

Table 2: Series summed by the method based on Borel transformation and mapping for the zero $\tilde{g}^*$ of the $\beta(g)$ function and the exponents γ and ν in the ϕ_3^4 field theory

However, since the series in ε are shorter than the series at fixed dimension 3, the apparent errors are larger. Finally, as already emphasized, the comparison between the different results is a check of the consistency of QFT methods combined with the summation procedures.

4.5 Numerical estimates of critical exponents

Fixed dimension 3. Table 3 displays the results obtained from summed perturbation series at fixed dimension 3. The last exponent $\theta = \omega\nu$ characterizes corrections to scaling in the temperature variable.

Note that shorter series have been generated in dimension 2 (five loops). Because the series are short and the fixed coupling constant larger, the apparent errors are large, but the results are consistent with exact $N = 1$ results.

The ε-expansion. In Table 4, we give the results coming from the summed ε-expansion for $\varepsilon = 2$ and compare them with exact results.

We see in this table that the agreement for $N = 0$ and $N = 1$ QFT and lattice models is satisfactory. We feel justified, therefore, in using a summation procedure of the ε-expansion which automatically incorporates the $d = 2$, $\varepsilon = 2$ values. Note, however, that in both cases, the identification of ω remains a problem.

Table 5 then displays the results for $\varepsilon = 1$, both for the ε series (free) and a modified ε series where the $d = 2$ results are imposed (bc).

Discussion. One can now compare the two sets of results coming from the perturbation series at fixed dimension, and the ε-expansion. First let us emphasize that the agreement is quite spectacular, although the apparent errors of the ε-expansion are in general larger because the series are shorter. Moreover, the agreement has improved with longer series.

The best agreement is found for the exponents ν and β. On the other hand, the values of η coming from the ε-expansion are systematically larger by about

N	0	1	2	3
$\tilde{g}^*$	1.413 ± 0.006	1.411 ± 0.004	1.403 ± 0.003	1.390 ± 0.004
g^*	26.63 ± 0.11	23.64 ± 0.07	21.16 ± 0.05	19.06 ± 0.05
γ	1.1596 ± 0.0020	1.2396 ± 0.0013	1.3169 ± 0.0020	1.3895 ± 0.0050
ν	0.5882 ± 0.0011	0.6304 ± 0.0013	0.6703 ± 0.0015	0.7073 ± 0.0035
η	0.0284 ± 0.0025	0.0335 ± 0.0025	0.0354 ± 0.0025	0.0355 ± 0.0025
β	0.3024 ± 0.0008	0.3258 ± 0.0014	0.3470 ± 0.0016	0.3662 ± 0.0025
α	0.235 ± 0.003	0.109 ± 0.004	-0.011 ± 0.004	-0.122 ± 0.010
ω	0.812 ± 0.016	0.799 ± 0.011	0.789 ± 0.011	0.782 ± 0.0013
$\theta = \omega\nu$	0.478 ± 0.010	0.504 ± 0.008	0.529 ± 0.009	0.553 ± 0.012

Table 3: Estimates of critical exponents in the $O(N)$ symmetric $(\phi^2)^2_3$ field theory

	γ	ν	η	β	ω
$N = 0$	1.39 ± 0.04	0.76 ± 0.03	0.21 ± 0.05	0.065 ± 0.015	1.7 ± 0.2
Exact	1.34375	0.75	$0.2083 \cdots$	$0.0781 \cdots$	?
$N = 1$	1.73 ± 0.06	0.99 ± 0.04	0.26 ± 0.05	0.120 ± 0.015	1.6 ± 0.2
Ising	1.75	1.	0.25	0.125	$1.33\ldots?$

Table 4: Critical exponents in the ϕ^4_2 field theory from the ε-expansion

N	0	1	2	3
γ(free) γ (bc)	1.1575 ± 0.0060 1.1571 ± 0.0030	1.2355 ± 0.0050 1.2380 ± 0.0050	1.3110 ± 0.0070 1.317	1.3820 ± 0.0090 1.392
ν (free) ν (bc)	0.5875 ± 0.0025 0.5878 ± 0.0011	0.6290 ± 0.0025 0.6305 ± 0.0025	0.6680 ± 0.0035 0.671	0.7045 ± 0.0055 0.708
η (free) η (bc)	0.0300 ± 0.0050 0.0315 ± 0.0035	0.0360 ± 0.0050 0.0365 ± 0.0050	0.0380 ± 0.0050 0.0370	0.0375 ± 0.0045 0.0355
β (free) β (bc)	0.3025 ± 0.0025 0.3032 ± 0.0014	0.3257 ± 0.0025 0.3265 ± 0.0015	0.3465 ± 0.0035	0.3655 ± 0.0035
ω	0.828 ± 0.023	0.814 ± 0.018	0.802 ± 0.018	0.794 ± 0.018
θ	0.486 ± 0.016	0.512 ± 0.013	0.536 ± 0.015	0.559 ± 0.017

Table 5: Critical exponents in the $(\phi^2)^2_3$ field theory from the ε-expansion

3×10^{-3}, though the error bars always overlap. The corresponding effect is observed on γ.

Comparison with lattice model estimates. The N-vector with nearest-neighbour interactions has been studied on various lattices. Most of the results for critical exponents come from the analysis of high temperature (HT) series expansion by different types of ratio methods, Padé approximants or differential approximants. Some results have also been obtained from low temperature expansions, computer calculations using stochastic methods, and in low dimensions, transfer matrix methods. Table 6 tries to give an idea of the agreement between lattice and QFT results. A historical remark is here in order: the agreement between both types of theoretical results has improved as the HT series became longer which is of course encouraging. The main reason is that, in the analysis of longer HT series, it has become possible to take into account the influence of confluent singularities due to corrections to the leading power law behaviour, as predicted by the RG. The effect of this improvement has been specially spectacular for the exponents γ and ν of the 3D Ising model.

The obvious conclusion is that one observes no systematic differences. In particular, the agreement is extremely good in the case of the Ising model where the HT series are the most accurate. To the best of our knowledge, the N-vector lattice models and the $(\phi^2)^2$ field theory belong to the same universality class.

Critical exponents from experiments. We have discussed the N-vector model in the ferromagnetic language, even though most of our experimental knowledge comes from physical systems that are non-magnetic, but belong to the universality class

N	0	1	2	3
γ	1.1575 ±0.0006	1.2385± 0.0025	1.322±0.005	1.400±0.006
ν	0.5877±0.0006	0.631±0.002	0.674±0.003	0.710±0.006
α	0.237±0.002	0.103±0.005	-0.022±0.009	-0.133±0.018
β	0.3028±0.0012	0.329±0.009	0.350±0.007	0.365±0.012
θ	0.56±0.03	0.53±0.04	0.60± 0.08	0.54 ±0.10

Table 6: Critical exponents in the N-vector model on the lattice

	γ	ν	β	α	θ
(a)	1.236 ± 0.008	0.625 ± 0.010	0.325 ± 0.005	0.112 ± 0.005	0.50 ± 0.03
(b)	1.23–1.25	0.625 ± 0.006	0.316–0.327	0.107 ± 0.006	0.50 ± 0.03
(c)	1.25 ± 0.01	0.64 ± 0.01	0.328 ± 0.009	0.112 ± 0.007	

Table 7: Critical exponents in fluids and antiferromagnets

of the N-vector model. The case $N = 0$ describes the statistical properties of long polymers, that is, long non-intersecting chains or self-avoiding walks. The case $N = 1$ (Ising-like systems) describes liquid–vapour transitions in classical fluids, critical binary fluids and uniaxial antiferromagnets. The helium superfluid transition corresponds to $N = 2$. Finally, for $N = 3$, the experimental information comes from ferromagnetic systems.

Critical exponents and polymers. In the case of polymers, only the exponent ν is easily accessible. The best results are

$$\nu = 0.586 \pm 0.004 ,$$

in excellent agreement with the RG result.

Ising-like systems $N = 1$. Table 7 gives a survey of the experimental situation for critical binary fluids (a), liquid–vapour transition in classical fluids (b), and antiferromagnets (c). For binary mixtures, we quote a weighted world average. In the case of the liquid–vapour transition, we quote a range of experimental results rather than statistical errors for all exponents but ν, the reason being that the values depend much on the method of analysis of the experimental data. The agreement with RG results is clearly impressive.

Helium superfluid transition, $N = 2$. The helium transition allows measurements very close to T_c and this explains the remarkably precise determination of the

γ	ν	β	α	θ
1.40 ± 0.03	0.700–0.725	0.35 ± 0.03	-0.09– -0.012	0.54 ± 0.10

Table 8: Ferromagnetic systems

critical exponents α and ν. The order parameter, however, is not directly accessible in helium. Most recent reported values are

$$\nu = 0.6705 \pm 0.0006\,, \quad \nu = 0.6708 \pm 0.0004 \quad \text{and} \quad \alpha = -0.01285 \pm 0.00038\,.$$

The agreement with RG values is quite remarkable but the precision of ν is now a challenge to field theory.

Ferromagnetic systems, $N = 3$. Finally, Table 8 displays some results concerning magnetic systems.

Conclusion and prospects. If one takes into account all data (critical exponents, equation of state, amplitude ratios,...) one is forced to conclude that the RG predictions are remarkably consistent with the whole experimental and lattice information available. Considering the variety of experimental situations, this is a spectacular confirmation of the RG ideas and the concept of universality.

The current effort goes in several directions. First, improve the precision of critical exponents, in particular trying to complete the seven loop calculation in three dimensions, which is a very demanding problem from the point of view of computer algebra and numerical integration: it involves calculating about 3500 Feynman diagrams, each of them being given *a priori* by a 21-dimensional integral (a few Feynman diagrams are displayed in Figure 2). After a large number of tricks have been used the number of integrations can be reduced (Figure 3).

Critical exponents are only the simplest universal quantities, but many other universal quantities are worth calculating, like the equation of state, in particular for $N > 1$ in 3 dimensions, $N = 1$ at higher orders in the ε expansion, or the two-point correlation function.

Then, other models with more than one coupling are also of interest, like the model with cubic anisotropy, which has been investigated.

These efforts are paralleled by similar efforts using HT series and simulations.

Much interesting work has been done in recent years using the functional RG equations expanded in the form of a derivative expansion. The main problem there is that it is difficult to go beyond the simplest approximation, and thus difficult to assess the reliability of the results which are obtained. In the future efforts to improve the approximation should be undertaken.

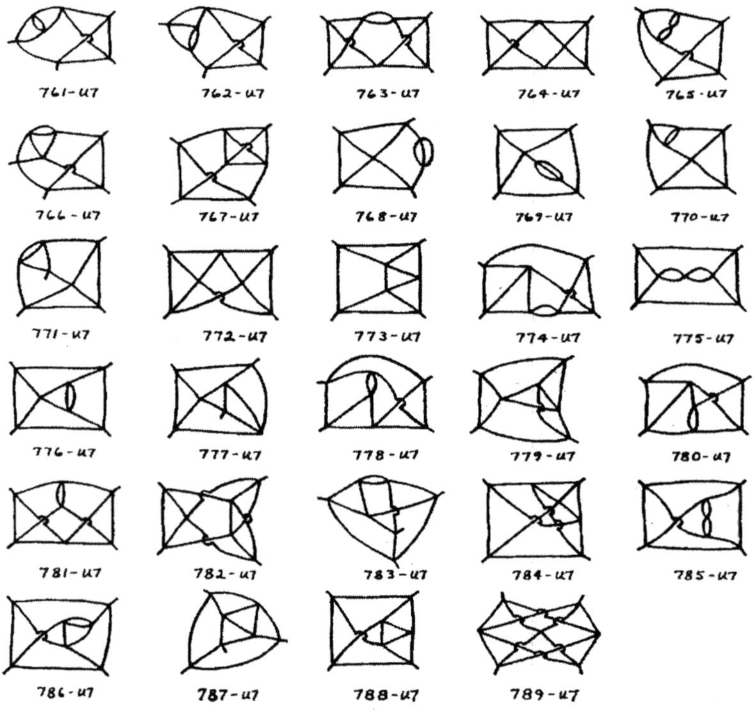

Figure 2: A few seven-loop diagrams.

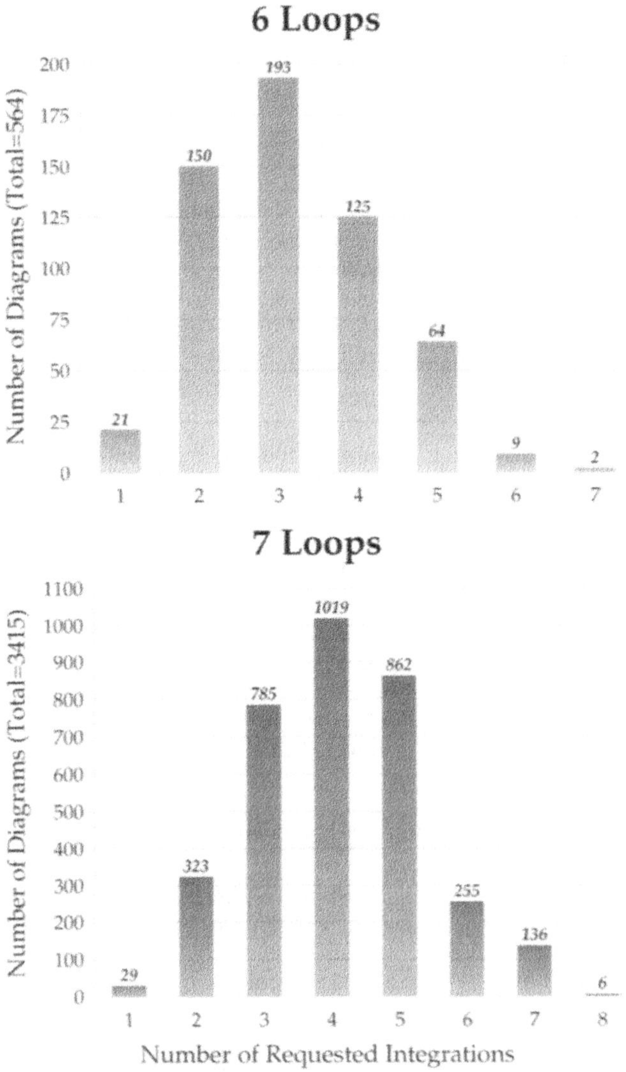

Figure 3: Number of remaining integrations after many tricks have been used (B.G. Nickel, R. Guida, P. Ribeca).

Bibliographical Notes

We give here only a short bibliography.

Many interesting details and references concerning the early history of Quantum Electrodynamics and divergences can be found in
S. Weinberg, *The Theory of Quantum Fields*, vol. 1, chap. 1, Cambridge 1995 (Cambridge Univ. Press).

A review of the situation after the discovery of the Standard Model Standard is found in
Methods in Field Theory, Les Houches 1975, R. Balian et J. Zinn-Justin eds., (North-Holland, Amsterdam 1976); C. Itzykson and J.B. Zuber, *Quantum Field Theory*, (McGraw-Hill, New-York 1980).

For a presentation of RG ideas applied to critical phenomena see
L.P. Kadanoff, *Physics* **2** (1966) 263; K.G. Wilson and J. Kogut, *Phys. Rep.* **12C** (1974) 75,
and the contributions to
Phase Transitions and Critical Phenomena, vol. 6, C. Domb et M.S. Green eds. (Academic Press, London 1976).

In particular the contribution
Field Theory Approach to Critical Phenomena par E. Brézin, J.C. Le Guillou and J. Zinn-Justin,
describes the application quantum field theory methods to the proof of scaling laws and the calculation of universal de quantities.

The RG equations, as written in section 3.1, have been first presented in
J. Zinn-Justin, Cargèse Summer School 1973, Saclay preprint SPhT-T73/049, (Oct. 1973).

The Borel summation of section 4.4 has been first applied on the ε-expansion in
J.C. Le Guillou and J. Zinn-Justin, *J. Physique Lett. (Paris)* **46** (1985) L137; *J. Physique (Paris)* **48** (1987) 19; *ibidem* **50** (1989) 1365.

The use of perturbation series at fixed dimension has been advocated by
G. Parisi, *Cargèse Lectures 1973*, published in *J. Stat. Phys.* **23** (1980) 49.

The calculation of the series expansion for the RG functions has been initiated by Nickel and reported (together with an estimate of $N = 1$ exponents) in
G.A. Baker, B.G. Nickel, M.S. Green and D.I. Meiron, *Phys. Rev. Lett.* **36** (1976) 1351; B.G. Nickel, D.I.Meiron, G.B. Baker, *Univ. of Guelph Report* 1977.

First precise estimates based on the method of section 4.4 were published in
J.C. Le Guillou and J. Zinn-Justin, *Phys. Rev. Lett.* **39** (1977) 95; *Phys. Rev.* **B21** (1980) 3976.

The updated 3D results presented in Section 4.5 are taken from
R. Guida and J. Zinn-Justin, *J. Phys. A* **31** (1998) 8103, cond-mat/9803240,
and obtained from corrected ε series, and seven-loop 3D terms for γ and η reported in
D.B. Murray and B.G. Nickel, unpublished Guelph University report (1991).

Series and estimates for equation of state and ratios of amplitude are reported in
C. Bagnuls and C. Bervillier, *Phys. Rev.* **B24** (1981) 1226; *J. Phys. A* **19** (1986) L85; C. Bervillier, *Phys. Rev.* **B34** (1986) 8141; C. Bagnuls, C. Bervillier, D.I. Meiron and B.G. Nickel, *Phys. Rev.* **B35** (1987) 3585; R. Schloms and V. Dohm, *Phys. Rev.* **B42** (1990) 6142; H. J. Krause, R. Schloms and V. Dohm, *Z. Physik* **B79** (1990) 287; F.J. Halfkann and V. Dohm, *Z. Phys.* **B89** (1992) 79; G. Münster, J. Heitger, *Nucl. Phys.* **B424** [FS] (1994) 582; R. Guida and J. Zinn-Justin, *Nucl. Phys.* **B489** [FS] (1997) 626; S.A. Larin, M. Moennigmann, M. Stroesser and V. Dohm, *Phys. Rev.* **B58** (1998) 3394; M. Stroesser, S.A. Larin and V. Dohm, *Nucl. Phys.* **B540** (1999) 654.

Functional RG equations have been introduced in :
F.J. Wegner and A. Houghton, *Phys. Rev.* **A8** (1973) 401; F.J. Wegner, *J. Physics* **C7** (1974) 2098; K.G. Wilson and J. Kogut, *Phys. Rep.* **12C** (1974) 75.

For recent applications of such ideas see for example
M. Bonini, M. D'Attanasio and G. Marchesini, *Nucl. Phys.* **B409** (1993) 441, *ibidem* B444 (1995) 602; J. Adams, J. Berges, S. Bornholdt, F. Freire, N. Tetradis and C. Wetterich, *Mod. Phys. Lett.* **A10** (1995) 2367; T.R. Morris, *Prog. Theor. Phys. Suppl.* **131** (1998) 395, hep-th/9802039,

Jean Zinn-Justin
Service de Physique Théorique[1]
CEA/Saclay
F-91191 Gif-sur-Yvette Cedex
France

[1]Laboratoire de la Direction des Sciences de la Matière du Commissariat à l'Energie Atomique, Unité de recherche associée au CNRS

Poincaré Seminar 2002, 241 – 264

Poincaré Seminar 2002

Symétries Galoisiennes et Renormalisation

Alain Connes

1 Introduction

La renormalisation est sans doute l'un des procédés les plus élaborés pour obtenir des quantités numériques signifiantes à partir d'expressions mathématiques a-priori dépourvues de sens. A ce titre elle est fascinante autant pour le physicien que pour le mathématicien. La profondeur de ses origines en théorie des champs et la précision avec laquelle elle est corroborée par l'expérience en font l'un des joyaux de la physique théorique. Pour le mathématicien épris de sens, mais non corseté par la rigueur, les explications données jusqu'à présent butaient toujours sur le sens conceptuel de la partie proprement calculatoire, celle qui est utilisée par exemple en électrodynamique quantique et ne tombe pas sous la coupe des "théories asymptotiquement libres" auxquelles la théorie constructive peut prétendre avoir donné un statut mathématique satisfaisant. Cet état de fait a changé récemment et cet exposé se propose de donner la signification conceptuelle des calculs effectués par les physiciens dans la théorie de la renormalisation grâce à mon travail sur la renormalisation en collaboration avec Dirk Kreimer et la relation que nous avons établie entre renormalisation et problème de Riemann–Hilbert.

Le résultat clef est l'identité entre le procédé récursif utilisé par les physiciens et les formules mathématiques qui résolvent un lacet à valeurs dans un groupe pronilpotent G en un produit d'un lacet holomorphe par un lacet anti-holomorphe. La signification géométrique de cette décomposition (de Riemann–Hilbert, Birkhoff ou Wiener–Hopf) provient directement de la théorie des fibrés holomorphes de groupe structural G sur la sphère de Riemann S^2. Dans la renormalisation perturbative, les points de la sphère S^2 sont les dimensions complexes parmi lesquelles la dimension d de l'espace-temps est un point privilégié. Le problème étant que dans les théories physiquement intéressantes les quantités à calculer conspirent pour diverger précisément au point d. On peut organiser ces quantités comme le développement de Taylor d'un difféomorphisme $g \in G$ et donner un sens à $g = g(z)$ en remplaçant dans les formules la dimension d par une valeur complexe $z \neq d$. Le procédé de renormalisation acquiert alors la signification suivante : la valeur cherchée $g \in G$ n'est autre que la valeur $g_+(d)$ en d de la partie holomorphe de la décomposition de Riemann–Hilbert $g(z) = g_-^{-1}(z)g_+(z)$ du lacet $g(z)$. La nature exacte du groupe G impliqué dans la renormalisation a été clarifiée par les étapes essentielles suivantes. La première est la découverte due à Dirk Kreimer de la structure d'algèbre de Hopf secrètement présente dans les formules récursives de Bogoliubov Parasiuk Hepp et Zimmermann.

La seconde qui est le point de départ de notre collaboration est la similitude entre l'algèbre de Hopf des arbres enracinés de Dirk et une algèbre de Hopf que j'avais introduite avec Henri Moscovici pour organiser les calculs très complexes de géométrie noncommutative. Ceci nous a conduit avec Dirk à définir une algèbre de Hopf directement en termes de graphes de Feynman et à lui appliquer le théorème de Milnor–Moore pour en déduire une algèbre de Lie et un groupe de Lie pronilpotent G, analogue du groupe des difféomorphismes formels.

Enfin la troisième étape cruciale est la construction d'une action du groupe G sur les constantes de couplage de la théorie physique. Ceci permet de relever le groupe de renormalisation comme un sous-groupe à un paramètre du groupe G et de montrer directement que les développements polaires des divergences sont entièrement déterminés par leurs résidus.

Le problème de Riemann–Hilbert joue un rôle clef dans la théorie de Galois différentielle, il est donc naturel d'interpréter en termes Galoisiens l'ambiguïté que le groupe de renormalisation introduit dans les théories physiques. La dernière section contient l'esquisse d'une telle interprétation.

Nous commencerons cette section par une introduction très élémentaire à la théorie de Galois pour les équations algébriques, en passant par un beau problème de géométrie plane.

Nous montrerons ensuite le rôle que le groupe de renormalisation devrait jouer pour comprendre la composante connexe du groupe des classes d'idèles de la théorie du corps de classe comme un groupe de Galois. Cette idée s'appuie à la fois sur l'analogie entre la théorie des facteurs et la théorie de Brauer pour un corps local et sur la présence implicite en théorie des champs d'un "corps de constantes" plus élaboré que le corps $\mathbb{C}$ des nombres complexes. En fait les calculs des physiciens regorgent d'exemples de "constantes" telles les constantes de couplage g des interactions (électromagnétiques, faibles et fortes) qui n'ont de "constantes" que le nom. Elles dépendent en réalité du niveau d'énergie μ auquel les expériences sont réalisées et sont des fonctions $g(\mu)$, de sorte que les physiciens des hautes énergies étendent implicitement le "corps des constantes" avec lequel ils travaillent, passant du corps $\mathbb{C}$ des scalaires à un corps de fonctions $g(\mu)$. Le groupe d'automorphismes de ce corps engendré par $\mu\partial/\partial\mu$ est le groupe d'ambiguïté de la théorie physique.

2 Renormalisation, position du problème

La motivation physique de la renormalisation est très claire et remonte aux travaux de Green au dix-neuvième siècle sur l'hydrodynamique. Pour prendre un exemple simple* si l'on calcule l'accélération initiale d'une balle de ping-pong plongée à quelques mètres sous l'eau, l'on obtient en appliquant la loi de Newton $F = ma$ et la poussée d'Archimède $F = (M - m)g$, où m est la masse inerte, et M la

* *voir le cours de théorie des champs de Sidney Coleman*

masse d'eau occupée, une accélération initiale de l'ordre de $11g$!† En réalité, si l'on réalise l'expérience, l'accélération est de l'ordre de $2g$. En fait la présence du fluide autour de la balle oblige à corriger la valeur m de la masse inerte dans la loi de Newton et à la remplacer par une "masse effective" qui en l'occurrence vaut $m+\frac{1}{2}M$. Dans cet exemple, l'on peut bien sur déterminer la masse m en pesant la balle de ping-pong hors de l'eau, mais il n'en va pas de même pour un electron dans le champ electromagnétique, dont il est impossible de l'extraire. De plus le calcul montre que, pour une particule ponctuelle comme le demande la relativité, la correction qui valait $\frac{1}{2}M$ ci-dessus est infinie.

Vers 1947 les physiciens ont réussi à utiliser la distinction entre les deux masses qui apparaissent ci-dessus et plus généralement le concept de quantité physique "effective" pour éliminer les quantités infinies qui apparaissent en théorie des champs quantiques (voir [16] pour un aperçu historique).

Une théorie des champs en d dimensions est donnée par une fonctionnelle d'action classique

$$S(A)=\int \mathcal{L}(A)\,d^dx \tag{1}$$

où A désigne un champ classique et le Lagrangien est de la forme,

$$\mathcal{L}(A)=(\partial A)^2/2-\frac{m^2}{2}A^2+\mathcal{L}_{\text{int}}(A) \tag{2}$$

où $\mathcal{L}_{\text{int}}(A)$ est un polynome en A.

On peut décrire la théorie par les fonctions de Green,

$$G_N(x_1,\ldots,x_N)=\langle 0\,|T\,\phi(x_1)\ldots\phi(x_N)|\,0\,\rangle \tag{3}$$

où le symbole T signifie que les champs quantiques $\phi(x_j)$'s sont écrits à temps croissant de droite à gauche.

L'amplitude de probabilité d'une configuration classique A est donnée par,

$$e^{i\,\frac{S(A)}{\hbar}} \tag{4}$$

et si l'on pouvait ignorer les problèmes de renormalisation, l'on pourrait calculer les fonctions de Green grâce à la formule

$$G_N(x_1,\ldots,x_N)=\mathcal{N}\int e^{i\,\frac{S(A)}{\hbar}}\,A(x_1)\ldots A(x_N)\,[dA] \tag{5}$$

où $\mathcal{N}$ est un facteur de normalisation requis par,

$$\langle 0\,|\,0\,\rangle=1\,. \tag{6}$$

† *La balle pèse* $m=2{,}7$ *grammes et a un diamètre de* 4 *cm de sorte que* $M=33{,}5$ *grammes*

L'on pourrait alors calculer l'intégrale fonctionnelle (5) en théorie des perturbations, en traitant le terme $\mathcal{L}_{\text{int}}$ de (2) comme une perturbation, le Lagrangien libre étant,

$$\mathcal{L}_0(A) = (\partial A)^2/2 - \frac{m^2}{2} A^2 , \tag{7}$$

de sorte que,

$$S(A) = S_0(A) + S_{\text{int}}(A) \tag{8}$$

où l'action libre S_0 définit une mesure Gaussienne $\exp(i\, S_0(A))\, [dA] = d\mu$.
On obtient alors le développement perturbatif des fonctions de Green sous la forme,

$$G_N(x_1, \dots, x_N) = \left(\sum_{n=0}^{\infty} i^n/n! \int A(x_1)\dots A(x_N)\,(S_{\text{int}}(A))^n\, d\mu\right)$$
$$\left(\sum_{n=0}^{\infty} i^n/n! \int S_{\text{int}}(A)^n\, d\mu\right)^{-1} . \tag{9}$$

Les termes de ce développement s'obtiennent en intégrant par partie sous la Gaussienne. Cela engendre un grand nombre de termes $U(\Gamma)$, où les paramètrès Γ sont les graphes de Feynman Γ, i.e. des graphes dont les sommets correspondent aux termes du Lagrangien de la théorie. En règle générale les valeurs des termes $U(\Gamma)$ sont données par des intégrales divergentes. Les divergences les plus importantes sont causées par la présence dans le domaine d'intégration de moments de taille arbitrairement grande. La technique de renormalisation consiste d'abord à "régulariser" ces intégrales divergentes par exemple en introduisant un paramètre de "cutoff" Λ et en se restreignant à la portion correspondante du domaine d'intégration. Les intégrales sont alors finies, mais continuent bien entendu à diverger quand $\Lambda \to \infty$. On établit ensuite une dépendance entre les termes du Lagrangien et Λ pour que les choses s'arrangent et que les résultats ayant un sens physique deviennent finis! Dans le cas particulier des théories asymptotiquement libres, la forme explicite de la dépendance entre les constantes nues et le paramètre de régularisation Λ a permis dans des cas très importants ([19],[17]) de mener à bien le programme de la théorie constructive des champs ([20]).
Décrivons maintenant en détail la technique de renormalisation perturbative. Pour faire les choses systématiquement, on rajoute un "contreterme" $C(\Gamma)$ au Lagrangien de départ $\mathcal{L}$, chaque fois que l'on rencontre un diagramme divergent Γ, dans le but d'annuler la divergence correspondante. Pour les théories "renormalisables", les contre-termes dont on a besoin sont tous déja des termes du Lagrangien $\mathcal{L}$ et ces contorsions peuvent s'interpreter à partir de l'inobservabilité des quantités numériques qui apparaissent dans $\mathcal{L}$, par opposition aux quantités physiques qui, elles, doivent rester finies.
La principale complication dans cette procédure vient de l'existence de nombreux graphes Γ pour lesquels les divergences de $U(\Gamma)$ ne sont pas locales. La raison étant

que ces graphes possèdent déja des sous-graphes dont les divergences doivent être prises en compte avant d'aller plus loin. La méthode combinatoire précise, due à Bogoliubov–Parasiuk–Hepp et Zimmermann ([2]) consiste d'abord à "préparer" le graphe Γ en remplaçant $U(\Gamma)$ par l'expression formelle,

$$\overline{R}(\Gamma) = U(\Gamma) + \sum_{\gamma \subset \Gamma} C(\gamma) U(\Gamma/\gamma) \tag{10}$$

où γ varie parmis tous les sous-graphes divergents. On montre alors que le calcul des divergences du graphe "préparé" ne donne que des expressions locales, qui pour les théories renormalisables se trouvent déja dans le Lagrangien $\mathcal{L}$.

3 L'algèbre de Hopf des graphes de Feynman

Dirk Kreimer a eu l'idée remarquable en 97 ([23]) d'utiliser la formule (10) pour définir le coproduit d'une algèbre de Hopf.
En tant qu'algèbre $\mathcal{H}$ est l'algèbre commutative libre engendrée par les graphes "une particule irréductibles" (1PI) ‡
Elle admet ainsi une base indexée par les graphes Γ unions disjointes de graphes 1PI.

$$\Gamma = \bigcup_{j=1}^{n} \Gamma_j \, . \tag{11}$$

Le produit dans $\mathcal{H}$ est donné par l'union disjointe,

$$\Gamma \cdot \Gamma' = \Gamma \cup \Gamma' \, . \tag{12}$$

Pour définir le coproduit,

$$\Delta : \mathcal{H} \to \mathcal{H} \otimes \mathcal{H} \tag{13}$$

il suffit de le donner sur les graphes 1PI, on a

$$\Delta\Gamma = \Gamma \otimes 1 + 1 \otimes \Gamma + \sum_{\gamma \subset \Gamma} \gamma_{(i)} \otimes \Gamma/\gamma_{(i)} \tag{14}$$

Ici γ est un sous-ensemble (non vide et de complémentaire non-vide) $\gamma \subset \Gamma^{(1)}$ de l'ensemble $\Gamma^{(1)}$ des faces internes de Γ dont les composantes connexes γ' vérifient des conditions d'admissibilité détaillées dans la référence [12].

Le coproduit Δ defini par (14) sur les graphes 1PI se prolonge de manière unique en un homomorphisme de $\mathcal{H}$ dans $\mathcal{H} \otimes \mathcal{H}$. On a alors ([23],[12])

Théorème *Le couple* $(\mathcal{H},\Delta)$ *est une algèbre de Hopf.*

‡ *Un graphe de Feynman* Γ *est "une particule irreductible" (1PI) si il est connexe et le reste apres avoir enlevé n'importe laquelle de ses faces*

4 L'algèbre de Lie des graphes, le groupe G et sa structure.

J'avais à la mème époque, dans les calculs de Géométrie Noncommutative de l'indice transverse pour les feuilletages, montré, avec Henri Moscovici, ([15]) que la compléxité extrême de ces calculs conduisait à introduire une algèbre de Hopf $\mathcal{H}_{cm}$, qui n'est ni commutative ni cocommutative mais est intimement reliée au groupe des difféomorphismes, dont l'algèbre de Lie apparait en appliquant le théorème de Milnor–Moore à une sous-algèbre commutative.
Aprés l'exposé de Dirk à l'IHES en février 98, nous avons tous les deux été intrigués par la similarité apparente entre ces deux algèbres de Hopf et notre collaboration a commencé par l'application du théorème de Milnor–Moore à l'algèbre de Hopf $\mathcal{H}$. Le théorème de Milnor–Moore montre qu'elle est duale de l'algèbre enveloppante d'une algèbre de Lie graduée $\underline{G}$ dont une base est donnée par les graphes 1-particule irreductibles. Le crochet de Lie de deux graphes est obtenu par insertion d'un graphe dans l'autre. Le groupe de Lie correspondant G est le groupe des caractères de $\mathcal{H}$.
Nous avons ensuite analysé le groupe G et montré qu'il est produit semi-direct d'un groupe abélien par un groupe très relié au groupe des difféomorphismes des constantes de couplage sans dimension de la théorie des champs (voir section VII).

L'algèbre de Hopf $\mathcal{H}$ admet plusieurs graduations naturelles. Il suffit de donner le degré des graphes 1PI puis de poser en général,

$$\deg(\Gamma_1 \ldots \Gamma_e) = \sum \deg(\Gamma_j) \,, \quad \deg(1) = 0 \tag{15}$$

On doit vérifier que,

$$\deg(\gamma) + \deg(\Gamma/\gamma) = \deg(\Gamma) \tag{16}$$

pour tout sous-graphe admissible γ.
Les deux graduations les plus naturelles sont

$$I(\Gamma) = \text{nombre de faces internes } \Gamma \tag{17}$$

et

$$v(\Gamma) = V(\Gamma) - 1 = \text{nombre de sommets } \Gamma - 1 \,. \tag{18}$$

On a aussi la combinaison importante

$$L = I - v = I - V + 1 \tag{19}$$

qui est le nombre de boucles du graphe.
Soit G un graphe 1PI avec n faces externes indexées par $i \in \{1, \ldots, n\}$, on spécifie sa structure externe en donnant une distribution σ definie sur un espace convenable de fonctions test $\mathcal{S}$ sur

$$\Big\{(p_i)_{i=1,\ldots,n} \; ; \; \sum p_i = 0\Big\} = E_G \,. \tag{20}$$

Ainsi σ est une forme linéaire continue,

$$\sigma : \mathcal{S}\,(E) \to \mathbb{C}\,. \tag{21}$$

A un graphe Γ de structure externe σ correspond un élément de $\mathcal{H}$ et on a

$$\delta_{(\Gamma,\lambda_1\sigma_1+\lambda_2\sigma_2)} = \lambda_1\,\delta_{(\Gamma,\sigma_1)} + \lambda_2\,\delta_{(\Gamma,\sigma_2)}\,. \tag{22}$$

Nous appliquons alors le théorème de Milnor–Moore à l'algèbre de Hopf bigraduée $\mathcal{H}$.

Ce théorème donne une structure d'algèbre de lie sur,

$$\bigoplus_{\Gamma} \mathcal{S}\,(E_\Gamma) = L \tag{23}$$

où pour chaque graphe 1PI Γ, on définit $\mathcal{S}\,(E_\Gamma)$ comme dans (20). Soit $X \in L$ et soit Z_X la forme linéaire sur $\mathcal{H}$ donnée, sur les monomes Γ, par

$$\langle \Gamma, Z_X\rangle = 0 \tag{24}$$

sauf si Γ est connexe et 1PI, et dans ce cas par,

$$\langle \Gamma, Z_X\rangle = \langle \sigma_\Gamma, X_\Gamma\rangle \tag{25}$$

où σ_Γ est la distribution qui donne la structure externe de Γ et X_Γ la composante correspondante de X. Par construction Z_X est un caractère infinitésimal de $\mathcal{H}$ ainsi que les commutateurs,

$$[Z_{X_1}, Z_{X_2}] = Z_{X_1}\,Z_{X_2} - Z_{X_2}\,Z_{X_1}\,. \tag{26}$$

Le produit étant obtenu par transposition du coproduit de $\mathcal{H}$, i.e. par

$$\langle Z_1\,Z_2, \Gamma\rangle = \langle Z_1 \otimes Z_2, \Delta\,\Gamma\rangle\,. \tag{27}$$

Soient Γ_j, $j = 1{,}2$ des graphes 1PI et $\varphi_j \in \mathcal{S}\,(E_{\Gamma_j})$ les fonctions test correspondantes.

Pour $i \in \{0{,}1\}$, soit $n_i\,(\Gamma_1,\Gamma_2;\Gamma)$ le nombre de sous-graphes de Γ isomorphes à Γ_1 et tels que

$$\Gamma/\Gamma_1(i) \simeq \Gamma_2\,. \tag{28}$$

Soit (Γ,φ) l'élément de L associé à $\varphi \in \mathcal{S}\,(E_\Gamma)$, le crochet de Lie de (Γ_1,φ_1) et (Γ_2,φ_2) est donné par,

$$\sum_{\Gamma,i} \sigma_i\,(\varphi_1)\;n_i\,(\Gamma_1,\Gamma_2;\Gamma)\,(\Gamma,\varphi_2) - \sigma_i\,(\varphi_2)\;n_i\,(\Gamma_2,\Gamma_1;\Gamma)\,(\Gamma,\varphi_1)\,. \tag{29}$$

Théorème ([12]) *L'algèbre de Lie L est produit semi-direct d'une algèbre de Lie Abelienne L_0 par L_c où L_c admet une base canonique indéxée par les graphes $\Gamma^{(i)}$ avec*

$$[\Gamma,\Gamma'] = \sum_{v} \Gamma \circ_v \Gamma' - \sum_{v'} \Gamma' \circ_{v'} \Gamma$$

où $\Gamma \circ_v \Gamma'$ est obtenu en greffant Γ' sur Γ en v.

5 Renormalisation et problème de Riemann–Hilbert

Le problème de Riemann–Hilbert vient du 21eme problème de Hilbert qu'il formulait ainsi;

> "Montrer qu'il existe toujours une équation différentielle Fuchsienne linéaire de singularités et monodromies données."

Sous cette forme il admet une réponse positive due à Plemelj et Birkhoff (cf. [1] pour un exposé détaillé). Quand on le reformule pour les systèmes linéaires de la forme,

$$y'(z) = A(z)\, y(z) \ , \ A(z) = \sum_{\alpha \in S} \frac{A_\alpha}{z - \alpha}\,, \tag{30}$$

où S est l'ensemble fini donné des singularités, $\infty \notin S$, et les A_α sont des matrices complexes telles que

$$\sum A_\alpha = 0 \tag{31}$$

pour éviter les singularités à ∞, la réponse n'est pas toujours positive [3], mais la solution existe quand les matrices de monodromie M_α sont suffisamment proches de 1. On peut alors l'écrire explicitement sous la forme d'une série de polylogarithmes [24].

Une autre formulation du problème de Riemann–Hilbert, intimement reliée à la classification des fibrés vectoriels holomorphes sur la sphère de Riemann $P_1(\mathbb{C})$, est en termes de la décomposition de Birkhoff

$$\gamma\,(z) = \gamma_-(z)^{-1}\,\gamma_+(z) \qquad z \in C \tag{32}$$

où $C \subset P_1(\mathbb{C})$ désigne une courbe simple, C_- la composante connexe du complément de C contenant $\infty \notin C$ et C_+ la composante bornée.

Les trois lacets γ et $\gamma_\pm$ sont à valeurs dans $\mathrm{GL}_n(\mathbb{C})$,

$$\gamma\,(z) \in G = \mathrm{GL}_n(\mathbb{C}) \qquad \forall\, z \in \mathbb{C} \tag{33}$$

et $\gamma_\pm$ sont les valeurs au bord d'applications holomorphes

$$\gamma_\pm : C_\pm \to \mathrm{GL}_n(\mathbb{C})\,. \tag{34}$$

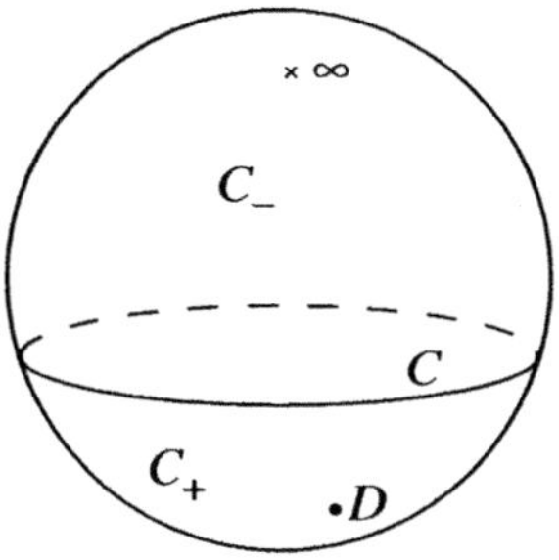

Figure 1

La condition $\gamma_-(\infty) = 1$ assure l'unicité de la décomposition (32) si elle existe.

L'existence de la décomposition de Birkhoff (32) est équivalente à l'annulation,

$$c_1(L_j) = 0 \tag{35}$$

des nombres de Chern $n_j = c_1(L_j)$ des fibrés en droites holomorphes de la décomposition de Birkhoff-Grothendieck,

$$E = \oplus L_j \tag{36}$$

où E est le fibré vectoriel holomorphe sur $P_1(\mathbb{C})$ associé à γ, i.e. d'espace total:

$$(C_+ \times \mathbb{C}^n) \cup_\gamma (C_- \times \mathbb{C}^n) . \tag{37}$$

La discussion ci-dessus pour $G = \mathrm{GL}_n(\mathbb{C})$ s'étend aux groupes de Lie complexes arbitraires.

Quand G est un groupe de Lie complexe nilpotent et simplement connexe l'existence (et l'unicité) de la décomposition de Birkhoff (32) est vraie pour tout γ. Quand le lacet $\gamma : C \to G$ se prolonge en un lacet holomorphe: $C_+ \to G$, la décomposition de Birkhoff est donnée par $\gamma_+ = \gamma$, $\gamma_- = 1$. En général, pour $z \in C_+$ l'évaluation,

$$\gamma \to \gamma_+(z) \in G \tag{38}$$

donne un principe naturel pour extraire une valeur finie à partir de l'expression singulière $\gamma(z)$. Cette extraction de partie finie est une division par la partie polaire pour un lacet méromorphe γ en prenant pour C un cercle infinitésimal centré en z_0.

Soit G un groupe de Lie complexe pro-nilpotent, $\mathcal{H}$ son algèbre de Hopf de coordonnées (graduée). Rappelons la traduction entre langages algébriques et géométriques, en désignant par $\mathcal{R}$ l'anneau des fonctions méromorphes, $\mathcal{R}_-$ le sous anneau des polynomes en $(z - z_0)^{-1}$ et $\mathcal{R}_+$ celui des fonctions régulières en z_0,

Homomorphismes de $\mathcal{H} \to \mathcal{R}$	Lacets de C à valeurs dans G
$\phi(\mathcal{H}) \subset \mathcal{R}_-$	γ se prolonge en une application holomorphe de $\mathbb{C}\backslash\{z_0\} \to G$ avec $\gamma(\infty) = 1$.
$\phi(\mathcal{H}) \subset \mathcal{R}_+$	γ se prolonge en une application holomorphe définie en $z = z_0$.
$\phi = \phi_1 \star \phi_2$	$\gamma(z) = \gamma_1(z)\gamma_2(z),\ \forall\, z \in \mathbf{C}$.
$\phi \circ S$	$z \to \gamma(z)^{-1}$.

(5.1)

Pour $X \in \mathcal{H}$ notons le coproduit sous la forme

$$\Delta(X) = X \otimes 1 + 1 \otimes X + \sum X' \otimes X''$$

La décomposition de Birkhoff d'un lacet s'obtient de manière récursive grâce au théorème suivant,

Théorème ([12]) *Soit $\phi : \mathcal{H} \to \mathcal{R}$, un homomorphisme d'algèbres. La décomposition de Birkhoff du lacet correspondant est donnée de manière récursive par les égalités,*

$$\phi_-(X) = -T\left(\phi(X) + \sum \phi_-(X')\phi(X'')\right)$$

$$\phi_+(X) = \phi(X) + \phi_-(X) + \sum \phi_-(X')\phi(X'').$$

Ici T désigne la projection sur $\mathcal{R}_-$ parallèlement à $\mathcal{R}_+$.
La clef de notre travail avec Dirk Kreimer réside dans l'identité entre ces formules et celles qui gouvernent la combinatoire des calculs de graphes. Nous avons déja vu la formule qui définit la préparation d'un graphe,

$$\overline{R}(\Gamma) = U(\Gamma) + \sum_{\gamma \subset \Gamma} C(\gamma) U(\Gamma/\gamma) \tag{39}$$

Celle qui donne le contreterme $C(\Gamma)$ est alors,

$$C(\Gamma) = -T(\overline{R}(\Gamma)) = -T\left(U(\Gamma) + \sum_{\gamma \subset \Gamma} C(\gamma) U(\Gamma/\gamma)\right) \tag{40}$$

et celle qui donne la valeur renormalisée du graphe est,

$$R(\Gamma) = \overline{R}(\Gamma) + C(\Gamma) = U(\Gamma) + C(\Gamma) + \sum_{\gamma \subset \Gamma} C(\gamma) U(\Gamma/\gamma) \tag{41}$$

Il est alors clair en posant $\phi = U$, $\phi_- = C$, $\phi_+ = R$ que ces équations sont identiques à celles du théorème donnant la construction récursive de la décomposition de Birkhoff.

Décrivons plus en détails ce résultat. Etant donnée une théorie renormalisable en dimension D la théorie nonrenormalisée donne en utilisant la régularisation dimensionnelle un lacet γ d'éléments du groupe G associé à la théorie dans la section IV. Le paramètre z du lacet $\gamma(z)$ est une variable complexe et $\gamma(z)$ est méromorphe dans un voisinage de D. Notre résultat principal est que la théorie renormalisée est donnée par l'évaluation à $z = D$ de la partie nonsingulière γ_+ de la décomposition de Birkhoff ,

$$\gamma(z) = \gamma_-(z)^{-1}\,\gamma_+(z) \tag{42}$$

de γ.

Les règles de Feynman et la régularisation dimensionnelle associent un nombre,

$$U_\Gamma(p_1,\ldots,p_N) = \int d^d k_1 \ldots d^d k_L \; I_\Gamma(p_1,\ldots,p_N,k_1,\ldots,k_L) \tag{43}$$

à chaque graphe Γ. Nous les utilisons en métrique Euclidienne pour éviter les facteurs imaginaires.

Pour respecter les dimensions physiques des quantités impliquées quand on écrit ces règles en dimension d, il faut introduire une unité de masse μ et remplacer partout la constante de couplage par $\mu^{3-d/2}\,g$. On normalise ainsi les calculs par,

$$U(\Gamma) = g^{(2-N)}\,\mu^{-B}\,\langle\sigma,U_\Gamma\rangle \tag{44}$$

où $B = B(d)$ est la dimension de $\langle\sigma,U_\Gamma\rangle$.

On étend la définition (44) aux réunions disjointes de graphes 1PI Γ_j par,

$$U(\Gamma = \cup\Gamma_j) = \Pi\, U(\Gamma_j)\,. \tag{45}$$

Le résultat principal est alors le suivant:

Théorème ([12]) a) *Il existe une unique application méromorphe $\gamma(z) \in G$, $z \in \mathbb{C}$, $z \neq D$ dont les Γ-coordonnées sont données par $U(\Gamma)_{d=z}$.*

b) *La valeur renormalisée d'une observable physique $\mathcal{O}$ est obtenue en remplaçant $\gamma(D)$ dans le développement perturbatif de $\mathcal{O}$ par $\gamma_+(D)$ où*

$$\gamma(z) = \gamma_-(z)^{-1}\,\gamma_+(z)$$

est la décomposition de Birkhoff du lacet $\gamma(z)$ relativement à un cercle infinitésimal autour de D.

6 Le groupe de renormalisation

Montrons comment le groupe de renormalisation apparait très simplement de notre point de vue.
Comme nous l'avons vu ci-dessus, la régularisation dimensionnelle implique le choix arbitraire d'une unité de masse μ et l'on constate d'abord que la partie singulière de la décomposition de Riemann–Hilbert de γ est en fait indépendante de ce choix. Il en résulte une contrainte très forte sur cette partie singulière et le groupe de renormalisation s'en déduit immédiatement. Nous en déduisons également une formule explicite pour l'action nue. On montre d'abord, en se limitant à la théorie φ_6^3 pour simplifier les notations, que bien que le lacet $\gamma(d)$ dépende du choix de l'unité de masse μ,

$$\mu \to \gamma_\mu(d)\,, \tag{46}$$

la partie singulière γ_{μ^-} de sa décomposition de Birkhoff,

$$\gamma_\mu(d) = \gamma_{\mu^-}(d)^{-1}\,\gamma_{\mu^+}(d) \tag{47}$$

est en fait indépendante de μ,

$$\frac{\partial}{\partial\mu}\,\gamma_{\mu^-}(d) = 0\,. \tag{48}$$

Cet énoncé découle immédiatement de l'analyse dimensionnelle.
De plus, par construction le groupe de Lie G est muni d'un groupe à un paramètre d'automorphismes,

$$\theta_t \in \operatorname{Aut} G\;,\quad t \in \mathbb{R}\,, \tag{49}$$

associé à la graduation de l'algèbre de Hopf $\mathcal{H}$ donnée par le nombre de boucles,

$$L(\Gamma) = \text{nombre de boucles } \Gamma \tag{50}$$

pour tout graphe 1PI Γ.
On a l'égalité

$$\gamma_{e^t\mu}(d) = \theta_{t\varepsilon}(\gamma_\mu(d)) \qquad \forall\, t \in \mathbb{R}\,,\ \varepsilon = D - d \tag{51}$$

Il en résulte que les lacets γ_μ associés à la théorie nonrenormalisée satisfont la propriété suivante: la partie singulière de leur décomposition de Birkhoff est inchangée par l'opération,

$$\gamma(\varepsilon) \to \theta_{t\varepsilon}(\gamma(\varepsilon))\,, \tag{52}$$

En d'autres termes, si l'on remplace $\gamma(\varepsilon)$ par $\theta_{t\varepsilon}(\gamma(\varepsilon))$ l'on ne modifie pas la partie singulière de sa décomposition de Birkhoff. On a posé

$$\varepsilon = D - d \in \mathbb{C}\backslash\{0\}\,. \tag{53}$$

Nous donnons une caractérisation complète des lacets $\gamma(\varepsilon) \in G$ vérifiant cette proprieté. Cette caractérisation n'implique que la partie singulière $\gamma_-(\varepsilon)$ qui vérifie par hypothèse,

$$\gamma_-(\varepsilon)\, \theta_{t\varepsilon}(\gamma_-(\varepsilon)^{-1}) \text{ est convergent pour } \varepsilon \to 0\,. \tag{54}$$

Il est facile de voir que la limite de (54) pour $\varepsilon \to 0$ définit un sous-groupe à un paramètre,

$$F_t \in G\,,\; t \in \mathbb{R} \tag{55}$$

et que le générateur $\beta = \left(\frac{\partial}{\partial t}\, F_t\right)_{t=0}$ de ce sous-groupe est relié au *résidu* de γ

$$\operatorname*{Res}_{\varepsilon=0} \gamma = -\left(\frac{\partial}{\partial u}\, \gamma_-\left(\frac{1}{u}\right)\right)_{u=0} \tag{56}$$

par l'équation,

$$\beta = Y \operatorname{Res} \gamma\,, \tag{57}$$

où $Y = \left(\frac{\partial}{\partial t}\, \theta_t\right)_{t=0}$ est la graduation.

Ceci est immédiat mais notre résultat ([13]) donne la formule explicite (59) qui exprime $\gamma_-(\varepsilon)$ en fonction de β. Introduisons le produit semi-direct de l'algèbre de Lie G (des éléments primitifs de $\mathcal{H}^*$) par la graduation. On a donc un élément Z_0 tel que

$$[Z_0, X] = Y(X) \qquad \forall\, X \in \operatorname{Lie} G\,. \tag{58}$$

La formule pour $\gamma_-(\varepsilon)$ est alors

$$\gamma_-(\varepsilon) = \lim_{t\to\infty} e^{-t\left(\frac{\beta}{\varepsilon} + Z_0\right)}\, e^{tZ_0}\,. \tag{59}$$

Les deux facteurs du terme de droite appartiennent au produit semi-direct du groupe G par sa graduation, mais leur rapport (59) appartient au groupe G.

Cette formule montre que toute la structure des divergences est uniquement déterminée par le résidu et donne une forme forte des relations de t'Hooft [21].

7 Le groupe G et les difféomorphismes

Bien entendu, on pourrait facilement objecter aux développements précédents en arguant que le mystère de la renormalisation n'est pas completement éclairci car le groupe G construit à partir des graphes de Feynman apparait également mysterieux. Cette critique est completement levée par la merveilleuse relation, basée sur la physique entre les algèbres de Hopf $\mathcal{H}$ des graphes de Feynman et celle, $\mathcal{H}_{cm}$ des difféomorphismes.

Nous montrons, dans le cas de masse nulle, que la formule qui donne la constante de couplage effective,

$$g_0 = \left(g + \sum g^{2\ell+1} \frac{\Gamma}{S(\Gamma)}\right)\left(1 - \sum g^{2\ell} \frac{\Gamma}{S(\Gamma)}\right)^{-3/2} \tag{60}$$

considérée comme une série formelle dans la variable g d'éléments de l'algèbre de Hopf $\mathcal{H}$, définit en fait un homomorphisme d'algèbres de Hopf de l'algèbre de Hopf $\mathcal{H}_{cm}$ des coordonnées sur le groupe des difféomorphismes formels de $\mathbb{C}$ tels que,

$$\varphi(0) = 0\,,\ \varphi'(0) = \mathrm{id} \tag{61}$$

vers l'algèbre de Hopf $\mathcal{H}$ de la théorie de masse nulle.
Il en résulte en transposant, une action formelle du groupe G sur la constante de couplage. Nous montrons en particulier que l'image par ρ de $\beta = Y\,\mathrm{Res}\,\gamma$ est la fonction β de la constante de couplage g.
Nous obtenons ainsi un corollaire du théorème principal qui se formule sans faire intervenir ni le groupe G ni l'algèbre de Hopf $\mathcal{H}$.

Théorème ([13]) *Considérons la constante de couplage effective nonrenormalisée* $g_{\mathrm{eff}}(\varepsilon)$ *comme une série formelle en* g *et soit* $g_{\mathrm{eff}}(\varepsilon) = g_{\mathrm{eff}_+}(\varepsilon)\,(g_{\mathrm{eff}_-}(\varepsilon))^{-1}$ *sa décomposition de Birkhoff (opposée) dans le groupe des difféomorphismes formels. Alors le lacet* $g_{\mathrm{eff}_-}(\varepsilon)$ *est la constante de couplage nue et* $g_{\mathrm{eff}_+}(0)$ *la constante de couplage renormalisée.*

Comme la décomposition de Birkhoff d'un lacet à valeurs dans le groupe des difféomorphismes (formels) est évidemment reliée à la classification des fibrés (non-vectoriels) holomorphes, ce résultat suggère qu'un tel fibré ayant pour base un voisinage de la dimension d de l'espace temps et pour fibre les valeurs (compléxifiées) des constantes de couplage devrait donner une interprétation géométrique de l'opération de renormalisation. Il faut tout de même noter que la décomposition de Birkhoff a lieu ici relativement à un cercle infinitésimal autour de d et qu'il s'agit de difféomor- phismes formels.
Les résultats ci-dessus montrent qu'au niveau des développements perturbatifs le procédé de renormalisation admet une interprétation géométrique simple grâce au groupe G et à la décomposition de Riemann–Hilbert. Le problème essentiel consiste à passer du développement perturbatif à la théorie non-perturbative, ce qui revient en termes de difféomorphismes à passer du developpement de Taylor à la formule globale.

8 Le groupe de renormalisation et la théorie de Galois aux places archimédiennes

Le problème de Riemann–Hilbert joue un rôle clef dans la théorie de Galois differentielle, il est donc naturel d'interpréter en termes Galoisiens l'ambiguïté que le groupe de renormalisation introduit dans les théories physiques. Cette section contient l'esquisse d'une telle interprétation. Nous montrerons en particulier le rôle que le groupe de renormalisation devrait jouer pour comprendre la composante connexe du groupe des classe d'idèles de la théorie du corps de classe comme un groupe de Galois.

Commençons par une introduction très élémentaire à la théorie de Galois pour les équations algébriques.
Si la technique de résolution des équations du second degré remonte à la plus haute Antiquité (Babyloniens, Egyptiens...), elle n'a pu être étendue au troisième degré que bien plus tard, et ne sera publiée par Girolamo (Jérôme) Cardano qu'en 1545 dans les chapitres 11 à 23 de son livre *Ars magna sive de regulis algebraicis*. Bien que cela n'ait pas été reconnu avant le dix-huitième siècle, la clef de la résolution par radicaux de l'équation générale du troisième degré, $x^3 + nx^2 + px + q = 0$, de racines a, b, c, est l'existence d'une fonction rationnelle $\alpha(a,b,c)$ de a, b, c, qui ne prend que deux déterminations différentes sous l'action des six permutations de a, b, c.
La méthode de Cardan revient à poser $\alpha = \big((1/3)(a+bj+cj^2)\big)^3$ où le nombre j est la première racine cubique de l'unité. La permutation circulaire transformant a en b, b en c et c en a laisse manifestement α inchangée et la seule autre détermination de la fonction α sous l'action des six permutations de a, b, c, est obtenue en transposant b et c par exemple, ce qui donne $\beta = \big((1/3)(a + cj + bj^2)\big)^3$.
Comme l'ensemble de ces deux nombres α et β est invariant par toutes les permutations de a, b, c, le polynôme du second degré dont α et β sont racines se calcule rationnellement en fonction des coefficients de l'équation initiale $x^3+nx^2+px+q = 0$: c'est $X^2 + 2qX - p^3 = (X + q + s)(X + q - s)$ où s est l'une des racines carrées de $p^3 + q^2$ et où l'on a réécrit l'équation initiale sous la forme équivalente $x^3 + 3px + 2q = 0$ débarrassée du terme du deuxième degré en effectuant une translation convenable des racines et où l'on a introduit les coefficients 2 et 3 pour simplifier les formules.
Un calcul simple montre alors que chacune des racines a, b et c, de l'équation initiale s'exprime comme somme de l'une des trois racines cubiques de α et de l'une des trois racines cubiques de β, ces deux choix étant liés par le fait que leur produit doit être impérativement égal à $-p$ (il n'y a donc que trois couples de choix de ces racines à prendre en compte, ce qui est rassurant, à la place des neuf possibilités que l'on aurait pu envisager *a priori*).
C'est à l'occasion de ces formules que l'utilisation des nombres complexes s'est imposée. En effet, même dans le cas où les trois racines sont réelles, il se peut que $p^3 + q^2$ soit négatif et que α et β soient nécessairement des nombres complexes.
Si la résolution des équations du troisième degré que nous venons d'exposer a été très longue à être mise au point (sans doute pour au moins l'un de ses cas particuliers entre 1500 et 1515 par Scipione del Ferro), celle du quatrième degré a été plus preste à la suivre puisqu'elle figure également dans l'*Ars magna* (chapitre 39) où Cardano l'attribue à son secrétaire Ludovico Ferrari qui l'aurait mise au point entre 1540 et 1545 (René Descartes en publiera une autre en 1637).
Ici encore, l'on peut partir d'un polynôme débarrassé d'un coefficient, annulé par translation, disons $X^4 + pX^2 + qX + r = (X - a)(X - b)(X - c)(X - d)$.

La fonction rationnelle $\alpha(a,b,c,d)$ la plus simple[§], ne prenant que trois déterminations différentes sous l'action des vingt-quatre permutations de a, b, c et d, est $\alpha = ab + cd$. Les deux autres déterminations sont $\beta = ac + bd$, $\gamma = ad + bc$. Ce sont donc les racines d'une équation du troisième degré dont les coefficients s'expriment rationnellement en fonction de p, q et r. Un calcul simple montre que le polynôme $(X-\alpha)(X-\beta)(X-\gamma)$ est égal à $X^3 - pX^2 - 4rX + (4pr - q^2)$. Il peut donc être décomposé comme on l'a vu plus haut pour en déduire α, β et γ; en fait, il suffit même de calculer l'une seulement de ces racines, disons α, pour pouvoir en déduire a, b, c et d (nous connaissons alors en effet la somme α et le produit r des deux nombres ab et cd, donc ces deux nombres eux-mêmes par une équation du second degré, et il ne reste plus qu'à exploiter les égalités $(a+b)+(c+d) = 0$ et $ab(c+d) + cd(a+b) = -q$ pour pouvoir en déduire $a+b$ et $c+d$, donc enfin a, b, c et d par une autre équation du second degré).
C'est à Joseph Louis Lagrange en 1770 et 1771 (publication en 1772, mais aussi, dans une moindre mesure, à Alexandre Vandermonde dans un mémoire publié en 1774 mais également rédigé vers 1770, ainsi qu'à Edward Waring dans ses *Meditationes algebricæ* de 1770 et à Francesco Malfatti) que l'on doit la mise en lumière du rôle fondamental des permutations sur les racines a, b, c... et sur les quantités auxiliaires α, β..., d'ailleurs aujourd'hui justement appelées "résolvantes de Lagrange".
Ces résolvantes ne sont pas uniques, et par exemple le choix $\alpha = (a+b-c-d)^2$ correspond à la solution de Descartes, mais elles fournissent la clef de toute les résolutions générales par radicaux. Il y en a une qui est particulièrement belle car elle est covariante pour le groupe affine, c'est à dire vérifie l'égalité,

$$\alpha(\lambda a + z, \lambda b + z, \lambda c + z, \lambda d + z) = \lambda\, \alpha(a,b,c,d) + z$$

et admet donc une interprétation géométrique. Elle est donnée algèbriquement par

$$\alpha = \frac{ad - bc}{a + d - b - c}$$

et correspond géométriquement (Figure 2) au point d'intersection des cercles circonscrits aux triangles ABJ et JCD où J désigne le point d'intersection des droites AC et BD.
J'ai rencontré récemment cette résolvante à propos du problème[¶] de l'étoile à cinq branches (Figure 3), dont elle permet une résolution algébrique que je laisse à la sagacité du lecteur.
L'étape suivante dans la théorie des équations algébriques est évidemment celle du cinquième degré. Descartes a certainement essayé et avec lui bien des chercheurs. Elle a toujours opposé des obstacles infranchissables, et nous savons depuis Abel et Galois, aux alentours de 1830, pourquoi cette quête était vaine.

§ *Voir [6] pour l'ubiquité de la symétrie en question, et son rôle dans l'organisation des tournois de football*

¶ *posé par le président Chinois Jiang Zemin à la délégation de mathématiciens venue à sa rencontre en l'an 2000.*

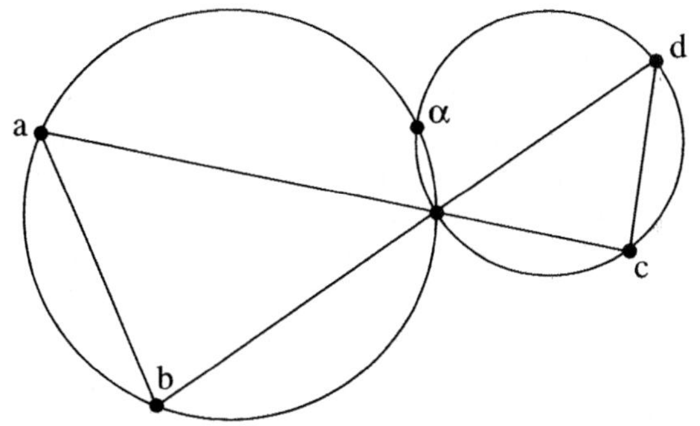

Figure 1: Le point α est fonction méromorphe et séparément homographique des quatre points A, B, C, D.

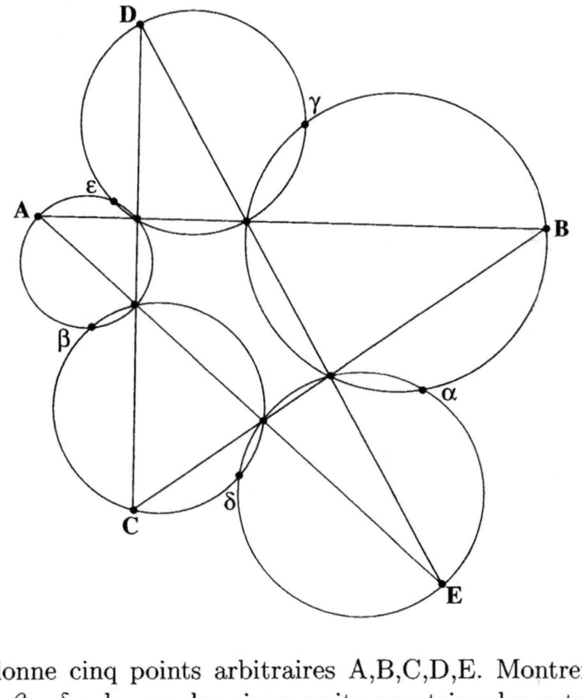

Figure 2: On donne cinq points arbitraires A,B,C,D,E. Montrer que les points d'intersection α,β,γ,δ,ϵ des cercles circonscrits aux triangles externes consécutifs de l'étoile sont situés sur un même cercle.

Descartes par exemple, persuadé qu'il n'existait pas de formule analogue à celle de Cardano, avait proposé en 1637, dans *La Géométrie*, une méthode graphique de résolution grâce à l'intersection de cercles et de cubiques qu'il avait inventées pour l'occasion. Entre 1799 et 1813 (date de l'édition de ses *Riflessioni intorno alla solutione delle equazioni algebraiche generali*), Paolo Ruffini a publié diverses tentatives de démonstrations, de plus en plus affinées, visant à établir l'impossiblité de résoudre l'équation générale du cinquième degré par radicaux. À toute fonction rationnelle des racines, il a eu l'idée juste d'associer le groupe des permutations de ces racines qui la laissent invariante, mais a cru à tort (d'après un rapport de Ludwig Sylow) que les radicaux intervenant dans la résolution de l'équation, comme les racines cubiques de α pour le degré trois, étaient nécessairement des fonctions rationnelles des racines.

Il faudra attendre 1824 pour que Niels Abel justifie l'intuition de Ruffini dans son *Mémoire sur les équations algébriques* et – après avoir cru trouver au contraire une méthode de résolution générale – prouve l'impossiblité de résoudre l'équation générale du cinquième degré par radicaux, en 1826 dans le *Mémoire sur une classe particulière d'équations résolubles algébriquement*, où il amorce une théorie générale qui ne s'épanouira que dans les écrits de Galois, vers 1830. Les travaux de Galois inaugurent une ère nouvelle des mathématiques, où les calculs font place à la réflexion sur leur potentialité, et les concepts, tels celui de groupe abstrait ou d'extension algébrique, occupent le devant de la scène.

L'idée lumineuse de Galois consiste d'abord à associer à une équation arbitraire un groupe de permutations qu'il définit de la manière suivante, [18]

> *Soit une équation donnée, dont a, b, c,... sont les m racines. Il y aura toujours un groupe de permutations des lettres a, b, c,...qui jouira de la propriété suivante:*
>
> *1) que toute fonction des racines, invariante par les substitutions de ce groupe, soit rationnellement connue;*
>
> *2) réciproquement, que toute fonction des racines, déterminée rationnellement, soit invariante par ces substitutions.*

puis à étudier comment ce groupe "d'ambiguïté" se trouve modifié par l'adjonction de quantités auxiliaires considérées comme "rationnelles". Ainsi, dans le cas de l'équation du quatrième degré, si l'on adjoint la quantité α obtenue en résolvant l'équation auxiliaire du troisième degré, l'on réduit le groupe d'ambiguïté au sous-groupe normal formé des quatre permutations (a,b,c,d), (b,a,d,c), (c,d,a,b), (d,c,b,a). Ce groupe est le produit de deux groupes à deux éléments et l'adjonction des solutions de deux équations du second degré suffit alors pour éliminer totalement l'ambiguïté, c'est à dire résoudre l'équation initiale.

Si l'on désigne par k le corps des "quantités rationnelles" et par K celui engendré par k et par toutes[‖] les racines de l'équation que l'on se propose de résoudre, le groupe de Galois, $G = \mathrm{Gal}(K : k)$ est le groupe des automorphismes de K qui fixent tous les éléments de k.

‖ *Il ne suffit pas d'adjoindre une seule de ces racines, il faut les adjoindre toutes*

L'impossibilité de réduire l'équation du cinquième degré à des équations de degré inférieur provient alors de la "simplicité" du groupe A_5 des soixantes permutations paires (produits d'un nombre pair de transpositions) des cinq racines a, b, c, d, e d'une telle équation. Un groupe abstrait fini est "simple" si l'on ne peut le réduire, par un homomorphisme non trivial, à un groupe plus petit. Le groupe A_5 est le plus petit groupe simple non commutatif et il apparaît très souvent en mathématiques.
J'en viens maintenant au rôle que le groupe de renormalisation devrait jouer pour comprendre la composante connexe du groupe des classes d'idèles de la théorie du corps de classe comme un groupe de Galois.
La théorie du corps de classe et sa généralisation aux groupes de Galois non commutatifs par le programme de Langlands constituent l'information la plus profonde que nous ayons sur le groupe de Galois des nombres algébriques. La source de ces théories est la loi de réciprocité quadratique qui joue un rôle central dans l'histoire de la théorie des nombres. Elle est démontrée en 1801 par Gauss dans ses " Disquisitiones " mais son énoncé était déjà connu d'Euler et de Legendre. La loi de réciprocité exprime, étant donnés deux nombres premiers p et q, une symétrie entre p et q dans la résolution de l'équation $x^2 = p$ modulo q. Elle montre par exemple que pour savoir si l'équation $x^2 = 5$ admet une solution modulo un nombre premier q il suffit de connaître la valeur de q modulo 5 ce que donne le dernier chiffre de q dans son développement décimal, (par exemple 19 et 1999, ou 7 et 1997 donnent le même résultat) de sorte que les nombres premiers ainsi sélectionnés se répartissent en classes. Il a fallu plus d'un siècle pour comprendre conceptuellement la loi de réciprocité quadratique dont Gauss avait donné plusieurs démonstrations, sous la forme de la théorie du corps de classe qui permet de calculer à partir de classes de nombres idéaux le groupe de Galois de l'extension Abelienne maximale d'un corps de nombres.
La généralisation conceptuelle de la notion de corps de nombres est celle de corps global. Un corps k est *global* si c'est un sous-corps discret cocompact d'un anneau localement compact (non discret) semi-simple et commutatif A. (Cf. Iwasawa *Ann. of Math.* **57** (1953).) L'anneau topologique A est alors canoniquement associé à k et s'appelle l'anneau des Adèles de k, on a,

$$A = \prod_{\text{res}} k_v \,, \tag{62}$$

où le produit est le produit restreint des corps locaux k_v indéxés par les places de k. Les k_v sont les corps localement compacts obtenus comme complétions de k de même que l'on obtient les nombres réels en complétant les rationnels.
Quand la caractéristique de k est $p > 1$ i.e. quand k est un corps de fonctions sur $\mathbb{F}_q$, on a

$$k \subset k_{\text{un}} \subset k_{\text{ab}} \subset k_{\text{sep}} \subset \overline{k} \,, \tag{63}$$

où $\overline{k}$ désigne une cloture algébrique de k, k_{sep} la cloture algébrique séparable, k_{ab} l'extension abélienne maximale et k_{un}, extension non ramifiée maximale, est obtenue en adjoignant à k les racines de l'unité d'ordre premier à p.

On définit le groupe de Weil W_k comme le sous-groupe de $\mathrm{Gal}(k_{\mathrm{ab}} : k)$ formé par les automorphismes de $(k_{\mathrm{ab}} : k)$ qui induisent sur k_{un} une puissance entière de l'automorphisme de "Frobenius", θ,

$$\theta(\mu) = \mu^q \qquad \forall\, \mu \text{ racine de l'unité d'ordre premier à } p. \tag{64}$$

Le résultat principal de la théorie du corps de classe global est l'existence d'un isomorphisme canonique,

$$W_k \simeq C_k = GL_1(A)/GL_1(k)\,, \tag{65}$$

de groupes localement compacts.
Quand k est de caractéristique nulle, i.e. un corps de nombres, on a un isomorphisme canonique,

$$\mathrm{Gal}(k_{\mathrm{ab}} : k) \simeq C_k/D_k\,, \tag{66}$$

où D_k désigne la composante connexe de l'élément neutre dans le groupe des classes d'idèles $C_k = GL_1(A)/GL_1(k)$, mais à cause des places Archimédiennes de k l'on n'a pas d'interprétation de C_k analogue au cas des corps de fonctions. Citons A. Weil [27],

"La recherche d'une interprétation pour C_k si k est un corps de nombres, analogue en quelque manière à l'interprétation par un groupe de Galois quand k est un corps de fonctions, me semble constituer l'un des problèmes fondamentaux de la théorie des nombres à l'heure actuelle ; il se peut qu'une telle interprétation renferme la clef de l'hypothèse de Riemann ...".

Cela signifie qu'aux places Archimédiennes (i.e. aux complétions de k qui donnent soit les nombres réels soit les nombres complexes), il devrait y avoir un groupe continu de symétries secrètement présent.
Mon intérêt pour ce problème vient de mon travail sur la classification des facteurs qui indiquait clairement que l'on possédait là l'analogue de la théorie de Brauer qui est l'une des clefs de la théorie du corps de classe local.
Les groupes de Galois sont par construction des limites projectives de groupes finis attachés à des extensions finies. Pour obtenir des groupes connexes il faut évidemment relaxer cette condition de finitude, qui est la même que la restriction en théorie de Brauer aux algèbres simples centrales de dimension finie. Comme ce sont les places Archimédiennes de k qui sont à l'origine de la composante connexe D_k, il est naturel de considérer la question préliminaire suivante,

"Existe-t-il une théorie de Brauer non-triviale d'algèbres simples centrales sur $\mathbb{C}$."

J'ai montré dans [5] que la classification des facteurs *approximativement finis* sur $\mathbb{C}$ donnait une réponse satisfaisante à cette question. Ils sont classifiés par leur module,

$$\mathrm{Mod}(M) \underset{\sim}{\subset} \mathbb{R}_+^*\,, \tag{67}$$

qui est un sous-groupe (virtuel) fermé de $\mathbb{R}_+^*$.

Ce groupe joue un rôle analogue dans le cas Archimédien au module des algèbres simples centrales sur un corps local nonarchimédien. Dans ce dernier cas le module se définit trés simplement par l'action du groupe multiplicatif d'une algèbre simple centrale sur la mesure de Haar du groupe additif. La définition de Mod(M) pour les facteurs est beaucoup plus élaborée, mais reste basée sur l'action du groupe $\mathbb{R}_+^*$ de changement d'échelle.
Pour poursuivre l'analogie avec la théorie de Brauer où le lien avec le groupe de Galois s'obtient par la construction d'algèbres simples centrales comme produits croisés d'un corps par un groupe d'automorphismes, le pas suivant consiste à trouver des exemples naturels de construction de facteurs comme produits croisés d'un corps K, extension transcendante de $\mathbb{C}$ par un groupe d'automorphismes. Dans nos recherches sur les variétés sphériques noncommutatives [7], avec M. Dubois-Violette, l'algèbre de Sklyanin ([26]) est apparue comme solution en dimension 3, du problème de classification formulé dans [14]. La représentation "régulière" de cette algèbre engendre une algèbre de von-Neumann intégrale directe de facteurs approximativement finis de type II_1, tous isomorphes au facteur hyperfini R. Les homomorphismes correspondants de l'algèbre de Sklyanin ([26]) vers le facteur R se factorisent miraculeusement à travers le produit croisé du corps K_q des fonctions elliptiques, où le module $q = e^{2\pi i\tau}$ est réel, par l'automorphisme de translation par un nombre réel (mais en général irrationnel). On obtient ainsi le facteur R comme produit croisé de K_q par un sous-groupe du groupe de Galois, en parfaite analogie avec la construction des algèbres simples centrales sur un corps local. Il reste à obtenir une construction semblable et naturelle du facteur R_∞ de type III_1.
Il est sans doute prématuré d'essayer d'identifier le corps K correspondant, qui devrait jouer le rôle de l'extension nonramifiée maximale $\mathbb{C}_{un}$ de $\mathbb{C}$ et être doté d'une action naturelle du groupe multiplicatif $\mathbb{R}_+^*$.
Le rôle du corps K en physique des hautes énergies devrait être relié à l'observation suivante concernant les "constantes" qui interviennent en théorie des champs. En fait les calculs des physiciens regorgent d'exemples de "constantes" telles les constantes de couplage g des interactions (électromagnétiques, faibles et fortes) qui n'ont de "constantes" que le nom. Elles dépendent en réalité du niveau d'énergie μ auquel les expériences sont réalisées et sont donc des fonctions $g(\mu)$. Ainsi les physiciens des hautes énergies étendent implicitement le "corps des constantes" avec lequel ils travaillent, passant du corps $\mathbb{C}$ des scalaires à un corps de fonctions $g(\mu)$. Le générateur du groupe de renormalisation est simplement $\mu\partial/\partial\mu$.
L'on peut mettre l'exemple plus simple du corps K_q des fonctions elliptiques sous la même forme en passant aux fonctions loxodromiques, c'est à dire en posant $\mu = e^{2\pi i z}$ de sorte que la première périodicité (en $z \to z+1$) est automatique alors que la deuxième s'écrit $g(q\,\mu) = g(\mu)$. Le groupe des automorphismes de la courbe elliptique est alors lui aussi engendré par $\mu\partial/\partial\mu$.
Les points fixes du groupe de renormalisation sont les scalaires ordinaires, mais il se pourrait que la physique quantique conspire pour nous empêcher d'espérer une théorie qui englobe toute la physique des particules et soit construite comme point fixe du groupe de renormalisation. Les interactions fortes sont asymptotiquement

libres et l'on peut les analyser à très hautes énergies en utilisant les points fixes du groupe de renormalisation, mais la présence du secteur électrodynamique montre qu'il est vain de vouloir s'en tenir à de tels points fixes pour décrire une théorie qui incorpore l'ensemble des forces observées. Le problème est le même dans le domaine infrarouge où les rôles des interactions fortes et électrodynamiques sont inversés.

Il est bien connu des physiciens que le groupe de renormalisation joue le rôle d'un groupe d'ambiguïté, l'on ne peut distinguer entre elles deux théories physiques qui appartiennent à la même orbite de ce groupe, ce qui nous ramène à Galois dont la "théorie de l'ambiguïté" allait bien au delà des équations algébriques.

Citons Émile Picard (voir [25]) qui dans sa préface aux œuvres complètes d'Evariste Galois écrivait,

"Il aurait édifié dans ses parties essentielles, la théorie des fonctions algébriques d'une variable telle que nous la connaissons aujourd'hui. Les méditations de Galois portèrent encore plus loin; il termine sa lettre en parlant de l'application à l'analyse transcendante de la théorie de l'ambiguïté. On devine à peu près ce qu'il entend par là, et sur ce terrain qui, comme il le dit est immense, il reste encore aujourd'hui bien des découvertes à faire".

Références

[1] A. Beauville, Monodromie des systèmes différentiels linéaires à pôles simples sur la sphère de Riemann, Séminaire Bourbaki, 45ème année, 1992–1993, n.**765**.

[2] N. N. Bogoliubov, O. Parasiuk *Acta Math.* **97**, 227 (1957);
K. Hepp, *Commun. Math. Phys.* **2**, 301 (1966).

[3] A. Bolibruch, Fuchsian systems with reducible monodromy and the Riemann–Hilbert problem, *Lecture Notes in Math.* **1520**, 139–155 (1992).

[4] J. Collins, Renormalization, Cambridge monographs in math. physics, Cambridge University Press (1984).

[5] A. Connes, Noncommutative Geometry and the Riemann Zeta Function, Mathematics: Frontiers and perspectives, IMU 2000 volume.

[6] A. Connes, Symétries, de Galois au monde quantique, Volume en l'honneur de Louis Michel (à paraître).

[7] A. Connes, M. Dubois-Violette, Noncommutative finite-dimensional manifolds. I. Spherical manifolds and related examples. Math QA/0107070

[8] A. Connes, D. Kreimer, *Lett. Math. Phys.* **48**, 85 (1999); hep-th/9904044.

[9] A. Connes, D. Kreimer, *Commun. Math. Phys.* **199**, 203 (1998); hep-th/9808042.

[10] A. Connes, D. Kreimer, Hopf algebras, renormalization and Noncommuative geometry, *Comm. Math. Phys.* **199**, 203–242 (1998).

[11] A. Connes, D. Kreimer, Renormalization in Quantum Field Theory and the Riemann–Hilbert problem, *J. High Energy Phys.* **09**, 024 (1999).

[12] A. Connes, D. Kreimer, Renormalization in Quantum Field Theory and the Riemann–Hilbert problem I, *Comm. Math. Phys.* **210** (2000).

[13] A. Connes, D. Kreimer, Renormalization in quantum field theory and the Riemann–Hilbert problem II, the β function, diffeomorphisms and the renormalization group; hep-th/0003188, *Comm. Math. Phys.* (2001).

[14] A. Connes, G. Landi, Noncommutative manifolds, the instanton algebra and isospectral deformations, Math. QA/0011194.

[15] A. Connes, H. Moscovici, Hopf algebras, cyclic cohomology and the transverse index theorem, *Comm. Math. Phys.* **198**, 199–246 (1998).

[16] M. Dresden, Renormalization in historical perspective – The first stage, in *Renormalization*, ed. L. Brown, Springer-Verlag, New York, Berlin, Heidelberg (1994).

[17] J. Feldman, J. Magnen, V. Rivasseau, R. Seneor, Massive Gross–Neveu model: a rigorous perturbative construction, *Phys. Rev. Lett.* **54**, (1985).

[18] É. Galois, Écrits et mémoires mathématiques d'Évariste Galois. Gauthier Villars, Paris, London (1962).

[19] K. Gawedzki, A. Kupiainen, Exact renormalization of the Gross–Neveu model of quantum fields, *Phys. Rev. Lett.* **54** (1985).

[20] J. Glimm, A. Jaffe, Quantum Physics, Springer Verlag, New York, Berlin, Heidelberg (1987).

[21] G. 't Hooft, *Nuclear Physics* **B 61**, 455 (1973).

[22] T. Krajewski, R. Wulkenhaar, *Eur. Phys. J.* **C7**, 697–708 (1999); hep-th/9805098.

[23] D. Kreimer, *Adv. Theor. Math. Phys.* **2.2**, 303 (1998); q-alg/9707029.

[24] I. Lappo-Danilevskii, Mémoires sur la théorie des systèmes des équations différentielles linéaires, Chelsea, New York (1953).

[25] J. P. Ramis, Séries divergentes et théories asymptotiques, Panoramas et synthèses, *S. M. F.* (1993), Tome **121**.

[26] E.K. Sklyanin, Some algebraic structures connected with the Yang–Baxter equation, *Func. Anal. Appl.* **16**, , 263–270 (1982).

[27] A. Weil, Sur la théorie du corps de classe, *J. Math. Soc. Japan* **3**, (1951).

Alain Connes
Collège de France
3, rue Ulm
75005 Paris
et
I.H.E.S.
35, route de Chartres
91440 Bures-sur-Yvette

Poincaré Seminar 2002, 265 – 309

Poincaré Seminar 2002

The Anomalous Magnetic Moments of the Electron and the Muon

Marc Knecht

1 Introduction

In February 2001, the Muon (g-2) Collaboration of the E821 experiment at the Brookhaven AGS released a new value of the anomalous magnetic moment of the muon, measured with an unprecedented accuracy of 1.3 ppm. This announcement has caused quite some excitement in the particle physics community. Indeed, this experimental value was claimed to show a deviation of 2.6 σ with one of the most accurate evaluation of the anomalous magnetic moment of the muon within the standard model. It was subsequently shown that a sign error in one of the theoretical contributions was responsible for a sizable part of this discrepancy, which eventually only amounted to 1.6 σ. However, this event had the merit to draw the attention to the fact that low energy but high precision experiments represent real potentialities, complementary to the high energy accelerator programs, for evidencing possible new degrees of freedom, supersymmetry or whatever else, beyond those described by the standard model of electromagnetic, weak, and strong interactions.

Clearly, in order for theory to match such an accurate measurement (in the meantime, the relative error has even been further reduced, to 0.7 ppm), calculations in the standard model have to be pushed to their very limits. The difficulty is not only one of having to compute higher orders in perturbation theory, but also to correctly take into account strong interaction contributions involving low-energy scales, where non perturbative effects are important, and which therefore represent a real theoretical challenge.

The purpose of this account is to give an overview of the main features of the theoretical calculations that have been done in order to obtain accurate predictions for the anomalous magnetic moments of the electron and of the muon within the standard model. There exist several excellent reviews of the subject, which the interested reader may consult. As far as the situation up to 1990 is concerned, the collection of articles published in Ref. [1] offers a wealth of information, on both theory and experiment. A very useful account of earlier theoretical work is presented in Ref. [2]. Among the more recent reviews, Refs. [3, 4, 5, 6] are most informative. I shall not touch on the subject of the study of new physics scenarios which might offer an explanation for a possible deviation between the standard model prediction of the magnetic moment of the muon and its experimental value. For this aspect, I refer the reader to [7] and to the articles quoted therein, or to [8].

2 General considerations

In the context of relativistic quantum mechanics, the interaction of a pointlike spin one-half particle of charge e_ℓ and mass m_ℓ with an external electromagnetic field $\mathcal{A}_\mu(x)$ is described by the Dirac equation with the minimal coupling prescription,

$$i\hbar\,\frac{\partial\psi}{\partial t} \;=\; \left[c\boldsymbol{\alpha}\cdot\left(-i\hbar\boldsymbol{\nabla} - \frac{e_\ell}{c}\boldsymbol{\mathcal{A}}\right) + \beta m_\ell c^2 + e_\ell\mathcal{A}_0\right]\psi\,. \tag{2.1}$$

In the non relativistic limit, this reduces to the Pauli equation for the two-component spinor φ describing the large components of the Dirac spinor ψ,

$$i\hbar\,\frac{\partial\varphi}{\partial t} \;=\; \left[\frac{(-i\hbar\boldsymbol{\nabla} - (e_\ell/c)\boldsymbol{\mathcal{A}})^2}{2m_\ell} \;-\; \frac{e_\ell\hbar}{2m_\ell c}\,\boldsymbol{\sigma}\cdot\mathrm{B} \,+\, e_\ell\mathcal{A}_0\right]\varphi\,. \tag{2.2}$$

As is well known, this equation amounts to associate with the particle's spin a magnetic moment

$$\mathrm{M}_s \;=\; g_\ell\left(\frac{e_\ell}{2m_\ell c}\right)\mathrm{S}\,,\;\; \mathrm{S} = \hbar\,\frac{\boldsymbol{\sigma}}{2}\,, \tag{2.3}$$

with a gyromagnetic ratio predicted to be $g_\ell = 2$.

In the context of quantum field theory, the response to an external electromagnetic field is described by the matrix element of the electromagnetic current [1] $\mathcal{J}^\rho$ [spin projections and Dirac indices are not written explicitly]

$$\langle \ell^-(p')|\mathcal{J}^\rho(0)|\ell^-(p)\rangle \;=\; \bar{\mathrm{u}}(p')\Gamma^\rho(p',p)\mathrm{u}(p)\,, \tag{2.4}$$

with $[k_\mu \equiv p'_\mu - p_\mu]$

$$\Gamma^\rho(p',p) \;=\; F_1(k^2)\gamma^\rho \;+\; \frac{i}{2m_\ell}\,F_2(k^2)\sigma^{\rho\nu}k_\nu \;-\; F_3(k^2)\gamma_5\sigma^{\rho\nu}k_\nu\,. \tag{2.5}$$

This expression of the matrix element $\langle \ell^-(p')|\mathcal{J}^\rho(0)|\ell^-(p)\rangle$ is the most general that follows from Lorentz invariance, the Dirac equation for the two spinors, $(\not{p} - m)\mathrm{u}(p) = 0$, $\bar{\mathrm{u}}(p')(\not{p}' - m) = 0$, and the conservation of the electromagnetic current, $(\partial\cdot\mathcal{J})(x) = 0$. The two first form factors, $F_1(k^2)$ and $F_2(k^2)$, are known as the Dirac (or electric) form factor and the Pauli (or magnetic) form

[1]In the standard model, $\mathcal{J}^\rho$ denotes the total electromagnetic current, with the contributions of all the charged elementary fields in presence, leptons, quarks, electroweak gauge bosons,...

factor, respectively. Since the electric charge operator $\mathcal{Q}$ is given, in units of the charge e_ℓ, by

$$\mathcal{Q} = \int d\mathrm{x}\, \mathcal{J}_0(x^0, \mathrm{x}) \,, \tag{2.6}$$

the form factor $F_1(k^2)$ satisfies the normalization condition $F_1(0) = 1$. The presence of the form factor $F_3(k^2)$ requires both parity and time reversal invariance to be broken. It is therefore absent if only electromagnetic interactions are considered. On the other hand, in the standard model, the weak interactions violate both parity and time reversal symmetry, so that they may induce such a form factor.

The analytic structure of these form factors is dictated by general properties of quantum field theory [causality, analyticity, and crossing symmetry]. They are real functions of k^2 in the spacelike region $k^2 < 0$. In the timelike region, they become complex, with a cut starting at $k^2 > 4m_\ell^2$. At $k^2 = 0$, they describe the residue of the s-channel pole in the S-matrix element for elastic $\ell^+\ell^-$ scattering.

At tree level, i.e. in the classical limit, one finds

$$F_1^{\text{tree}}(k^2) = 1\,,\ F_2^{\text{tree}}(k^2) = 0\,,\ F_3^{\text{tree}}(k^2) = 0\,. \tag{2.7}$$

In order to obtain non zero values for $F_2(k^2)$ and $F_3(k^2)$ already at tree level, the interaction of the Dirac field with the photon field $\mathcal{A}_\mu$ would have to depart from the minimal coupling prescription. For instance, the modification [$\mathcal{F}_{\mu\nu} = \partial_\mu \mathcal{A}_\nu - \partial_\nu \mathcal{A}_\mu$, $\mathcal{J}^\rho = \overline{\psi}\gamma^\rho\psi$]

$$\int d^4x \mathcal{L}_{\text{int}} = -\frac{e_\ell}{c}\int d^4x \mathcal{J}^\rho \mathcal{A}_\rho \ \rightarrow$$

$$\rightarrow \int d^4x \widehat{\mathcal{L}}_{\text{int}} = -\frac{e_\ell}{c}\int \left[\mathcal{J}^\rho \mathcal{A}_\rho + \frac{\hbar}{4m_\ell} a_\ell \overline{\psi}\sigma_{\mu\nu}\psi\mathcal{F}^{\mu\nu} + \frac{\hbar}{2e_\ell} d_\ell \overline{\psi} i\gamma_5 \sigma_{\mu\nu}\psi\mathcal{F}^{\mu\nu}\right]$$

$$= -\frac{e_\ell}{c}\int d^4x \widehat{\mathcal{J}}^\rho \mathcal{A}_\rho\,, \tag{2.8}$$

with [2]

$$\widehat{\mathcal{J}}_\rho = \mathcal{J}_\rho - \frac{\hbar}{2m_\ell} a_\ell \partial^\mu\left(\overline{\psi}\sigma_{\mu\rho}\psi\right) - \frac{\hbar}{d_\ell} e_\ell \partial^\mu\left(\overline{\psi} i\gamma_5\sigma_{\mu\rho}\psi\right), \tag{2.9}$$

leads to

$$\widehat{F}_1^{\text{tree}}(k^2) = 1\,,\ \widehat{F}_2^{\text{tree}}(k^2) = a_\ell\,,\ \widehat{F}_3^{\text{tree}}(k^2) = d_\ell/e_\ell\,. \tag{2.10}$$

The equation satisfied by the Dirac spinor ψ then reads

$$i\hbar\frac{\partial\psi}{\partial t} = \Big[c\boldsymbol{\alpha}\cdot\left(-i\hbar\boldsymbol{\nabla} - \frac{e_\ell}{c}\boldsymbol{A}\right) + \beta m_\ell c^2 + e_\ell A_0$$

$$+\frac{e_\ell\hbar}{2m_\ell} a_\ell\beta\left(i\boldsymbol{\alpha}\cdot\mathrm{E} - \boldsymbol{\Sigma}\cdot\mathrm{B}\right) - \hbar d_\ell \beta\left(\boldsymbol{\Sigma}\cdot\mathrm{E} + i\boldsymbol{\alpha}\cdot\mathrm{B}\right)\Big]\psi\,, \tag{2.11}$$

[2] The current $\widehat{\mathcal{J}}^\rho$ is still a conserved four-vector, therefore the matrix element $\langle \ell^-(p')|\widehat{\mathcal{J}}^\rho(0)|\ell^-(p)\rangle$ also takes the form (2.4), (2.5), with appropriate form factors $\widehat{F}_i(k^2)$.

and the corresponding non relativistic limit becomes [3]

$$i\hbar\frac{\partial\varphi}{\partial t} = \left[\frac{(-i\hbar\boldsymbol{\nabla} - (e_\ell/c)\boldsymbol{\mathcal{A}})^2}{2m_\ell} - \frac{e_\ell\hbar}{2m_\ell c}(1+a_\ell)\boldsymbol{\sigma}\cdot\mathrm{B}\cdot\mathrm{E} + e_\ell\mathcal{A}_0 + \cdots\right]\varphi. \tag{2.12}$$

Thus the coupling constant a_ℓ induces a shift in the gyromagnetic factor, $g_\ell = 2(1+a_\ell)$, while d_ℓ gives rise to an electric dipole moment. The modification (2.8) of the interaction with the photon field introduces two arbitrary constants, and both terms produces a *non renormalizable* interaction. Non constant values of the form factors could be generated at tree level upon introducing [9] additional non renormalizable couplings, involving derivatives of the external field of the type $\Box^n\mathcal{A}_\mu$, which preserve the gauge invariance of the corresponding field equation satisfied by ψ. In a renormalizable framework, like QED or the standard model, calculable non vanishing values for $F_2(k^2)$ and $F_3(k^2)$ are generated by the loop corrections. In particular, the latter will likewise induce an *anomalous magnetic moment*

$$a_\ell = \frac{1}{2}(g_\ell - 2) = F_2(0) \tag{2.13}$$

and an electric dipole moment $d_\ell = e_\ell F_3(0)$.

If we consider only the electromagnetic and the strong interactions, the current $\mathcal{J}^\rho$ is gauge invariant, and the two form factors symmetry $F_1(k^2)$ and $F_2(k^2)$ do not depend on the gauges chosen in order to quantize the photon and the gluon gauge fields. This is no longer the case if the weak interactions are included as well, since $\mathcal{J}^\rho$ now transforms under a weak gauge transformation, and the corresponding form factors in general depend on the gauge choices. As we have already mentioned above, the zero momentum transfer values $F_i(0)$, $i = 1,2,3$ describe a physical S-matrix element. To the extent that the perturbative S-matrix of the standard model does not depend on the gauge fixing parameters to any order of the renormalized perturbation expansion, the quantities $F_i(0)$ should define *bona fide* gauge-fixing independent observables.

The computation of $\Gamma_\rho(p',p)$ is often a tedious task, especially if higher loop contributions are considered. It is therefore useful to concentrate the efforts on computing the form factor of interest, e.g. $F_2(k^2)$ in the case of the anomalous magnetic moment. This can be achieved upon projecting out the different form factors [10, 11] using the following general expression [4]

$$F_i(k^2) = \mathrm{tr}\left[\Lambda_i^\rho(p',p)(\not p' + m_\ell)\Gamma_\rho(p',p)(\not p + m_\ell)\right], \tag{2.14}$$

[3] Terms involving the gradients of the external fields **E** and **B** or terms nonlinear in these fields are not shown.

[4] From now on, I most of the time use the system of units where $\hbar = 1$, $c = 1$.

with

$$\begin{aligned}
\Lambda_1^\rho(p',p) &= \frac{1}{4}\frac{1}{k^2-4m_\ell^2}\gamma^\rho + \frac{3m_\ell}{2}\frac{1}{(k^2-4m_\ell^2)^2}(p'+p)^\rho \\
\Lambda_2^\rho(p',p) &= -\frac{m_\ell^2}{k^2}\frac{1}{k^2-4m_\ell^2}\gamma^\rho - \frac{m_\ell}{k^2}\frac{k^2+2m_\ell^2}{(k^2-4m_\ell^2)^2}(p'+p)^{\ rho} \\
\Lambda_3^\rho(p',p) &= -\frac{i}{2k^2}\frac{1}{k^2-4m_\ell^2}\gamma_5(p'+p)^\rho . \qquad (2.15)
\end{aligned}$$

For $k \to 0$, one has

$$\Lambda_2^\rho(p,p') = \frac{1}{4k^2}\Big[\gamma^\rho - \frac{1}{m_\ell}\Big(1+\frac{k^2}{m_\ell^2}\Big)(p+\frac{1}{2}k)^\rho + \cdots\Big], \qquad (2.16)$$

and

$$(\not p+m_\ell)\Lambda_2^\rho(p,p')(\not p'+m_\ell) = \frac{1}{4}(\not p+m_\ell)\Big[-\frac{k^\rho}{k^2}+(\gamma^\rho-\frac{p^\rho}{m_\ell})\frac{\not k}{k^2}+\cdots\Big]. \qquad (2.17)$$

The last expression behaves as $\sim 1/k$ as the external photon four momentum k_μ vanishes, so that one may worry about the finiteness of $F_2(0)$ obtained upon using Eq. (2.14). This problem is solved by the fact that $\Gamma^\rho(p',p)$ satisfies the Ward identity

$$(p'-p)_\rho\Gamma^\rho(p',p) = 0, \qquad (2.18)$$

following from the conservation of the electromagnetic current. Therefore, the identity

$$\Gamma^\rho(p',p) = -k_\sigma\frac{\partial}{\partial k_\rho}\Gamma^\sigma(p',p) \qquad (2.19)$$

provides the additional power of k which ensures a finite result as $k_\mu \to 0$.

The presence of three different interactions in the standard model naturally leads one to consider the following decomposition of the anomalous magnetic moment a_ℓ:

$$a_\ell = a_\ell^{\text{QED}} + a_\ell^{\text{had}} + a_\ell^{\text{weak}} . \qquad (2.20)$$

By a_ℓ^{QED}, I denote all the contributions which arise from loops involving only virtual photons and leptons. Among these, it is useful to distinguish those which involve only the same lepton flavour ℓ for which we wish to compute the anomalous magnetic moment, and those which involve loops with leptons of different flavours, denoted collectively as ℓ' [$\alpha \equiv e^2/4\pi$],

$$a_\ell^{\text{QED}} = \sum_{n\geq 1} A_n\left(\frac{\alpha}{\pi}\right)^n + \sum_{n\geq 2} B_n(\ell,\ell')\left(\frac{\alpha}{\pi}\right)^n . \qquad (2.21)$$

The second type of contribution, a_ℓ^{had} involves also quark loops. Their contribution is far from being limited to the short distance scales, and a_ℓ^{had} is an intrinsically non perturbative quantity. From a theoretical point of view, this represents a serious difficulty. Finally, at some level of precision, the weak interactions can no longer be ignored, and contributions of virtual Higgs or massive gauge boson degrees of freedom induce the third component a_ℓ^{weak}. Of course, starting from the two loop level, a hadronic contribution to a_ℓ^{weak} will also be present. The remaining of this presentation is devoted to a detailed discussion of these various contributions.

Before starting this guided tour of the anomalous magnetic moments of the massive charged leptons of the standard model, it is useful to keep in mind a few simple and elementary considerations:

- The anomalous magnetic moment is a dimensionless quantity. Therefore, the coefficients A_n above are *universal*, i.e. they do not depend on the flavour of the lepton whose anomalous magnetic moment we wish to evaluate.

- The contributions to a_ℓ of degrees of freedom corresponding to a typical scale $M \gg m_\ell$ decouple [12], i.e. they are *suppressed* by powers of m_ℓ/M.[5]

- The contributions to a_ℓ originating from light degrees of freedom, characterized by a typical scale $m \ll m_\ell$ are *enhanced* by powers of $\ln(m_\ell/m)$. At a given order, the logarithmic terms that do not vanish as $m_\ell/m \to 0$ can often be computed from the knowledge of the lesser order terms and of the β function through the renormalization group equations [15, 16, 17, 18].

These general properties already allow to draw a few elementary conclusions. The electron being the lightest charged lepton, its anomalous magnetic moment is dominantly determined by the values of the coefficients A_n. The first contribution of other degrees of freedom comes from graphs involving, say, at least one muon loop, which occurs first at the two-loop level, and is of the order of $(m_e^2/m_\mu)^2(\alpha/\pi)^2 \sim 10^{-10}$. The hadronic effects, i.e. "quark and gluon loops", characterized by a scale of $\sim$ 1 GeV, or effects of degrees of freedom beyond the standard model, which may appear at some high scale M, will be felt more strongly, by a considerable factor $(m_\mu/m_e)^2 \sim 40\,000$, in a_μ than in a_e. Thus, a_e is well suited for testing the validity of QED at higher orders, whereas a_μ is more appropriate for detecting new physics. If we follow this line reasoning, a_τ would even be better suited for finding evidence of degrees of freedom beyond the standard model. Unfortunately, the very short lifetime of the τ lepton $[\tau_\tau \sim 3\times 10^{-13} s]$ makes a sufficiently accurate measurement of a_τ impossible at present.

[5]In the presence of the weak interactions, this statement has to be reconsidered, since the necessity for the cancellation of the $SU(2) \times U(1)$ gauge anomalies transforms the decoupling of, say, a single heavy fermion in a given generation, into a somewhat subtle issue [13, 14].

3 Brief overview of the experimental situation

3.1 Measurements of the magnetic moment of the electron

The first indication that the gyromagnetic factor of the electron is different from the value $g_e = 2$ predicted by the Dirac theory came from the precision measurement of hyperfine splitting in hydrogen and deuterium [19]. The first measurement of the gyromagnetic factor of free electrons was performed in 1958 [20], with a precision of 3.6%. The situation began to improve with the introduction of experimental setups based on the Penning trap. Some of the successive values obtained over a period of forty years are shown in Table 3.1. Technical improvements, eventually allowing for the trapping of a single electron or positron, produced, in the course of time, an enormous increase in precision which, starting from a few percents, went through the ppm [parts per million] levels, before culminating at 4 ppb [parts per billion] [21] in the last of a series of experiments performed at the University of Washington in Seattle. The same experiment has also produced a measurement of the magnetic moment of the positron with the same accuracy, thus providing a test of CPT invariance at the level of 10^{-12},

$$g_{e^-}/g_{e^+} \;=\; 1 \,+\, (0.5 \pm 2.1) \times 10^{-12} . \tag{3.1}$$

An extensive survey of the literature and a detailed description of the various experimental aspects can be found in [22]. The earlier experiments are reviewed in [23].

0.001 19(5)	4.2%	[24]
0.001 165(11)	1%	[25]
0.001 116(40)	3.6%	[20]
0.001 160 9(2 4)	2 100 ppm	[26]
0.001 159 622(27)	23 ppm	[27]
0.001 159 660(300)	258 ppm	[28]
0.001 159 657 7(3 5)	3 ppm	[29]
0.001 159 652 41(20)	172 ppb	[30]
0.001 159 652 188 4(4 3)	4 ppb	[21]

Table 1: Some experimental determinations of the electron's anomalous magnetic moment a_e with the corresponding relative precision.

3.2 Measurements of the magnetic moment of the muon

The anomalous magnetic moment of the muon has also been the subject of quite a few experiments. The very short lifetime of the muon, $\tau_\mu = (2.19703 \pm 0.00004) \times 10^{-6} s$, makes it necessary to proceed in a completely different way in order to attain a high precision. The experiments conducted at CERN during the years 1968-1977 used a muon storage ring [for details, see [31] and references quoted therein]. The more recent experiments at the AGS in Brookhaven are based on the same concept. Pions are produced by sending a proton beam on a target. The pions subsequently decay into longitudinally polarized muons, which are captured inside a storage ring, where they follow a circular orbit in the presence of both a uniform magnetic field and a quadrupole electric field, the latter serving the purpose of focusing the muon beam. The difference between the spin precession frequency and the orbit, or synchrotron, frequency is given by

$$\boldsymbol{\omega}_s - \boldsymbol{\omega}_c = \frac{e}{m_\mu c} \left\{ a_\mu \mathrm{B} - \left[a_\mu - \frac{1}{1-\gamma^2} \right] \boldsymbol{\beta} \wedge \mathrm{E} \right\} . \tag{3.2}$$

Therefore, if the Lorentz factor γ is tuned to its "magic" value $\gamma = \sqrt{1 + 1/a_\mu} = 29.3$, the measurement of $\omega_s - \omega_c$ and of the magnetic field B allows to determine a_μ. The spin direction of the muon is determined by detecting the electrons or positrons produced in the decay of the muons with an energy greater than some threshold energy E_t. The number of electrons detected decreases exponentially in time, with a time constant set by the muon's lifetime, and is modulated by the frequency $\omega_s - \omega_c$,

$$N_e(t) = N_0 e^{-t/\tau_\mu} \{1 + A \cos[(\omega_s - \omega_c)t + \phi]\} . \tag{3.3}$$

0.001 166 16(31)	265 ppm	[32]
0.001 165 895(27)	23 ppm	[33]
0.001 165 911(11)	10 ppm	[34]
0.001 165 925(15)	13 ppm	[35]
0.001 165 9191(59)	5 ppm	[36]
0.001 165 920 2(16)	1.3 ppm	[37]
0.001 165 920 3(8)	0.7 ppm	[38]

Table 2: Determinations of the anomalous magnetic moment of the positively charged muon from the storage ring experiments conducted at the CERN PS and at the BNL AGS.

Several experimental results for the anomalous magnetic moment of the positively charged muon, obtained at the CERN PS or, more recently, at the BNL AGS, are recorded in Table 3.2. Notice that the relative errors are measured in ppm units, to be contrasted with the ppb level of accuracy achieved in the electron case. The four last values in Table 3.2 were obtained by the E821 experiment at BNL. They show a remarkable stability and a steady increase in precision, and now completely dominate the world average value. Further data, for negatively charged muons [6] are presently being analyzed. The aim of the Brookhaven Muon (g - 2) Collaboration is to reach a precision of 0.35 ppm, but this will depend on whether the experiment will receive financial support to collect more data or not.

3.3 Experimental bounds on the anomalous magnetic moment of the τ lepton

As already mentioned, the very short lifetime of the τ precludes a measurement of its anomalous magnetic moment following any of the techniques described above. Indirect access to a_τ is provided by the reaction $e^+e^- \rightarrow \tau^+\tau^-\gamma$. The results obtained by OPAL [39] and L3 [40] at LEP only provide very loose bounds,

$$\begin{aligned} -0.052 < a_\tau < 0.058 \ (95\% C.L.) \\ -0.068 < a_\tau < 0.065 \ (95\% C.L.)\,, \end{aligned} \tag{3.4}$$

respectively.
We shall now turn towards theory, in order to see how the standard model predictions compare with these experimental values. Only the cases of the electron and of the muon will be treated in some detail. The theoretical aspects as far as the anomalous magnetic moment of the τ are concerned are discussed in [41].

4 The anomalous magnetic moment of the electron

We start with the anomalous magnetic moment of the lightest charged lepton, the electron. Since the electron mass m_e is much smaller than any other mass scale present in the standard model, the mass independent part of $a_e^{\rm QED}$ dominates its value. As mentioned before, non vanishing contributions appear at the level of the loop diagrams shown in Fig. 1.

4.1 The lowest order contribution

The one loop diagram gives

$$\begin{aligned} \Gamma^\rho(p',p)\big|_{\rm 1loop} &= (-ie)^2 \int \frac{d^4q}{(2\pi)^4}\gamma_\mu(\not p' + \not q + m_e)\gamma^\rho(\not p + \not q + m_e)\gamma^\mu \\ &\times \frac{i}{(p'+q)^2 - m_e^2}\,\frac{i}{(p+q)^2 - m_e^2}\,\frac{(-i)}{q^2}\,. \end{aligned} \tag{4.1}$$

[6] The CERN experiment had also measured $a_{\mu^-} = 0.001\,165\,937(12)$ with a 10 ppm accuracy, giving the average value $a_\mu = 0.001\,165\,924(8.5)$, with an accuracy of 7 ppm.

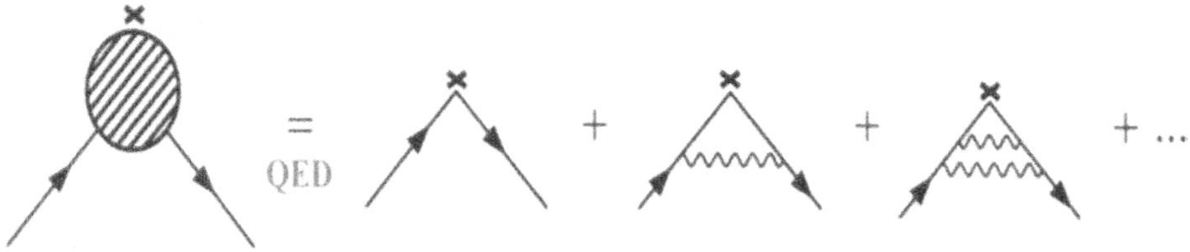

Figure 1: The perturbative expansion of $\Gamma^{\rho}(p',p)$ in single flavour QED. The tree graph gives $F_1 = 1$, $F_2 = F_3 = 0$. The one loop vertex correction graph gives the coefficient A_1 in Eq. (2.21). The cross denotes the insertion of the external field.

The form factor $F_2(k^2)$ is obtained by using Eqs. (2.14) and (2.15) and, upon evaluating the corresponding trace of Dirac matrices, one finds

$$\begin{aligned} F_2(k^2)\big|_{1\text{loop}} &= ie^2 \frac{32m_e^2}{k^2(k^2-4m_e^2)^2} \int \frac{d^4q}{(2\pi)^4} \frac{1}{(p'+q)^2-m_e^2} \frac{1}{(p+q)^2-m_e^2} \frac{1}{q^2} \\ &\times \left[-3k^2(p\cdot q)^2 + 2k^2m_e^2(p\cdot q) + k^2m_e^2q^2 - m_e^2(k\cdot q)^2\right]. \end{aligned} \tag{4.2}$$

Then follow the usual steps of introducing two Feynman parameters, of performing a trivial change of variables and a symmetric integration over the loop momentum q, so that one arrives at

$$\begin{aligned} F_2(k^2)\big|_{1\text{loop}} &= ie^2 \frac{64m_e^2}{(k^2-4m_e^2)^2} \int_0^1 dx \int_0^1 dy \int \frac{d^4q}{(2\pi)^4} \frac{1}{(q^2-\mathcal{R}^2)^3} \\ &\times \left[2x(1-x)m_e^4 - \frac{3}{4}x^2y^2(k^2)^2 + m_e^2k^2x\left(3xy - y + \frac{1}{2}x\right)\right] \\ &= \frac{e^2}{\pi^2} \frac{2m_e^2}{(k^2-4m_e^2)^2} \int_0^1 dx \int_0^1 dy \frac{1}{\mathcal{R}^2} \\ &\times \left[2x(1-x)m_e^4 - \frac{3}{4}x^2y^2(k^2)^2 + m_e^2k^2x\left(3xy - y + \frac{1}{2}x\right)\right], \end{aligned} \tag{4.3}$$

with

$$\mathcal{R}^2 = x^2y(1-y)(2m_e^2-k^2) + x^2y^2m_e^2 + x^2(1-y)^2m_e^2. \tag{4.4}$$

As expected, the limit $k^2 \to 0$ can be taken without problem, and gives

$$a_e\big|_{1\text{loop}} \equiv F_2(0)\big|_{1\text{loop}} = \frac{1}{2}\frac{\alpha}{\pi}. \tag{4.5}$$

Let us stress that although the integral (4.1) diverges, we have obtained a finite result for $F_2(k^2)$, and hence for a_e, without introducing any regularization. This is of course expected, since a divergence in, say, $F_2(0)$ would require that a counterterm of the form given by the second term in $\widehat{\mathcal{L}}_{\text{int}}$, see Eq. (2.8), be introduced. This would in turn spoil the renormalizability of the theory. In fact, as is well known, the divergence lies in $F_1(0)$, and is absorbed into the renormalization of the electron's charge.

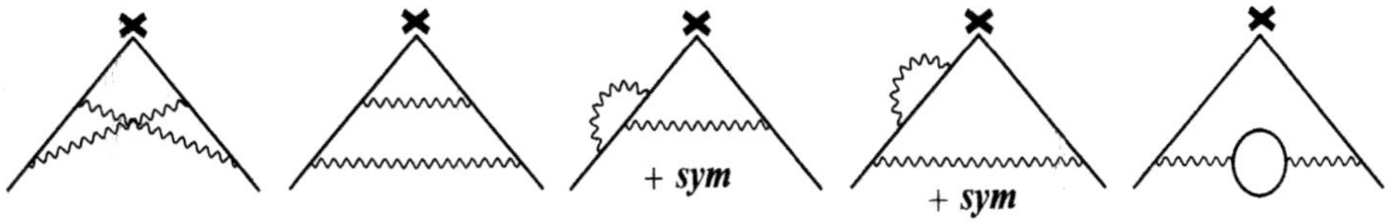

Figure 2: The Feynman diagrams which contribute to the coefficient A_2 in Eq. (2.21).

4.2 Higher order mass independent corrections

The previous calculation is rather straightforward and amounts to the result

$$A_1 = \frac{1}{2} \tag{4.6}$$

first obtained by Schwinger [42]. Schwinger's calculation was soon followed by a computation of A_2 [43], which requires the evaluation of 7 graphs, representing five distinct topologies, and shown in Fig. 2. Historically, the result of Ref. [43] was important, because it provided the first explicit example of the realization of the renormalization program of QED at two loops. However, the value for A_2 was not given correctly. The correct expression of the second order mass independent contribution was derived in [44, 45, 46] (see also [47, 48]) and reads [7]

$$\begin{aligned} A_2 &= \frac{197}{144} + \left(\frac{1}{2} - 3\ln 2\right)\zeta(2) + \frac{3}{4}\zeta(3) \\ &= -0.328\,478\,965... \end{aligned} \tag{4.7}$$

with $\zeta(p) = \sum_{n=1}^{\infty} 1/n^p$, $\zeta(2) = \pi^2/6$. The occurrence of transcendental numbers like zeta functions or polylogarithms is a general feature of higher order calculations in perturbative quantum field theory. The pattern of these transcendentals in perturbation theory has also been put in relationship with other mathematical structures, like knot theory.

The analytic evaluation of the three-loop mass independent contribution to the anomalous magnetic moment required quite some time, and is mainly due to the

[7] Actually, the experimental result of Ref. [25] disagreed with the value $A_2 = -2.973$ obtained in [43], and prompted theoreticians to reconsider the calculation. The result obtained by the authors of Refs. [44, 45, 46] reconciled theory with experiment.

dedication of E. Remiddi and his coworkers during the period 1969–1996. There are now 72 diagrams to consider, involving many different topologies, see Fig. 3.

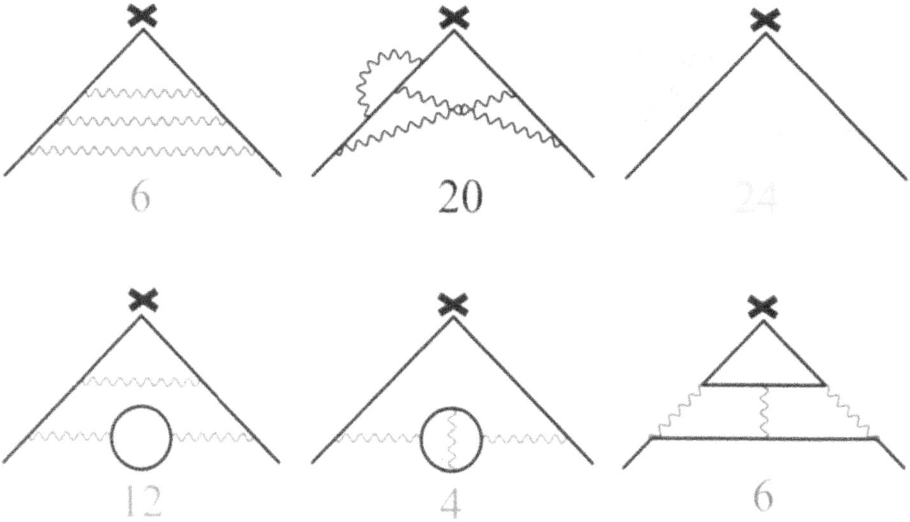

Figure 3: The 72 Feynman diagrams which make up the coefficient A_3 in Eq. (2.21).

The calculation was completed [49] in 1996, with the analytical evaluation of a last class of diagrams, the non planar "triple cross" topologies. The result reads [8]

$$\begin{aligned} A_3 &= \frac{87}{72}\pi^2\zeta(3) - \frac{215}{24}\zeta(5) + \frac{100}{3}\left[\left(a_4 + \frac{1}{24}\ln^4 2\right) - \frac{1}{24}\pi^2\ln^2 2\right] \\ &\quad - \frac{239}{2160}\pi^4 + \frac{139}{18}\zeta(3) - \frac{298}{9}\pi^2\ln 2 + \frac{17101}{810}\pi^2 + \frac{28259}{5184} \\ &= 1.181\,241\,456... \end{aligned} \tag{4.8}$$

where [9] $a_p = \sum_{n=1}^{\infty} \frac{1}{2^n n^p}$. The numerical value extracted from the exact analytical expression given above can be improved to any desired order of precision.

[8]The completion of this three-loop program can be followed through Refs. [50]–[55] and [49]. A description of the technical aspects related to this work and an account of its status up to 1990, with references to the corresponding literature, are given in Ref. [56].

[9]The first three values are known to be $a_1 = \ln 2$, $a_2 = \mathrm{Li}_2(1/2) = (\zeta(2) - \ln^2 2)/2$, $a_3 = \frac{7}{8}\zeta(3) - \frac{1}{2}\zeta(2)\ln 2 + \frac{1}{6}\ln^3 2$ [56].

In parallel to these analytical calculations, numerical methods for the evaluation of the higher order contributions were also developed, in particular by Kinoshita and his collaborators (for details, see [57]). The numerical evaluation of the full set of three loop diagrams was achieved in several steps [58]-[64]. The value quoted in [64] is $A_3 = 1.195(26)$, where the error comes from the numerical procedure. In comparison, let us quote the value [65, 57] $A_3 = 1.176\,11\,(42)$ obtained if only a subset of 21 three loop diagrams out of the original set of 72 is evaluated numerically, relying on the analytical results for the remaining 51 ones, and recall the value $A_3 = 1.181\,241\,456...$ obtained from the full analytical evaluation. The error induced on a_e due to the numerical uncertainty in the second, more accurate, value is still $\Delta(a_e) = 5.3 \times 10^{-12}$, whereas the experimental error is only $\Delta(a_e)|_{\rm exp} = 4.3 \times 10^{-12}$. This discussion shows that the analytical evaluations of higher loop contributions to the anomalous magnetic moment of the electron have a strong practical interest as far as the precision of the theoretical prediction is concerned, and which goes well beyond the mere intellectual satisfaction and technical skills involved in these calculations. [10]

At the four loop level, there are 891 diagrams to consider. Clearly, only a few of them have been evaluated analytically [66, 67]. The complete numerical evaluation of the whole set gave [65] $A_4 = -1.434(138)$. The development of computers allowed subsequent reanalyzes to be more accurate, i.e. $A_4 = -1.557(70)$ [68], while the "latest of [these] constantly improving values" is [4]

$$A_4 = -1.509\,8(38\,4)\,. \tag{4.9}$$

Needless to say, so far the five loop contribution A_5 is unknown territory. On the other hand, $(\alpha/\pi)^5 \sim 7 \times 10^{-14}$, so that one may reasonably expect that, in view of the present experimental situation, its knowledge is not yet required.

4.3 Mass dependent QED corrections

We now turn to the QED loop contributions to the electron's anomalous magnetic moment involving the heavier leptons, μ and τ. The lowest order contribution of this type occurs at the two loop level, $\mathcal{O}(\alpha^2)$, and corresponds to a heavy lepton vacuum polarization insertion in the one loop vertex graph, cf. Fig. 4. Quite generally, the contribution to a_ℓ arising from the insertion, into the one loop vertex correction, of a vacuum polarization graph due to a loop of lepton ℓ', reads

[10] It is only fair to point out that the numerical values that are quoted here correspond to those given in the original references. It is to be expected that they would improve if today's numerical possibilities were used.

[69, 70] [11]

$$B_2(\ell,\ell') = \frac{1}{3}\int_{4m_{\ell'}^2}^{\infty} dt\sqrt{1-\frac{4m_{\ell'}^2}{t}}\,\frac{t+2m_{\ell'}^2}{t^2}\int_0^1 dx\,\frac{x^2(1-x)}{x^2+(1-x)\frac{t}{m_\ell^2}}\,. \tag{4.10}$$

If $m_{\ell'} \gg m_\ell$, the second integrand can be approximated by $x^2 m_\ell^2/t$, and one obtains [72]

$$B_2(\ell,\ell') = \frac{1}{45}\left(\frac{m_\ell}{m_{\ell'}}\right)^2 + \mathcal{O}\left[\left(\frac{m_\ell}{m_{\ell'}}\right)^3\right], \quad m_{\ell'} \gg m_\ell\,. \tag{4.11}$$

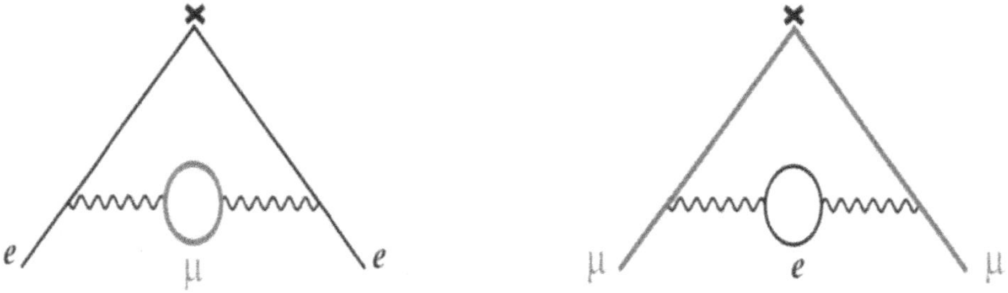

Figure 4: The insertion of a muon vacuum polarization loop into the electron vertex correction (left) or of an electron vacuum polarization loop into the muon vertex correction (right).

Numerically, this translates into [3] [m_e = 0.51099907(15) MeV, m_μ/m_e = 206.768273(24), $m_\tau = 1\,777.05(26)$]

$$\begin{aligned} B_2(e,\mu) &= 5.197\times10^{-7} \\ B_2(e,\tau) &= 1.838\times10^{-9}\,. \end{aligned} \tag{4.12}$$

For later use, it is interesting to briefly discuss the structure of Eq. (4.10). The quantity which appears under the integral is related to the cross section for the scattering of a $\ell^+\ell^-$ pair into a pair $(\ell')^+(\ell')^-$ *at lowest order in QED,*

$$\sigma_{\text{QED}}^{(\ell^+\ell^-\to(\ell')^+(\ell')^-)}(s) = \frac{4\pi\alpha^2}{3s^2}\sqrt{1-\frac{4m_{\ell'}^2}{s}}\,(s+2m_{\ell'}^2)\,, \tag{4.13}$$

[11] A trivial change of variable on t brings the expression (4.10) into the form given in [69, 70]. Furthermore, the analytical result obtained upon performing the double integration is available in [71].

so that

$$B_2(\ell;\ell') = \frac{1}{3}\int_{4m_{\ell'}^2}^{\infty} dt K(t) R^{(\ell')}(t)\,, \tag{4.14}$$

where

$$K(t) = \int_0^1 dx \, \frac{x^2(1-x)}{x^2+(1-x)\frac{t}{m_\ell^2}}\,, \tag{4.15}$$

and $R^{(\ell')}(t)$ is the *lowest order* QED cross section $\sigma_{\rm QED}^{(\ell^+\ell^-\to(\ell')^+(\ell')^-)}(s)$ divided by the asymptotic form of the cross section of the reaction $e^+e^- \to \mu^+\mu^-$ for $s \gg m_\mu^2$, $\sigma_\infty^{(e^+e^-\to\mu^+\mu^-)}(s) = \frac{4\pi\alpha^2}{3s}$.

The three loop contributions with different lepton flavours in the loops are also known analytically [73, 74]. It is convenient to distinguish three classes of diagrams. The first group contains all the diagrams with one or two vacuum polarization insertion involving the same lepton, μ or τ, of the type shown in Fig. 5. The second group consists of the leptonic light-by-light scattering insertion diagrams, Fig. 6. Finally, since there are three flavours of massive leptons in the standard model, one has also the possibility of having graphs with two heavy lepton vacuum polarization insertions, one made of a muon loop, the other of a τ loop. This gives

$$\begin{aligned} B_3(e,\ell) = {} & B_3^{(\rm v.p.)}(e;\mu) + B_3^{(\rm v.p.)}(e;\tau) + B_3^{(\rm L\times L)}(e;\mu) \\ & + B_3^{(\rm L\times L)}(e;\tau) + B_3^{(\rm v.p.)}(e;\mu,\tau)\,. \end{aligned} \tag{4.16}$$

The analytical expression for $B_3^{(\rm v.p.)}(e;\mu)$ can be found in Ref. [73], whereas [74] gives the corresponding result for $B_3^{(\rm L\times L)}(e;\mu)$. For practical purposes, it is both sufficient and more convenient to use their expansions in powers of m_e/m_μ,

$$\begin{aligned} B_3^{(\rm v.p.)}(e;\mu) &= \left(\frac{m_e}{m_\mu}\right)^2\left[-\frac{23}{135}\ln\left(\frac{m_\mu}{m_e}\right) - \frac{2}{45}\pi^2 + \frac{10117}{24300}\right] \\ &+ \left(\frac{m_e}{m_\mu}\right)^4\left[\frac{19}{2520}\ln^2\left(\frac{m_\mu}{m_e}\right) - \frac{14233}{132300}\ln\left(\frac{m_\mu}{m_e}\right) + \frac{49}{768}\zeta(3)\right. \\ &- \left.\frac{11}{945}\pi^2 + \frac{2976691}{296352000}\right] \\ &+ \mathcal{O}\left[\left(\frac{m_e}{m_\mu}\right)^6\right] \\ &= -0.000\,021\,768... \end{aligned} \tag{4.17}$$

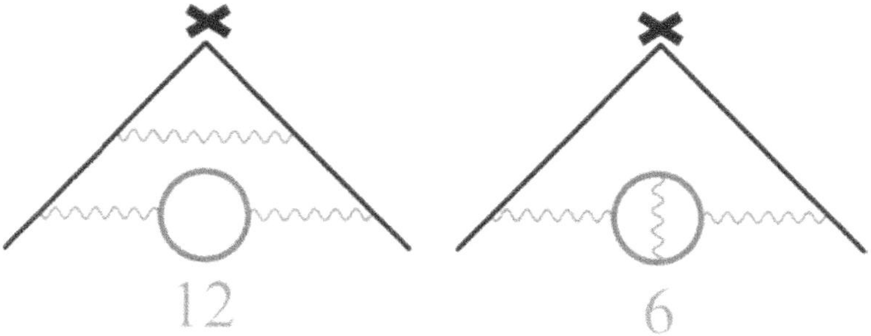

Figure 5: Three loop QED corrections with insertion of a heavy lepton vacuum polarization which make up the coefficient $B_3^{(\text{v.p.})}(e;\mu)$.

and [74]

$$
\begin{aligned}
B_3^{(\text{L}\times\text{L})}(e;\mu) &= \left(\frac{m_e}{m_\mu}\right)^2 \left[\frac{3}{2}\zeta(3) - \frac{19}{16}\right] \\
&+ \left(\frac{m_e}{m_\mu}\right)^4 \left[-\frac{161}{810}\ln^2\left(\frac{m_\mu}{m_e}\right) - \frac{16189}{48600}\ln\left(\frac{m_\mu}{m_e}\right) + \frac{13}{18}\zeta(3)\right. \\
&- \left.\frac{161}{9720}\pi^2 - \frac{831931}{972000}\right] \\
&+ \mathcal{O}\left[\left(\frac{m_e}{m_\mu}\right)^6\right] \\
&= 0.000\,014\,394\,5\ldots
\end{aligned}
\tag{4.18}
$$

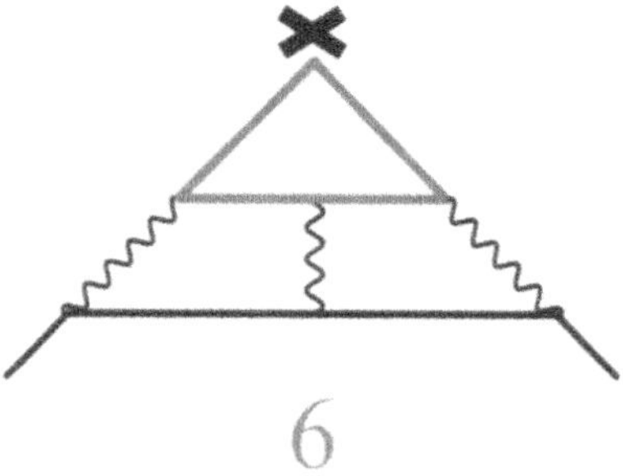

Figure 6: The three loop QED correction with the insertion of a heavy lepton light-by-light scattering subgraph, corresponding to the coefficient $B_3^{(\text{L}\times\text{L})}(e;\mu)$.

The expressions for $B_3^{(\rm v.p.)}(e;\tau)$ and $B_3^{(\rm L\times L)}(e;\tau)$ follow upon replacing the muon mass m_μ by m_τ. This again gives a suppression factor $(m_\mu/m_\tau)^2$, which makes these contributions negligible at the present level of precision. For the same reason, $B_3^{(\rm v.p.)}(e;\mu,\tau)$ can also be discarded.

4.4 Other contributions to a_e

In order to make the discussion of the standard model contributions to a_e complete, there remains to mention the hadronic and weak components, $a_e^{\rm had}$ and $a_e^{\rm weak}$, respectively. Their features will be discussed in detail below, in the context of the anomalous magnetic moment of the muon. I therefore only quote the numerical values [12]

$$a_e^{\rm had} = 1.67(3) \times 10^{-12}, \tag{4.19}$$

and [75]

$$a_e^{\rm weak} = 0.030 \times 10^{-12} \tag{4.20}$$

4.5 Comparison with experiment and determination of α

Summing up the various contributions discussed so far gives the standard model prediction [3, 4, 7]

$$a_e^{\rm SM} = 0.5\,\frac{\alpha}{\pi} - 0.328\,478\,444\,00\left(\frac{\alpha}{\pi}\right)^2 + 1.181\,234\,017\left(\frac{\alpha}{\pi}\right)^3 - 1.509\,8(38\,4)\left(\frac{\alpha}{\pi}\right)^4 + 1.70\times 10^{-12}. \tag{4.21}$$

In order to obtain a number that can be compared to the experimental result, a sufficiently accurate determination of the fine structure constant α is required. The best available measurement of the latter comes from the quantum Hall effect [76],

$$\alpha^{-1}(qH) = 137.036\,003\,00(2\,70) \tag{4.22}$$

and leads to

$$a_e^{\rm SM}(qH) = 0.001\,159\,652\,153\,5(24\,0), \tag{4.23}$$

about six times less accurate than the latest experimental value [21]

$$a_e^{\rm exp} = 0.001\,159\,652\,188\,4(4\,3). \tag{4.24}$$

On the other hand, if one excludes other contributions to a_e than those from the standard model considered so far, and believes that all theoretical errors are under control, then the above value of $a_e^{\rm exp}$ provides the best determination of α to date,

$$\alpha^{-1}(a_e) = 137.035\,999\,58(52). \tag{4.25}$$

[12] I reproduce here the values given in [3, 4], except for the fact that I have taken into account the changes in the value of the hadronic light-by-light contribution to a_μ, see below, for which I take $a_\mu^{(\rm L\times L)} = +8(4)\times 10^{-10}$, and which translates into $a_e^{(\rm L\times L)} \sim a_\mu^{(\rm L\times L)}(m_e/m_\mu)^2 = 0.02\times 10^{-12}$.

5 The anomalous magnetic moment of the muon

In this section, we discuss the theoretical aspects concerning the anomalous magnetic moment of the muon. Since the muon is much heavier than the electron, a_μ will be more sensitive to higher mass scales. In particular, it is a better probe for possible degrees of freedom beyond the standard model, like supersymmetry. The drawback, however, is that a_μ will also be more sensitive to the non perturbative strong interaction dynamics at the ~ 1 GeV scale.

5.1 QED contributions to a_μ

As already mentioned before, the mass independent QED contributions to a_μ are described by the same coefficients A_n as in the case of the electron. We therefore need only to discuss the coefficients $B_n(\mu;\ell')$ associated with the mass dependent corrections.

For $m_{\ell'} \ll m_\ell$, Eq. (4.10) gives [69, 70, 71]

$$B_2(\ell;\ell') = \frac{1}{3}\ln\left(\frac{m_\ell}{m_{\ell'}}\right) - \frac{25}{36} + \frac{3}{2}\frac{m_\ell}{m_{\ell'}}\zeta(2) - 4\left(\frac{m_\ell}{m_{\ell'}}\right)^2 \ln\left(\frac{m_\ell}{m_{\ell'}}\right) + 3\left(\frac{m_\ell}{m_{\ell'}}\right)^2 + \mathcal{O}\left[\left(\frac{m_\ell}{m_{\ell'}}\right)^3\right], \quad (5.1)$$

which translates into the numerical values [3]

$$B_2(\mu;e) = 1.094\,258\,294(37) \quad (5.2)$$

$$B_2(\mu;\tau) = 0.00\,078\,059(23)\,. \quad (5.3)$$

Although these numbers follow from an analytical expression, there are uncertainties attached to them, induced by those on the corresponding values of the ratios of the lepton masses.

The three loop QED corrections decompose as

$$B_3(\mu,\ell) = B_3^{(\text{v.p.})}(\mu;e) + B_3^{(\text{v.p.})}(\mu;\tau) + B_3^{(\text{L}\times\text{L})}(\mu;e) + B_3^{(\text{L}\times\text{L})}(\mu;\tau) + B_3^{(\text{v.p.})}(\mu;e,\tau)\,. \quad (5.4)$$

with [73, 74]

$$
\begin{aligned}
B_3^{(\text{v.p.})}(\mu; e) &= \frac{2}{9}\ln^2\left(\frac{m_\mu}{m_e}\right) + \left[\zeta(3) - \frac{2}{3}\pi^2\ln 2 + \frac{1}{9}\pi^2 + \frac{31}{27}\right]\ln\left(\frac{m_\mu}{m_e}\right) \\
&+ \frac{11}{216}\pi^4 - \frac{2}{9}\pi^2\ln^2 2 - \frac{8}{3}a_4 - \frac{1}{9}\ln^4 2 - 3\zeta(3) + \frac{5}{3}\pi^2\ln 2 - \frac{25}{18}\pi^2 + \frac{1075}{216} \\
&+ \frac{m_e}{m_\mu}\left[-\frac{13}{18}\pi^3 - \frac{16}{9}\pi^2\ln 2 + \frac{3199}{1080}\pi^2\right] \\
&+ \left(\frac{m_e}{m_\mu}\right)^2\left[\frac{10}{3}\ln^2\left(\frac{m_\mu}{m_e}\right) - \frac{11}{9}\ln\left(\frac{m_\mu}{m_e}\right) - \frac{14}{3}\pi^2\ln 2 - 2\zeta(3)\right. \\
&\left.+ \frac{49}{12}\pi^2 - \frac{131}{54}\right] \\
&+ \left(\frac{m_e}{m_\mu}\right)^3\left[\frac{4}{3}\pi^2\ln\left(\frac{m_\mu}{m_e}\right) + \frac{35}{12}\pi^3 - \frac{16}{3}\pi^2\ln 2 - \frac{5771}{1080}\pi^2\right] \\
&+ \left(\frac{m_e}{m_\mu}\right)^4\left[-\frac{25}{9}\ln^3\left(\frac{m_\mu}{m_e}\right) - \frac{1369}{180}\ln^2\left(\frac{m_\mu}{m_e}\right)\right. \\
&+ [-2\zeta(3) + 4\pi^2\ln 2 - \frac{269}{144}\pi^2 - \frac{7496}{675}]\ln\left(\frac{m_\mu}{m_e}\right) \\
&- \frac{43}{108}\pi^4 + \frac{8}{9}\pi^2\ln^2 2 + \frac{80}{3}a_4 + \frac{10}{9}\ln^4 2 - \frac{411}{32}\zeta(3) \\
&\left.+ \frac{89}{48}\pi^2\ln 2 - \frac{1061}{864}\pi^2 - \frac{274511}{54000}\right] \\
&+ \mathcal{O}\left[\left(\frac{m_e}{m_\mu}\right)^5\right],
\end{aligned}
\qquad (5.5)
$$

$$
\begin{aligned}
B_3^{(\mathrm{L\times L})}(\mu;e) &= \frac{2}{3}\pi^2 \ln\left(\frac{m_\mu}{m_e}\right) + \frac{59}{270}\pi^4 - 3\zeta(3) - \frac{10}{3}\pi^2 + \frac{2}{3} \\
&+ \frac{m_e}{m_\mu}\left[\frac{4}{3}\pi^2 \ln\left(\frac{m_\mu}{m_e}\right) - \frac{196}{3}\pi^2 \ln 2 + \frac{424}{9}\pi^2\right] \\
&+ \left(\frac{m_e}{m_\mu}\right)^2 \left[-\frac{2}{3}\ln^3\left(\frac{m_\mu}{m_e}\right) + (\frac{\pi^2}{9} - \frac{20}{3})\ln^2\left(\frac{m_\mu}{m_e}\right)\right. \\
&- \left[\frac{16}{135}\pi^4 + 4\zeta(3) - \frac{32}{9}\pi^2 + \frac{61}{3}\right]\ln\left(\frac{m_\mu}{m_e}\right) \\
&\left. + \frac{4}{3}\zeta(3)\pi^2 - \frac{61}{270}\pi^4 + 3\zeta(3) + \frac{25}{18}\pi^2 - \frac{283}{12}\right] \\
&+ \left(\frac{m_e}{m_\mu}\right)^3 \left[\frac{10}{9}\pi^2 \ln\left(\frac{m_\mu}{m_e}\right) - \frac{11}{9}\pi^2\right] \\
&+ \left(\frac{m_e}{m_\mu}\right)^4 \left[\frac{7}{9}\ln^3\left(\frac{m_\mu}{m_e}\right) + \frac{41}{18}\ln^2\left(\frac{m_\mu}{m_e}\right) + \frac{13}{9}\pi^2 \ln\left(\frac{m_\mu}{m_e}\right) + \frac{517}{108}\ln\left(\frac{m_\mu}{m_e}\right)\right. \\
&\left. + \frac{1}{2}\zeta(3) + \frac{191}{216}\pi^2 + \frac{13283}{2592}\right] + \mathcal{O}\left[\left(\frac{m_e}{m_\mu}\right)^5\right], \qquad (5.6)
\end{aligned}
$$

while $B_3^{(\mathrm{v.p.})}(\mu;\tau)$ and $B_3^{(\mathrm{L\times L})}(\mu;\tau)$ are derived from $B_3^{(\mathrm{v.p.})}(\mu;\tau)$ and from $B_3^{(\mathrm{L\times L})}(\mu;\tau)$, respectively, by trivial substitutions of the masses. Furthermore, the graphs with mixed vacuum polarization insertions, one electron loop, and one τ loop, are evaluated numerically using a dispersive integral [51, 73, 77].
Numerically, one obtains (we quote here the numerical values updated in [3])

$$
\begin{aligned}
B_3^{(\mathrm{v.p.})}(\mu;e) &= 1.920\,455\,1(2) \\
B_3^{(\mathrm{L\times L})}(\mu;e) &= 20.947\,924\,6(7) \\
B_3^{(\mathrm{v.p.})}(\mu;\tau) &= -0.001\,782\,2(4) \\
B_3^{(\mathrm{L\times L})}(\mu;\tau) &= 0.002\,142\,8(7) \\
B_3^{(\mathrm{v.p.})}(\mu;e,\tau) &= 0.000\,527\,6(2)\,. \qquad (5.7)
\end{aligned}
$$

Notice the large value of $B_3^{(\mathrm{L\times L})}(\mu;e)$, due to the occurrence of terms involving factors like
$\ln(m_\mu/m_e) \sim 5$ and powers of π.

5.2 Hadronic contributions to a_μ

On the level of Feynman diagrams, hadronic contributions arise through loops of virtual quarks and gluons. These loops also involve the soft scales, and therefore cannot be computed reliably in perturbative QCD. We shall decompose the hadronic contributions into three subsets: hadronic vacuum polarization insertions at order α^2, at order α^3, and hadronic light-by-light scattering,

$$a_\mu^{\text{had}} = a_\mu^{(\text{h.v.p.1})} + a_\mu^{(\text{h.v.p.2})} + a_\mu^{(\text{h.L}\times\text{L})} \tag{5.8}$$

5.2.1 Hadronic vacuum polarization

We first discuss $a_\mu^{(\text{h.v.p.1})}$, which arises at order $\mathcal{O}(\alpha^2)$ from the insertion of a single hadronic vacuum polarization into the lowest order vertex correction graph, see Fig. 7. The importance of this contribution to a_μ is known since long time [78, 79].

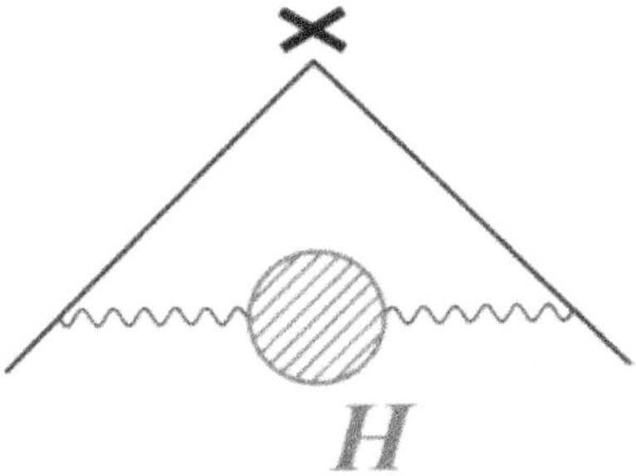

Figure 7: The insertion of the hadronic vacuum polarization into the one loop vertex correction, corresponding to $a_\mu^{(\text{h.v.p.1})}$.

There is a very convenient dispersive representation of this diagram, similar to Eq. (4.10)

$$\begin{aligned} a_\mu^{(\text{h.v.p.1})} &= \frac{\alpha}{\pi}\int_{4M_\pi^2}^\infty \frac{dt}{t}\,K(t)\,\frac{1}{\pi}\,\text{Im}\Pi(t) \\ &= \frac{1}{3}\left(\frac{\alpha}{\pi}\right)^2 \int_{4M_\pi^2}^\infty \frac{dt}{t}\,K(t)R^{\text{had}}(t)\,, \end{aligned} \tag{5.9}$$

Here, $\Pi(t)$ denotes the *hadronic* component of the vacuum polarization function, defined as [13]

$$(q_\mu q_\nu - q^2\eta_{\mu\nu})\Pi(Q^2) = i\int d^4x e^{iq\cdot x}\,\langle\Omega|\text{T}\{j_\mu(x)j_\nu(0)\}|\Omega\rangle\,, \tag{5.10}$$

with j_ρ the hadronic component of the electromagnetic current, $Q^2 = -q^2 \geq 0$ for q_μ spacelike, and $|\Omega\rangle$ the QCD vacuum. The function $K(t)$ was defined in Eq.

[13]Actually, $\Pi(t)$ defined this way has an ultraviolet divergence, produced by the QCD short distance singularity of the chronological product of the two currents. However, it only affects the real part of $\Pi(t)$. A renormalized, finite quantity is obtained by a single subtraction, $\Pi(t) - \Pi(0)$.

(4.15), and $R^{\rm had}(t)$ stands now for the cross section of $e^+e^- \to$ hadrons, *at lowest order in* α, divided by $\sigma_\infty^{(e^+e^-\to\mu^+\mu^-)}(s) = \frac{4\pi\alpha^2}{3s}$. A first principle computation of this strong interaction contribution is far beyond our present abilities to deal with the non perturbative aspects of confining gauge theories. This last relation is however very interesting because it expresses $a_\mu^{\rm (h.v.p.1)}$ through a quantity that can be measured experimentally. In this respect, two important properties of the function $K(t)$ deserve to be mentioned. First, it appears from the integral representation (4.15) that $K(t)$ is positive definite. Since $R_{e^+e^-}$ is also positive, one deduces that $a_\mu^{\rm (h.v.p.1)}$ itself is positive. Second, the function $K(t)$ decreases as $m_\mu^2/3t$ as t grows, so that it is indeed the low energy region which dominates the integral. Explicit evaluation of $a_\mu^{\rm (h.v.p.1)}$ using available data actually reveals that more than 80% of its value comes from energies below 1.4 GeV. Finally, the values obtained this way for $a_\mu^{\rm (h.v.p.1)}$ have evolved in time, as shown in Table 5.2.1. This evolution is mainly driven by the availability of more data, and is still going on, as the last entries of Table 5.2.1 show. In order to match the precision reached by the latest experimental measurement of a_μ, $a_\mu^{\rm (h.v.p.1)}$ needs to be known at $\sim$ 1%. Besides the very recent high quality e^+e^- data obtained by the BES Collaboration [80] in the region between 2 to 5 GeV, and by the CMD-2 collaboration [81] in the region dominated by the ρ resonance, the latest analyses sometimes also include or use, in the low-energy region, data obtained from hadronic decays of the τ by ALEPH [82], and, more recently, by CLEO [83]. We may notice from Table 5.2.1 that the precision obtained by using e^+e^- data alone has become comparable to the one achieved upon including the τ data. However, one of the latest analyses reveals a troubling discrepancy between the e^+e^- and τ evaluations. Additional work is certainly needed in order to resolve these problems. Further data are also expected in the future, from the KLOE experiment at the DAPHNE e^+e^- machine, or from the B factories BaBar and Belle. For additional comparative discussions and details of the various analyses, we refer the reader to the literature quoted in Table 5.2.1.

Let us briefly mention here that it is quite easy to estimate the order of magnitude of $a_\mu^{\rm (h.v.p.1)}$. For this purpose, it is convenient to introduce still another representation [93], which relates $a_\mu^{\rm (h.v.p.1)}$ to the hadronic Adler function $\mathcal{A}(Q^2)$, defined as [14]

$$\mathcal{A}(Q^2) = -Q^2\,\frac{\partial\Pi(Q^2)}{\partial Q^2} = \int_0^\infty dt\,\frac{Q^2}{(t+Q^2)^2}\,\frac{1}{\pi}\,{\rm Im}\Pi(t)\,, \tag{5.11}$$

by

$$a_\mu^{\rm (h.v.p.1)} = 2\pi^2\left(\frac{\alpha}{\pi}\right)^2\int_0^1\frac{dx}{x}(1-x)(2-x)\mathcal{A}\left(\frac{x^2}{1-x}\,m_\mu^2\right). \tag{5.12}$$

A simple representation of the hadronic Adler function can be obtained if one assumes that ${\rm Im}\Pi(t)$ is given by a single, zero width, vector meson pole, and,

[14]Unlike $\Pi(t)$ itself, $\mathcal{A}(Q^2)$ if free from ultraviolet divergences.

7024(153)	[84]	e^+e^-
7026(160)	[85]	e^+e^-
6950(150)	[86]	e^+e^-
7011(94)	[86]	τ, e^+e^-,
6951(75)	[87]	τ, e^+e^-, QCD
6924(62)	[88]	τ, e^+e^-, QCD
	[89]	QCD sum rules
7036(76)	[41]	τ, e^+e^-, QCD
7002(73)	[90]	e^+e^-, F_π
6974(105)	[91]	e^+e^-, incl. BES-II data
6847(70)	[92]	e^+e^-, incl. BES-II and CMD-2 data
7019(62)	[92]	τ, e^+e^-
6831(62)	[94]	e^+e^-

Table 3: Some of the recent evaluations of $a_\mu^{(\text{h.v.p.1})} \times 10^{-11}$ from e^+e^- and/or τ-decay data.

above a certain threshold s_0, by the QCD perturbative continuum contribution,

$$\frac{1}{\pi}\,\text{Im}\Pi(t) = \frac{2}{3}\,f_V^2 M_V^2 \delta(t - M_V^2) + \frac{2}{3}\,\frac{N_C}{12\pi^2}\,[1 + \mathcal{O}(\alpha_s)]\,\theta(t - s_0) \tag{5.13}$$

The justification [95] for this type of minimal hadronic ansatz can be found within the framework of the large-N_C limit [96, 97] of QCD, see Ref. [95] for a general discussion and a detailed study of this representation of the Adler function. The threshold s_0 for the onset of the continuum can be fixed from the property that there is no contribution in $1/Q^2$ in the short distance expansion of $\mathcal{A}(Q^2)$, which requires [95]

$$2f_V^2 M_V^2 = \frac{N_C}{12\pi^2}\,s_0\left(1 + \frac{3}{8}\,\frac{\alpha_s(s_0)}{\pi} + \mathcal{O}(\alpha_s^2)\right). \tag{5.14}$$

This then gives [98] $a_\mu^{(\text{h.v.p.1})} \sim (570 \pm 170) \times 10^{-10}$, which compares well with the more elaborate data based evaluations in Table 5.2.1, even though this simple estimate cannot claim to provide the required accuracy of about 1%.

We now come to the $\mathcal{O}(\alpha^3)$ corrections involving hadronic vacuum polarization subgraphs. Besides the contributions shown in Fig. 8, another one is obtained upon inserting a lepton loop in one of the two photon lines of the graph shown in

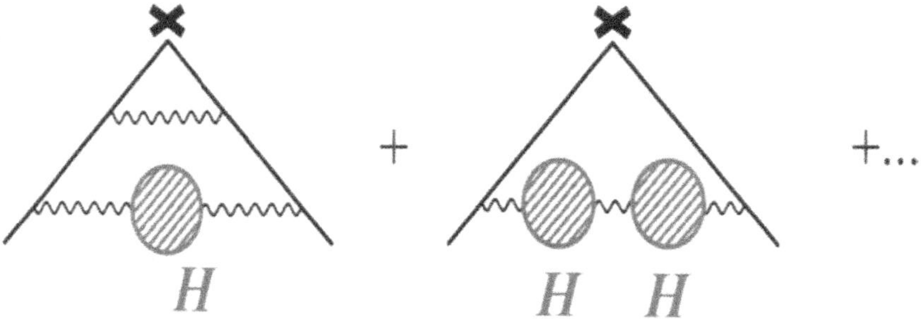

Figure 8: Higher order corrections containing the hadronic vacuum polarization contribution, corresponding to $a_\mu^{(\mathrm{h.v.p.2})}$.

Fig. 7. These can again be expressed in terms of R^{had} [99, 2, 77]

$$a_\mu^{(\mathrm{h.v.p.2})} = \frac{1}{3}\left(\frac{\alpha}{\pi}\right)^3 \int_{4M_\pi^2}^{\infty} \frac{dt}{t} K^{(2)}(t) R^{\mathrm{had}}(t) . \qquad (5.15)$$

Unlike $K(t)$, the function $K^{(2)}(t)$ is not positive definite, so that the sign of $a_\mu^{(\mathrm{h.v.p.2})}$ is not fixed on the basis of general considerations. The value obtained for this quantity is [77] $a_\mu^{(\mathrm{h.v.p.2})} \times 10^{11} = -101 \pm 6$.

5.2.2 Hadronic light-by-light scattering

We now discuss the so called hadronic light-by-light scattering graphs of Fig. 9. Actually, there is another $\mathcal{O}(\alpha^3)$ correction involving the amplitude for virtual light-by-light scattering, namely the one obtained by adding an additional photon line attached to the hadronic blob in Fig. 7. This contribution is usually included in the evaluations reported on in Table 5.2.1 [see the discussion in [92]], otherwise, it has been added. The reason for that is due to the fact that the measured e^+e^- data contain QED effects, and do not correspond to the cross section of $e^+e^- \to$ hadrons restricted to the *lowest order* in α. It is possible to compute and subtract away QED corrections involving the leptonic vertex, but there still remain radiative corrections between the final state hadrons, or which affect both the initial and the final states. These cannot be evaluated in a model independent way, and are not completely described by attaching a photon loop to the hadronic blob in Fig. 7.

Coming back to the diagram of Fig. 9, the contribution to $\Gamma_\rho(p',p)$ of relevance here is the matrix element, at lowest nonvanishing order in the fine structure constant α, of the light quark electromagnetic current

$$j_\rho(x) = \frac{2}{3}(\bar{u}\gamma_\rho u)(x) - \frac{1}{3}(\bar{d}\gamma_\rho d)(x) - \frac{1}{3}(\bar{s}\gamma_\rho s)(x) \qquad (5.16)$$

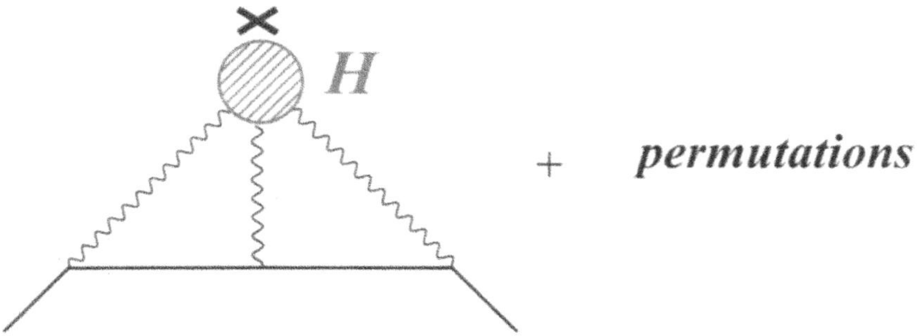

Figure 9: The hadronic light-by-light scattering graphs contributing to $a_\mu^{(\mathrm{h.L\times L})}$.

between μ^- states,

$$\begin{aligned}(-ie)\bar{\mathrm{u}}(p')\Gamma_\rho^{(\mathrm{h.L\times L})}(p',p)\mathrm{u}(p) &\equiv \langle\mu^-(p')|(ie)j_\rho(0)|\mu^-(p)\rangle \\ &= \int\frac{d^4q_1}{(2\pi)^4}\int\frac{d^4q_2}{(2\pi)^4}\frac{(-i)^3}{q_1^2\,q_2^2\,(q_1+q_2-k)^2} \\ &\times\frac{i}{(p'-q_1)^2-m^2}\,\frac{i}{(p'-q_1-q_2)^2-m^2} \\ &\times(-ie)^3\bar{\mathrm{u}}(p')\gamma^\mu(\not{p}'-\not{q}_1+m)\gamma^\nu(\not{p}'-\not{q}_1-\not{q}_2+m)\gamma^\lambda\mathrm{u}(p) \\ &\times(ie)^4\Pi_{\mu\nu\lambda\rho}(q_1,q_2,k-q_1-q_2)\,, \end{aligned} \tag{5.17}$$

with $k_\mu = (p'-p)_\mu$ and

$$\begin{aligned}\Pi_{\mu\nu\lambda\rho}(q_1,q_2,q_3) &= \int d^4x_1\int d^4x_2\int d^4x_3\,e^{i(q_1\cdot x_1+q_2\cdot x_2+q_3\cdot x_3)} \\ &\times\langle\,\Omega\,|\,\mathrm{T}\{j_\mu(x_1)j_\nu(x_2)j_\lambda(x_3)j_\rho(0)\}\,|\,\Omega\,\rangle\end{aligned} \tag{5.18}$$

the fourth-rank light quark hadronic vacuum-polarization tensor, $|\,\Omega\,\rangle$ denoting the QCD vacuum. Since the flavour diagonal current $j_\mu(x)$ is conserved, the tensor $\Pi_{\mu\nu\lambda\rho}(q_1,q_2,q_3)$ satisfies the Ward identities

$$\{q_1^\mu;q_2^\nu;q_3^\lambda;(q_1+q_2+q_3)^\rho\}\Pi_{\mu\nu\lambda\rho}(q_1,q_2,q_3) = 0\,. \tag{5.19}$$

This entails that [15]

$$\begin{aligned}&\bar{\mathrm{u}}(p')\Gamma_\rho^{(\mathrm{h.L\times L})}(p',p)\mathrm{u}(p) = \\ &\qquad\bar{\mathrm{u}}(p')\left[\gamma_\rho F_1^{(\mathrm{h.L\times L})}(k^2) + \frac{i}{2m}\sigma_{\rho\tau}k^\tau F_2^{(\mathrm{h.L\times L})}(k^2)\right]\mathrm{u}(p)\,,\end{aligned} \tag{5.20}$$

[15] We use the following conventions for Dirac's γ-matrices: $\{\gamma_\mu,\gamma_\nu\} = 2\eta_{\mu\nu}$, with $\eta_{\mu\nu}$ the flat Minkowski space metric of signature $(+---)$, $\sigma_{\mu\nu} = (i/2)[\gamma_\mu,\gamma_\nu]$, $\gamma_5 = i\gamma^0\gamma^1\gamma^2\gamma^3$, whereas the totally antisymmetric tensor $\varepsilon_{\mu\nu\rho\sigma}$ is chosen such that $\varepsilon_{0123} = +1$.

as well as $\Gamma_\rho^{(\mathrm{h.L\times L})}(p',p) = k^\tau \Gamma_{\rho\tau}^{(\mathrm{h.L\times L})}(p',p)$ with

$$\begin{aligned}
\bar{\mathrm{u}}(p')\Gamma_{\rho\sigma}^{(\mathrm{h.L\times L})}(p',p)\mathrm{u}(p) &= -ie^6 \int \frac{d^4q_1}{(2\pi)^4} \int \frac{d^4q_2}{(2\pi)^4} \frac{1}{q_1^2\, q_2^2\, (q_1+q_2-k)^2} \\
&\times \frac{1}{(p'-q_1)^2 - m^2} \frac{1}{(p'-q_1-q_2)^2 - m^2} \\
&\times \bar{\mathrm{u}}(p')\gamma^\mu(\not p' - \not q_1 + m)\gamma^\nu(\not p' - \not q_1 - \not q_2 + m)\gamma^\lambda \mathrm{u}(p) \\
&\times \frac{\partial}{\partial k^\rho} \Pi_{\mu\nu\lambda\sigma}(q_1, q_2, k - q_1 - q_2)\,. \qquad (5.21)
\end{aligned}$$

Following Ref. [58] and using the property $k^\rho k^\sigma \bar{\mathrm{u}}(p')\Gamma_{\rho\sigma}^{(\mathrm{h.L\times L})}(p',p)\mathrm{u}(p) = 0$, one deduces that $F_1^{(\mathrm{h.L\times L})}(0) = 0$ and that the hadronic light-by-light contribution to the muon anomalous magnetic moment is equal to

$$a_\mu^{(\mathrm{h.L\times L})} \equiv F_2^{(\mathrm{h.L\times L})}(0) = \frac{1}{48m} \mathrm{tr}\left\{ (\not p + m)[\gamma^\rho, \gamma^\sigma](\not p + m)\Gamma_{\rho\sigma}^{(\mathrm{h.L\times L})}(p,p) \right\}. \qquad (5.22)$$

This is about all we can say about the QCD four-point function $\Pi_{\mu\nu\lambda\rho}(q_1, q_2, q_3)$. Unlike the hadronic vacuum polarization function, there is no experimental data which would allow for an evaluation of $a_\mu^{(\mathrm{h.L\times L})}$. The existing estimates regarding this quantity therefore rely on specific models in order to account for the non perturbative QCD aspects. A few particular contributions can be identified, see Fig. 10. For instance, there is a contribution where the four photon lines are attached to a closed loop of charged mesons. The case of the charged pion loop with pointlike couplings is actually finite and contributes $\sim 4 \times 10^{-10}$ to a_μ [100]. If the coupling of charged pions to photons is modified by taking into account the effects of resonances like the ρ, this contribution is reduced by a factor varying between 3 [100, 102] and 10 [101], depending on the resonance model used. Another class of contributions consists of those involving resonance exchanges between photon pairs [100, 101, 102, 103]. Although here also the results depend on the models used, there is a constant feature that emerges from all the analyses that have been done: the contribution coming from the exchange of the pseudoscalars, π^0, η and η' gives practically the final result. Other contributions (charged pion loops, vector, scalar, and axial resonances,...) tend to cancel among themselves.

Some of the results obtained for $a_\mu^{(\mathrm{h.L\times L})} \times 10^{-11}$ have been gathered in Table 5.2.2. Leaving aside the first result [99, 2] shown there, which is affected by a bad numerical convergence [100], one notices that the sign of this contribution has changed twice. The first change resulted from a mistake in Ref. [100], that was corrected for in [101]. The minus sign that resulted was confirmed by an independent calculation, using the ENJL model, in Ref. [102]. A subsequent reanalysis [103] gave additional support to a negative result, while also getting better agreement with the value of Ref. [102].

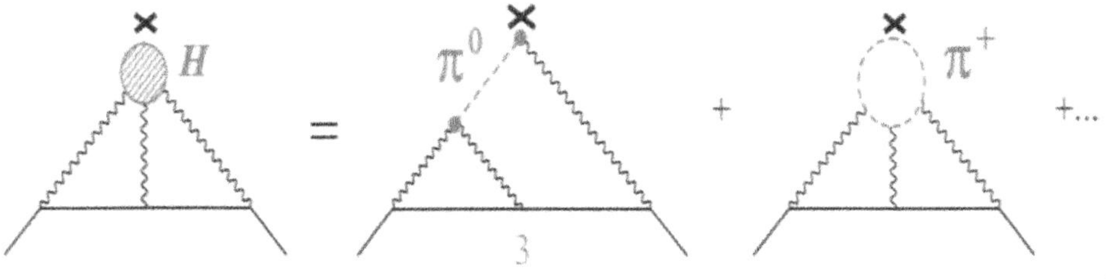

Figure 10: Some individual contributions to hadronic light-by-light scattering: the neutral pion pole and the charged pion loop. There are other contributions, not shown here.

-260(100)	constituent quark loop	[99, 2]
+60(4)	constituent quark loop	[100]
+49(5)	$\pi^{\pm}$loop, π^0 and resonance poles, $a_\mu^{(\text{h.L}\times\text{L};\pi^0)} = 65(6)$	[100]
-52(18)	$\pi^{\pm}$ loop, π^0 and resonance poles, and quark loop $a_\mu^{(\text{h.L}\times\text{L};\pi^0)} = -55.60(3)$	[101]
-92(32)	ENJL, $a_\mu^{(\text{h.L}\times\text{L};\pi^0+\eta+\eta')} = -85(13)$	[102]
-79.2(15.4)	$\pi^{\pm}$ loop, π^0 pole and quark loop, $a_\mu^{(\text{h.L}\times\text{L};\pi^0)} = -55.60(3)$	[103]
+83(12)	π^0, η and η' poles only	[104]
+89.6(15.4)	$\pi^{\pm}$ loop, π^0 pole and quark loop, $a_\mu^{(\text{h.L}\times\text{L};\pi^0)} = +55.60(3)$	[105]
+83(32)	ENJL, $a_\mu^{(\text{h.L}\times\text{L};\pi^0+\eta+\eta')} = 85(13)$	[106]

Table 4: Various evaluations of $a_\mu^{(\text{h.L}\times\text{L})} \times 10^{-11}$ and of the pion pole contribution $a_\mu^{(\text{h.L}\times\text{L};\pi^0)} \times 10^{-11}$.

Needless to say, these evaluations are based on heavy numerical work, which has the drawback of making the final results rather opaque to an intuitive understanding of the physics behind them. We [16] therefore decided to improve things on the analytical side, in order to achieve a better understanding of the relevant features that led to the previous results. Taking advantage of the observation that the pion pole contribution $a_\mu^{(\text{h.L}\times\text{L};\pi^0)}$ was found to dominate the final values obtained for $a_\mu^{(\text{h.L}\times\text{L})}$, we concentrated our efforts on that part, that I shall now describe in greater detail. For a detailed account on how the other contributions to $a_\mu^{(\text{h.L}\times\text{L})}$ arise, I refer the reader to the original works [100]–[103].

The contributions to $\Pi_{\mu\nu\lambda\rho}(q_1, q_2, q_3)$ arising from single neutral pion exchanges, see Fig. 11, read

[16] A. Nyffeler and myself, in Ref. [104].

$$\Pi^{(\pi^0)}_{\mu\nu\lambda\rho}(q_1,q_2,q_3) =$$
$$i\,\frac{\mathcal{F}_{\pi^0\gamma^*\gamma^*}(q_1^2,q_2^2)\;\mathcal{F}_{\pi^0\gamma^*\gamma^*}(q_3^2,(q_1+q_2+q_3)^2)}{(q_1+q_2)^2-M_\pi^2}\;\varepsilon_{\mu\nu\alpha\beta}\,q_1^\alpha q_2^\beta\;\varepsilon_{\lambda\rho\sigma\tau}\,q_3^\sigma(q_1+q_2)^\tau$$
$$+\,i\,\frac{\mathcal{F}_{\pi^0\gamma^*\gamma^*}(q_1^2,(q_1+q_2+q_3)^2)\;\mathcal{F}_{\pi^0\gamma^*\gamma^*}(q_2^2,q_3^2)}{(q_2+q_3)^2-M_\pi^2}\;\varepsilon_{\mu\rho\alpha\beta}\,q_1^\alpha(q_2+q_3)^\beta\;\varepsilon_{\nu\lambda\sigma\tau}\,q_2^\sigma q_3^\tau$$
$$+\,i\,\frac{\mathcal{F}_{\pi^0\gamma^*\gamma^*}(q_1^2,q_3^2)\;\mathcal{F}_{\pi^0\gamma^*\gamma^*}(q_2^2,(q_1+q_2+q_3)^2)}{(q_1+q_3)^2-M_\pi^2}\;\varepsilon_{\mu\lambda\alpha\beta}\,q_1^\alpha q_3^\beta\;\varepsilon_{\nu\rho\sigma\tau}\,q_2^\sigma(q_1+q_3)^\tau\,. \tag{5.23}$$

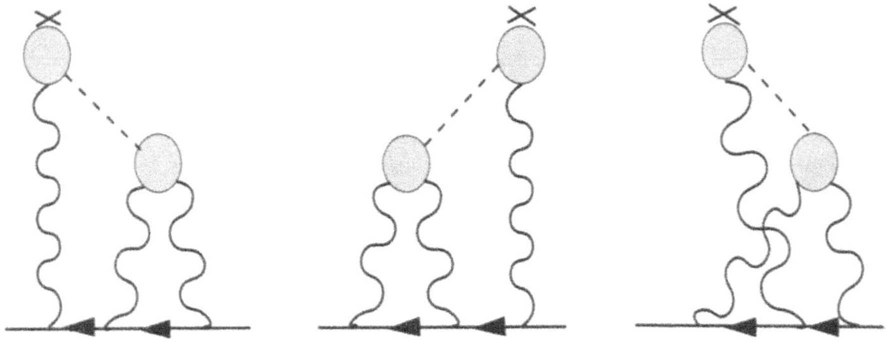

Figure 11: The pion-pole contributions to light-by-light scattering. The shaded blobs represent the form factor $\mathcal{F}_{\pi^0\gamma^*\gamma^*}$. The first and second graphs give rise to identical contributions, involving the function $T_1(q_1,q_2;p)$ in Eq. (5.25), whereas the third graph gives the contribution involving $T_2(q_1,q_2;p)$.

The form factor $\mathcal{F}_{\pi^0\gamma^*\gamma^*}(q_1^2,q_2^2)$, which corresponds to the shaded blobs in Fig. 11, is defined as

$$i\int d^4x e^{iq\cdot x}\langle\,\Omega|T\{j_\mu(x)j_\nu(0)\}|\pi^0(p)\rangle \;=\; \varepsilon_{\mu\nu\alpha\beta}\,q^\alpha p^\beta\,\mathcal{F}_{\pi^0\gamma^*\gamma^*}(q^2,(p-q)^2)\,, \tag{5.24}$$

with $\mathcal{F}_{\pi^0\gamma^*\gamma^*}(q_1^2,q_2^2)=\mathcal{F}_{\pi^0\gamma^*\gamma^*}(q_2^2,q_1^2)$. Inserting the expression (5.23) into (5.21) and computing the corresponding Dirac traces in Eq. (5.22), we obtain

$$a_\mu^{(\mathrm{h.L\times L};\pi^0)} =$$
$$e^6\int\frac{d^4q_1}{(2\pi)^4}\int\frac{d^4q_2}{(2\pi)^4}\,\frac{1}{q_1^2q_2^2(q_1+q_2)^2[(p+q_1)^2-m^2][(p-q_2)^2-m^2]}$$
$$\times\left[\frac{\mathcal{F}_{\pi^0\gamma^*\gamma^*}(q_1^2,(q_1+q_2)^2)\;\mathcal{F}_{\pi^0\gamma^*\gamma^*}(q_2^2,0)}{q_2^2-M_\pi^2}\;T_1(q_1,q_2;p)\right.$$
$$\left.+\;\frac{\mathcal{F}_{\pi^0\gamma^*\gamma^*}(q_1^2,q_2^2)\;\mathcal{F}_{\pi^0\gamma^*\gamma^*}((q_1+q_2)^2,0)}{(q_1+q_2)^2-M_\pi^2}\;T_2(q_1,q_2;p)\right]\,, \tag{5.25}$$

where $T_1(q_1, q_2; p)$ and $T_2(q_1, q_2; p)$ denote two polynomials in the invariants $p \cdot q_1$, $p \cdot q_2$, $q_1 \cdot q_2$. Their expressions can be found in Ref. [104]. The former arises from the two first diagrams shown in Fig. 11, which give identical contributions, while the latter corresponds to the third diagram on this same figure. At this stage, it should also be pointed out that the expression (5.23) does not, strictly speaking, represent the contribution arising from the pion pole only. The latter would require that the numerators in (5.23) be evaluated at the values of the momenta that correspond to the pole indicated by the corresponding denominators. For instance, the numerator of the term proportional to $T_1(q_1, q_2; p)$ in Eq. (5.25) should rather read $\mathcal{F}_{\pi^0\gamma^*\gamma^*}(q_1^2, (q_1^2 + 2q_1 \cdot q_2 + M_\pi^2)\ \mathcal{F}_{\pi^0\gamma^*\gamma^*}(M_\pi^2, 0)$ with $q_2^2 = M_\pi^2$. However, Eq. (5.25) corresponds to what previous authors have called the pion pole contribution, and for the sake of comparison I shall adopt the same definition.

From here on, information on the form factor $\mathcal{F}_{\pi^0\gamma^*\gamma^*}(q_1^2, q_2^2)$ is required in order to proceed. The simplest model for the form factor follows from the Wess–Zumino–Witten (WZW) term [107, 108] that describes the Adler–Bell–Jackiw anomaly [109, 110] in chiral perturbation theory. Since in this case the form factor is constant, one needs an ultraviolet cutoff, at least in the contribution to Eq. (5.25) involving T_1, the one involving T_2 gives a finite result even for a constant form factor [100]. Therefore, this model cannot be used for a reliable estimate, but at best serves only illustrative purposes in the present context.[17] Previous calculations [100, 101, 103] have also used the usual vector meson dominance form factor [see also Ref. [111]]. The expressions for the form factor $\mathcal{F}_{\pi^0\gamma^*\gamma^*}$ based on the ENJL model that have been used in Ref. [102] do not allow a straightforward analytical calculation of the loop integrals. However, compared with the results obtained in Refs. [100, 101, 103], the corresponding numerical estimates are rather close to the VMD case [within the error attributed to the model dependence]. Finally, representations of the form factor $\mathcal{F}_{\pi^0\gamma^*\gamma^*}$, based on the large-$N_C$ approximation to QCD and that takes into account constraints from chiral symmetry at low energies, and from the operator product expansion at short distances, have been discussed in Ref. [112]. They involve either one vector resonance [lowest meson dominance, LMD] or two vector resonances (LMD+V), see [112] for details. The four types of form factors just mentioned can be written in the form [F_π is the pion decay constant]

$$\mathcal{F}_{\pi^0\gamma^*\gamma^*}(q_1^2, q_2^2) = \frac{F_\pi}{3}\left[f(q_1^2) - \sum_{M_{V_i}} \frac{1}{q_2^2 - M_{V_i}^2} g_{M_{V_i}}(q_1^2) \right]. \tag{5.26}$$

For the VMD and LMD form factors, the sum in Eq. (5.26) reduces to a single term, and the corresponding function is denoted $g_{M_V}(q^2)$. It depends on the mass M_V of the vector resonance, which will be identified with the mass of the ρ meson. For

[17]In the context of an effective field theory approach, the pion pole with WZW vertices represents a chirally suppressed, but large-N_C dominant contribution, whereas the charged pion loop is dominant in the chiral expansion, but suppressed in the large-N_C limit.

our present purposes, it is enough to consider only these two last cases, along with the constant WZW form factor. The corresponding functions $f(q^2)$ and $g_{M_V}(q^2)$ are displayed in Table 5.2.2.

	$f(q^2)$	$g_{M_V}(q^2)$
WZW	$-\dfrac{N_C}{4\pi^2 F_\pi^2}$	0
VMD	0	$\dfrac{N_C}{4\pi^2 F_\pi^2}\dfrac{M_V^4}{q^2 - M_V^2}$
LMD	$\dfrac{1}{q^2 - M_V^2}$	$-\dfrac{q^2 + M_V^2 - c_V}{q^2 - M_V^2}$

Table 5: The functions $f(q^2)$ and $g_{M_V}(q^2)$ of Eq. (5.26) for the different form factors. N_C is the number of colors, taken equal to 3, and $F_\pi = 92.4$ MeV is the pion decay constant. Furthermore, $c_V = \frac{N_C}{4\pi^2}\frac{M_V^4}{F_\pi^2}$.

We may now come back to Eq. (5.25). With a representation of the form (5.26), the angular integrations can be performed, using for instance standard Gegenbauer polynomial techniques (hyperspherical approach), see Refs. [113, 114, 56]. This leads to a two-dimensional integral representation:

$$a_\mu^{(\mathrm{h.L\times L};\pi^0)} = \left(\frac{\alpha}{\pi}\right)^3 \left[a_\mu^{(\pi^0;1)} + a_\mu^{(\pi^0;2)}\right], \tag{5.27}$$

$$a_\mu^{(\pi^0;1)} = \int_0^\infty dQ_1 \int_0^\infty dQ_2 \left[w_{f_1}(Q_1,Q_2)\, f^{(1)}(Q_1^2,Q_2^2) + w_{g_1}(M_V,Q_1,Q_2)\, g_{M_V}^{(1)}(Q_1^2,Q_2^2)\right], \tag{5.28}$$

$$a_\mu^{(\pi^0;2)} = \int_0^\infty dQ_1 \int_0^\infty dQ_2 \left[\sum_{M=M_\pi,M_V} w_{g_2}(M,Q_1,Q_2)\, g_M^{(2)}(Q_1^2,Q_2^2)\right]. \tag{5.29}$$

The functions $f^{(1)}(Q_1^2,Q_2^2)$, $g_{M_V}^{(1)}(Q_1^2,Q_2^2)$, $g_{M_\pi}^{(2)}(Q_1^2,Q_2^2)$ and $g_{M_V}^{(2)}(Q_1^2,Q_2^2)$ are expressed in terms of the functions given in Table 5.2.2, see Ref. [104], where the universal [for the class of form factors that have a representation of the type shown in Eq. (5.26)] weight functions w in Eqs. (5.28) and (5.29) can also be found. The latter are plotted in Fig. 12.

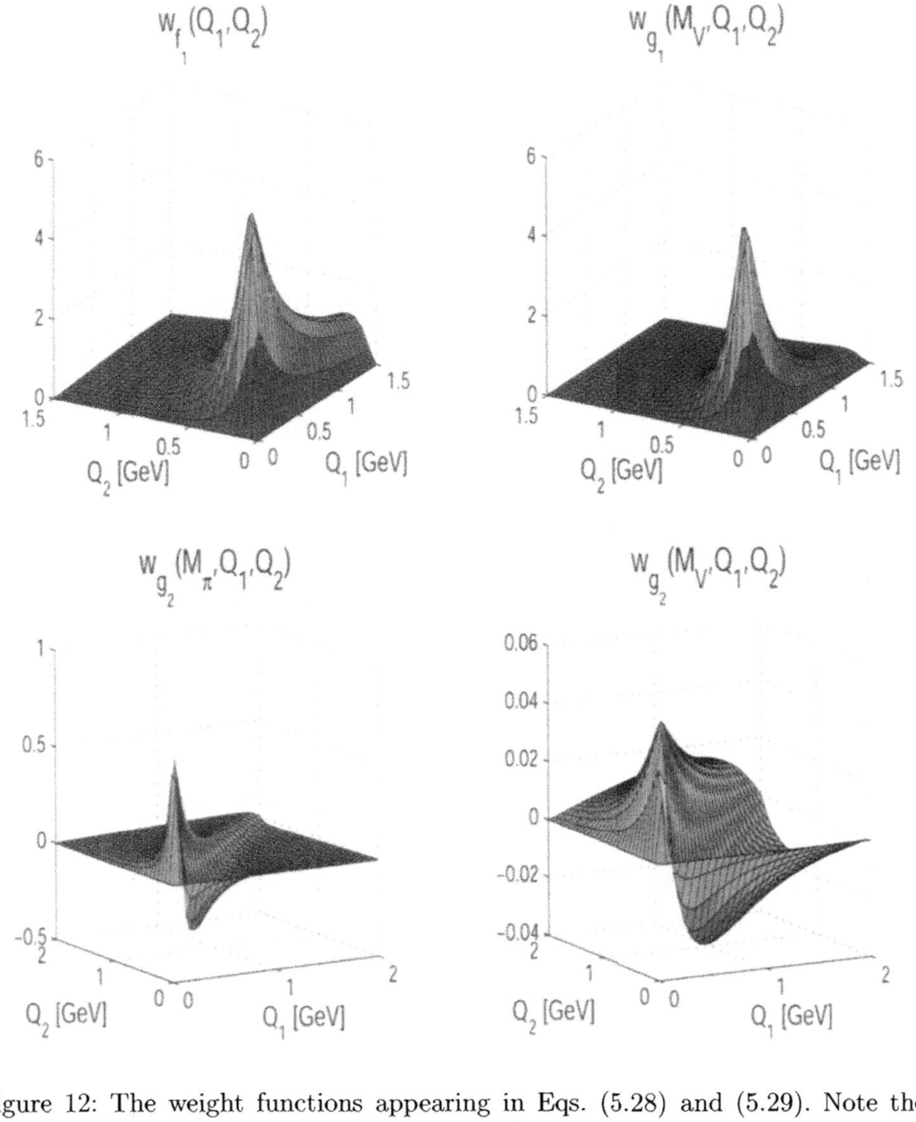

Figure 12: The weight functions appearing in Eqs. (5.28) and (5.29). Note the different ranges of Q_i in the subplots. The functions w_{f_1} and w_{g_1} are positive definite and peaked in the region $Q_1 \sim Q_2 \sim 0.5$ GeV. Note, however, the tail in w_{f_1} in the Q_1-direction for $Q_2 \sim 0.2$ GeV. The functions $w_{g_2}(M_\pi, Q_1, Q_2)$ and $w_{g_2}(M_V, Q_1, Q_2)$ take both signs, but their magnitudes remain small as compared to $w_{f_1}(Q_1, Q_2)$ and $w_{g_1}(M_V, Q_1, Q_2)$. We have used $M_V = M_\rho = 770$ MeV.

The functions w_{f_1} and w_{g_1} are positive and concentrated around momenta of the order of 0.5 GeV. This feature was already observed numerically in Ref. [102] by varying the upper bound of the integrals [an analogous analysis is contained in Ref. [101]]. Note, however, the tail in w_{f_1} in the Q_1 direction for $Q_2 \sim 0.2$ GeV. On the other hand, the function w_{g_2} has positive and negative contributions in that region, which will lead to a strong cancellation in the corresponding integrals, provided they are multiplied by a positive function composed of the form factors [see the numerical results below]. As can be seen from the plots, and checked analytically, the weight functions vanish for small momenta. Therefore, the integrals are infrared finite. The behaviors of the weight functions for large values of Q_1 and/or Q_2 can also be worked out analytically. From these, one can deduce that in the case of the WZW form factor, the corresponding, divergent, integral for $a_\mu^{(\pi^0;1)}$ behaves, as a function of the ultraviolet cut off Λ, as $a_\mu^{(\pi^0;1)} \sim \mathcal{C} \ln^2 \Lambda$, with [104]

$$\mathcal{C} = 3 \left(\frac{N_C}{12\pi} \right)^2 \left(\frac{m_\mu}{F_\pi} \right)^2 = 0.0248 . \tag{5.30}$$

The log-squared behavior follows from the general structure of the integral (5.28) for $a_\mu^{(\pi^0;1)}$ in the case of a constant form factor, as pointed out in [5]. The expression (5.30) of the coefficient $\mathcal{C}$ has been derived independently, in Ref. [115], through a renormalization group argument in the effective theory framework.

Form factor	$a_\mu^{(\pi^0;1)}$	$a_\mu^{(\pi^0;2)}$	$a_\mu^{\mathrm{h.L\times L};\pi^0} \times 10^{10}$
WZW	0.095	0.0020	12.2
VMD	0.044	0.0013	5.6
LMD	0.057	0.0014	7.3

Table 6: Results for the terms $a_\mu^{(\pi^0;1)}$, $a_\mu^{(\pi^0;2)}$ and for the pion exchange contribution to the anomalous magnetic moment $a_\mu^{\mathrm{h.L\times L};\pi^0}$ according to Eq. (5.27) for the different form factors considered. In the WZW model we used a cutoff of 1 GeV in the first contribution, whereas the second term is ultraviolet finite.

In the case of the other form factors, the integration over Q_1 and Q_2 is finite and can now be performed numerically. [18] Furthermore, since both the VMD and LMD model tend to the WZW constant form factor as $M_V \to \infty$, the results for $a_\mu^{(\pi^0;1)}$ in these models should scale as $\mathcal{C} \ln^2 M_V^2$ for a large resonance mass. This has been checked numerically, and the value of the coefficient $\mathcal{C}$ obtained that way was in perfect agreement with the value given in Eq. (5.30). The results of the integration over Q_1 and Q_2 are displayed in Table 5.2.2. They definitely show a sign difference when compared to those obtained in Refs. [100, 101, 103, 111], although in absolute value the numbers agree perfectly. After the results of Table

[18] In the case of the VMD form factor, an analytical result is now also available [116].

5.2.2 were made public [104], previous authors checked their calculations and soon discovered that they had made a sign mistake at some stage [105, 106]. Almost simultaneously, the results presented in Table 5.2.2 and in Refs. [104, 115] also received independent confirmations [117, 116].

The analysis of [104] leads to the following estimates

$$a_\mu^{\mathrm{h.L\times L};\pi^0} = 5.8(1.0) \times 10^{-10}\,, \tag{5.31}$$

and

$$a_e^{\mathrm{h.L\times L};\pi^0} = 5.1 \times 10^{-14}\,. \tag{5.32}$$

Taking into account the other contributions computed by previous authors, and adopting a conservative attitude towards the error to be ascribed to their model dependences, the total contribution to a_μ coming from the hadronic light-by-light scattering diagrams amounts to

$$a_\mu^{\mathrm{h.L\times L}} = 8(4) \times 10^{-10}\,. \tag{5.33}$$

5.3 Electroweak contributions to a_μ

Electroweak corrections to a_μ have been considered at the one and two loop levels. The one loop contributions, shown in Fig. 13, have been worked out some time ago, and read [118]-[122]

$$a_\mu^{\mathrm{W}(1)} = \frac{G_\mathrm{F}}{\sqrt{2}}\frac{m_\mu^2}{8\pi^2}\left[\frac{5}{3}+\frac{1}{3}\left(1-4\sin^2\theta_W\right)^2 + \mathcal{O}\left(\frac{m_\mu^2}{M_Z^2}\log\frac{M_Z^2}{m_\mu^2}\right)+\mathcal{O}\left(\frac{m_\mu^2}{M_H^2}\log\frac{M_H^2}{m_\mu^2}\right)\right]\,, \tag{5.34}$$

where the weak mixing angle is defined by $\sin^2\theta_W = 1 - M_W^2/M_Z^2$.

Numerically, with $G_\mathrm{F} = 1.16639(1) \times 10^{-5}\,\mathrm{GeV}^{-2}$ and $\sin^2\theta_W = 0.224$,

$$a_\mu^{\mathrm{W}(1)} = 19.48 \times 10^{-10}\,, \tag{5.35}$$

It is convenient to separate the two–loop electroweak contributions into two sets of Feynman graphs: those which contain closed fermion loops, which are denoted by $a_\mu^{\mathrm{EW}(2);\mathrm{f}}$, and the others, $a_\mu^{\mathrm{EW}(2);\mathrm{b}}$. In this notation, the electroweak contribution to the muon anomalous magnetic moment is

$$a_\mu^{\mathrm{EW}} = a_\mu^{\mathrm{W}(1)} + a_\mu^{\mathrm{EW}(2);\mathrm{f}} + a_\mu^{\mathrm{EW}(2);\mathrm{b}}\,. \tag{5.36}$$

I shall review the calculation of the two–loop contributions separately.

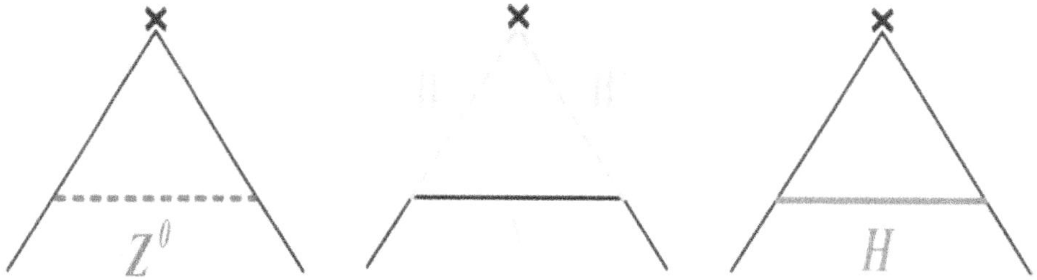

Figure 13: One loop weak interaction contributions to the anomalous magnetic moment.

5.3.1 Two loop bosonic contributions

The leading logarithmic terms of the two–loop electroweak bosonic corrections have been extracted using asymptotic expansion techniques, see e.g. Ref. [123]. In the approximation where $\sin^2\theta_W \to 0$ and $M_H \sim M_W$ these calculations simplify considerably and one obtains

$$a_\mu^{\mathrm{EW(2);b}} = \frac{G_{\mathrm{F}}}{\sqrt{2}}\frac{m_\mu^2}{8\pi^2}\frac{\alpha}{\pi} \times \left[-\frac{65}{9}\ln\frac{M_W^2}{m_\mu^2} + \mathcal{O}\left(\sin^2\theta_W \ln\frac{M_W^2}{m_\mu^2}\right)\right]. \tag{5.37}$$

In fact, these contributions have now been evaluated analytically, in a systematic expansion in powers of $\sin^2\theta_W$, up to $\mathcal{O}[(\sin^2\theta_W)^3]$, where $\ln\frac{M_W^2}{m_\mu^2}$ terms, $\ln\frac{M_H^2}{M_W^2}$ terms, $\frac{M_W^2}{M_H^2}\ln\frac{M_H^2}{M_W^2}$ terms, $\frac{M_W^2}{M_H^2}$ terms and constant terms are kept [75]. Using $\sin^2\theta_W = 0.224$ and $M_H = 250\,\mathrm{GeV}$, the authors of Ref. [75] find

$$a_\mu^{\mathrm{EW(2);b}} = \frac{G_{\mathrm{F}}}{\sqrt{2}}\frac{m_\mu^2}{8\pi^2}\frac{\alpha}{\pi} \times \left[-5.96\ln\frac{M_W^2}{m_\mu^2} + 0.19\right] = \frac{G_{\mathrm{F}}}{\sqrt{2}}\frac{m_\mu^2}{8\pi^2}\left(\frac{\alpha}{\pi}\right) \times (-79.3)\,, \tag{5.38}$$

showing, in retrospect, that the simple approximation in Eq. (5.37) is rather good.

5.3.2 Two loop fermionic contributions

The discussion of the two–loop electroweak fermionic corrections is more delicate. First, it contains a hadronic contribution. Next, because of the cancellation between lepton loops and quark loops in the electroweak $U(1)$ anomaly, one cannot separate hadronic effects from leptonic effects any longer. In fact, as discussed in Refs. [124, 125], it is this cancellation which eliminates some of the large logarithms which, incorrectly were kept in Ref. [126]. It is therefore appropriate to separate the two–loop electroweak fermionic corrections into two classes: One is the class arising from Feynman diagrams containing a lepton or a quark loop, with the external photon, a virtual photon and a virtual Z^0 attached to it, see Fig. 14.[19] The quark loop of course again represents non perturbative hadronic contributions which have to be evaluated using some model. This first class is denoted by $a_\mu^{\mathrm{EW}(2);\mathrm{f}}(\ell;q)$. It involves the QCD correlation function

$$W_{\mu\nu\rho}(q,k) = \int d^4x\, e^{iq\cdot x} \int d^4y\, e^{i(k-q)\cdot y} \langle\Omega\,|\mathrm{T}\{j_\mu(x)A_\nu^{(Z)}(y)j_\rho(0)\}|\Omega\rangle\,, \qquad (5.39)$$

with k the incoming external photon four-momentum associated with the classical external magnetic field. As previously, j_ρ denotes the hadronic part of the electromagnetic current, and $A_\rho^{(Z)}$ is the axial component of the current which couples the quarks to the Z^0 gauge boson. The other class is defined by the rest of the diagrams, where quark loops and lepton loops can be treated separately, and is called $a_\mu^{\mathrm{EW}(2);\mathrm{f}}(\text{residual})$.

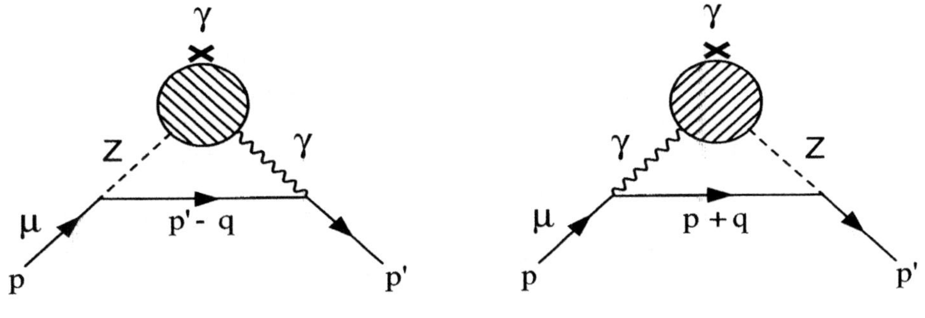

Figure 14: Graphs with hadronic contributions to $a_\mu^{\mathrm{EW}(2);\mathrm{f}}(\ell,q)$ and involving the QCD three point function $W_{\mu\nu\rho}(q,k)$.

The contribution from $a_\mu^{\mathrm{EW}(2);\mathrm{f}}(\text{residual})$ brings in factors of the ratio m_t^2/M_W^2. It has been estimated, to a very good approximation, in Ref. [125], with the result

$$a_\mu^{\mathrm{EW}(2);\mathrm{f}}(\text{residual}) = \frac{G_{\mathrm{F}}}{\sqrt{2}}\frac{m_\mu^2}{8\pi^2}\frac{\alpha}{\pi} \times \left[\frac{1}{2\sin^2\theta_W}\left(-\frac{5}{8}\frac{m_t^2}{M_W^2} - \log\frac{m_t^2}{M_W^2} - \frac{7}{3}\right) + \Delta_{\mathrm{Higgs}}\right], \qquad (5.40)$$

where Δ_{Higgs} denotes the contribution from diagrams with Higgs lines, which the authors of Ref. [125] estimate to be

$$\Delta_{\mathrm{Higgs}} = -5.5 \pm 3.7\,, \qquad (5.41)$$

[19] If one works in a renormalizable gauge, the contributions where the Z^0 is replaced by the neutral unphysical Higgs should also be included. The final result does not depend on the gauge fixing parameter ξ_Z, if one works in the class of 't Hooft gauges.

and therefore,

$$a_\mu^{\mathrm{EW}(2);\mathrm{f}}(\mathrm{residual}) = \frac{G_{\mathrm{F}}}{\sqrt{2}}\frac{m_\mu^2}{8\pi^2}\frac{\alpha}{\pi} \times [-21(4)]\,. \qquad (5.42)$$

Let us finally discuss the contributions to $a_\mu^{\mathrm{EW}(2);\mathrm{f}}(\ell; q)$. Here, it is convenient to treat the contributions from the three generations separately. The contribution from the third generation can be calculated in a straightforward way, with the result [124, 125]

$$\begin{aligned} a_\mu^{\mathrm{EW}(2);\mathrm{f}}(\tau; t, b) = \\ \frac{G_{\mathrm{F}}}{\sqrt{2}}\frac{m_\mu^2}{8\pi^2}\frac{\alpha}{\pi} \times \left[-3\ln\frac{M_Z^2}{m_\tau^2} - \ln\frac{M_Z^2}{m_b^2} - \frac{8}{3}\ln\frac{m_t^2}{M_Z^2} + \frac{8}{3} + \mathcal{O}\left(\frac{M_Z^2}{m_t^2}\ln\frac{m_t^2}{M_Z^2}\right)\right] \\ = \frac{G_{\mathrm{F}}}{\sqrt{2}}\frac{m_\mu^2}{8\pi^2}\frac{\alpha}{\pi} \times (-30.6)\,. \qquad (5.43) \end{aligned}$$

In fact the terms of $\mathcal{O}\left(\frac{M_Z^2}{m_t^2}\ln\frac{m_t^2}{M_Z^2}\right)$ and $\mathcal{O}\left(\frac{M_Z^2}{m_t^2}\right)$ have also been calculated in Ref. [125]. There are in principle QCD perturbative corrections to this estimate, which have not been calculated, but the result in Eq. (5.43) is good enough for the accuracy required at present. The contributions of the remaining charged standard model fermions involve the light quarks u and d, as well as the second generation s quark, for which non perturbative effects tied to the spontaneous breaking of chiral symmetry are important [124, 127]. The contributions from the first and second generation are thus most conveniently taken together, with the result

$$\begin{aligned} a_\mu^{\mathrm{EW}(2);\mathrm{f}}(e, \mu; u, d, s, c) = \frac{G_{\mathrm{F}}}{\sqrt{2}}\frac{m_\mu^2}{8\pi^2}\frac{\alpha}{\pi} \times \Bigg\{ -3\ln\frac{M_Z^2}{m_\mu^2} - \frac{5}{2} \\ -3\ln\frac{M_Z^2}{m_\mu^2} + 4\ln\frac{M_Z^2}{m_c^2} - \frac{11}{6} + \frac{8}{9}\pi^2 - 8 \\ + \left[\frac{4}{3}\ln\frac{M_Z^2}{m_\mu^2} + \frac{2}{3} + \mathcal{O}\left(\frac{m_\mu^2}{M_Z^2}\ln\frac{M_Z^2}{m_\mu^2}\right)\right] - 1.38(35) + 0.06(2)\Bigg\} \qquad (5.44) \\ = \frac{G_{\mathrm{F}}}{\sqrt{2}}\frac{m_\mu^2}{8\pi^2}\frac{\alpha}{\pi} \times [-34.5(4)]\,, \qquad (5.45) \end{aligned}$$

where the first line shows the result from the e loop and the second line the result from the μ loop and the c quark, which is treated as a heavy quark. The term between brackets in the third line is the one induced by the anomalous term in the hadronic three point function $W_{\mu\nu\rho}(q, k)$ The other contributions have been estimated on the basis of an approximation to the large-N_C limit of QCD, similar to the one discussed for the two-point function $\Pi(Q^2)$ after Eq. (5.12), see Ref. [127] for details.

The result in Eq. (5.44) for the contribution from the first and second generations of quarks and leptons is conceptually very different to the corresponding one proposed in Ref. [125],

$$a_{\mu}^{\mathrm{EW(2);f}}(\ell,q)(e,\mu;u,d,s,c) = \frac{G_{\mathrm{F}}}{\sqrt{2}}\frac{m_{\mu}^{2}}{8\pi^{2}}\frac{\alpha}{\pi}\left[-3\ln\frac{M_Z^2}{m_\mu^2}+4\ln\frac{M_Z^2}{m_u^2}-\ln\frac{M_Z^2}{m_d^2}-\frac{5}{2}\right.$$

$$\left. -6\ \acute{o}\ -3\ln\frac{M_Z^2}{m_\mu^2}+4\ln\frac{M_Z^2}{m_c^2}-\ln\frac{M_Z^2}{m_s^2}-\frac{11}{6}+\frac{8}{9}\pi^2-6\right] \tag{5.46}$$

$$= \frac{G_{\mathrm{F}}}{\sqrt{2}}\frac{m_{\mu}^{2}}{8\pi^{2}}\frac{\alpha}{\pi}\times(-31.9)\,. \tag{5.47}$$

where the light quarks are, *arbitrarily*, treated the same way as heavy quarks, with $m_u = m_d = 0.3\,\mathrm{GeV}$, and $m_s = 0.5\,\mathrm{GeV}$. Although, numerically, the two results turn out not to be too different, the result in Eq. (5.46) follows from an hadronic model which is in contradiction with basic properties of QCD. This is at the origin of the spurious cancellation of the $\ln M_Z$ terms in Eq. (5.46).

Putting together the numerical results in Eqs. (5.38), (5.42), (5.43) with the new result in Eq. (5.44), we finally obtain the value

$$a_{\mu}^{\mathrm{EW}} = \frac{G_{\mathrm{F}}}{\sqrt{2}}\frac{m_{\mu}^{2}}{8\pi^{2}}\left[\frac{5}{3}+\frac{1}{3}\left(1-4\sin^2\theta_W\right)^2-\left(\frac{\alpha}{\pi}\right)(165.4(4.0)\right] = 15.0(1)\times 10^{-10}\,, \tag{5.48}$$

which shows that the two–loop correction represents indeed a reduction of the one–loop result by an amount of 23%. The final error here does not include higher order electroweak effects [128].

5.4 Comparison with experiment

We may now put all the pieces together and obtain the value for a_μ predicted by the standard model. We have seen that in the case of the hadronic vacuum polarization contributions, the latest evaluation [92] shows a discrepancy between the value obtained exclusively from e^+e^- data and the value that arises if τ data are also included. This gives us the two possibilities

$$\begin{aligned} a_{\mu}^{\mathrm{SM}}(e^+e^-) &= (11\,659\,169.1 \pm 7.5 \pm 4.0 \pm 0.3)\times 10^{-10} \\ a_{\mu}^{\mathrm{SM}}(\tau) &= (11\,659\,186.3 \pm 6.2 \pm 4.0 \pm 0.3)\times 10^{-10}\,, \end{aligned} \tag{5.49}$$

where the first error comes from hadronic vacuum polarization, the second from hadronic light-by-light scattering, and the last from the QED and weak corrections. When compared to the present experimental average

$$a_{\mu}^{\mathrm{exp}} = (11\,659\,203 \pm 8)\times 10^{-10} \tag{5.50}$$

there results a difference,

$$\begin{aligned} a_\mu^{\text{exp}} - a_\mu^{\text{SM}}(e^+e^-) &= 33.9(11.2) \times 10^{-10}\,, \\ a_\mu^{\text{exp}} - a_\mu^{\text{SM}}(\tau) &= 16.7(10.7) \times 10^{-10}\,, \end{aligned}$$

which represents 3.0 and 1.6 standard deviations, respectively.

Although experiment and theory have now both reached the same level of accuracy, $\sim \pm 8 \times 10^{-10}$ or 0.7 ppm, the present discrepancy between the e^+e^- and τ based evaluations makes the interpretation of the above results a delicate issue as far as evidence for new physics is concerned. Other evaluations of comparable accuracy [88, 90, 41] cover a similar range of variation in the difference between experiment and theory. One possibility to come to a conclusion would be to have the experimental result still more accurate, so that even the difference $a_\mu^{\text{exp}} - a_\mu^{\text{SM}}(\tau)$ would become sufficiently significant. In this respect, it is certainly very important that the Brookhaven experiment is given the means to improve on the value of a_μ^{exp}, bringing its error down to $\sim \pm 4 \times 10^{-10}$ or 0.35 ppm. Furthermore, the value obtained for $a_\mu^{\text{SM}}(e^+e^-)$ relies strongly on the low-energy data obtained by the CMD-2 experiment, with none of the older data able to check them at the same level of precision. In this respect, the prospects for additional high statistics data in the future, either from KLOE or from BaBar, are most welcome. On the other hand, if the present discrepancy in the evaluations of the hadronic vacuum polarization finds a solution in the future, and if the experimental error is further reduced, by, say, a factor of two, then the theoretical uncertainty on the hadronic light-by-light scattering will constitute the next serious limitation on the theoretical side. It is certainly worthwhile to devote further efforts to a better understanding of this contribution, for instance by finding ways to feed more constraints with a direct link to QCD into the descriptions of the four-point function $\Pi_{\mu\nu\rho\sigma}(q_1, q_2, q_3)$.

6 Concluding remarks

With this review, I hope to have convinced the reader that the subject of the anomalous magnetic moments of the electron and of the muon is an exciting and fascinating topic. It provides a good example of mutual stimulation and strong interplay between experiment and theory.

The anomalous magnetic moment of the electron constitutes a very stringent test of QED and of the practical working of the framework of perturbatively renormalized quantum field theory at higher orders. It tests the validity of QED at very short distances, and provides at present the best determination of the fine structure constant.

The anomalous magnetic moment of the muon represents the best compromise between sensitivity to new degrees of freedom describing physics beyond the standard model and experimental feasibility. Important progress has been achieved on the experimental side during the last couple of years, with the results of the E821 collaboration at BNL. The experimental value of a_μ is now known with an accuracy of 0.7ppm. Hopefully, the Brookhaven experiment will be given the opportunity to reach its initial goal of achieving a measurement at the 0.35 ppm level.
As can be inferred from the examples mentioned in this text, the subject constitutes, from a theoretical point of view, a difficult and error prone topic, due to the technical difficulties encountered in the higher loop calculations. The theoretical predictions have reached a precision comparable to the experimental one, but unfortunately there appears a discrepancy between the most recent evaluations of the hadronic vacuum polarization according to whether τ data are taken into account or not. Hadronic contributions, especially from vacuum polarization and from light-by-light scattering, are responsible for the bulk part of the final uncertainty in the theoretical value a_μ^{SM}. Further efforts are needed in order to bring these aspects under better control.

Acknowledgments

I wish to thank A. Nyffeler, S. Peris, M. Perrottet, and E. de Rafael for stimulating and very pleasant collaborations, and for sharing many insights on this vast subject and on related topics. Most of the figures appearing in this text were kindly provided by M. Perrottet, to whom I am also most grateful for a careful and critical reading of the manuscript. Finally, I wish to thank B. Duplantier and V. Rivasseau for the invitation to give a presentation at the “Séminaire Poincaré”. This work is supported in part by the EC contract No. HPRN-CT-2002-00311 (EURIDICE).

References

[1] *Quantum Electrodynamics*, T. Kinoshita Ed., World Scientific Publishing Co. Pte. Ltd., 1990.

[2] J. Calmet, S. Narison, M. Perrottet and E. de Rafael, *Rev. Mod. Phys.* **49**, 21 (1977).

[3] A. Czarnecki and W. J. Marciano, *Nucl. Phys. B (Proc. Suppl.)* **76**, 245 (1998).

[4] V. W. Hughes and T. Kinoshita, *Rev. Mod. Phys.* **71**, S133 (1999).

[5] K. Melnikov, *Int. J. Mod. Phys. A* **16**, 4591 (2001).

[6] E. de Rafael, arXiv:hep-ph/0208251.

[7] A. Czarnecki and W. J. Marciano, *Phys. Rev. D* **64**, 013014 (2001).

[8] A list of recent papers on the subject can be found under the URL http://www.slac.stanford.edu/spires/find/hep/www?c=PRLTA,86,2227.

[9] L. L. Foldy, *Phys. Rev.* **87**, 688 (1952); *Rev. Mod. Phys.* **30**, 471 (1958).

[10] S. J. Brodsky and J. D. Sullivan, *Phys. Rev.* **156**, 1644 (1967).

[11] R. Barbieri, J. A. Mignaco and E. Remiddi, *Nuovo Cimento* **11A**, 824 (1972).

[12] T. Appelquist and J. Carazzone, *Phys. Rev. D* **11**, 2856 (1975).

[13] T. Sterling and M. J. Veltman, *Nucl. Phys.* **B 189**, 557 (1981).

[14] E. d'Hoker and E. Farhi, *Nucl. Phys.* **B 248**, 59, 77 (1984).

[15] T. Kinoshita, *Nuovo Cimento* **51B**, 140 (1967).

[16] B. E. Lautrup and E. de Rafael, *Nucl. Phys.* **B 70**, 317 (1974).

[17] E. de Rafael and J. L. Rosner, *Ann. Phys. (N.Y.)* **82**, 369 (1974).

[18] T. Kinoshita and W. J. Marciano, Theory of the Muon Anomalous Magnetic Moment, in [1], p. 419.

[19] J. E. Nafe, E. B. Nelson and I. I. Rabi, *Phys. Rev.* **71**, 914 (1947).

[20] H. G. Dehmelt, *Phys. Rev.* **109**, 381 (1958).

[21] R. S. Van Dyck, P. B. Schwinberg and H. G. Dehmelt, *Phys. Rev. Lett.* **59**, 26 (1987).

[22] R. S. Van Dyck, Anomalous Magnetic Moment of Single Electrons and Positrons: Experiment, in [1], p. 322.

[23] A. Rich and J. C. Wesley, *Rev. Mod. Phys.* **44**, 250 (1972).

[24] P. Kusch and H. M. Fowley, *Phys. Rev.* **72**, 1256 (1947).

[25] P. A. Franken and S. Liebes Jr., *Phys. Rev.* **104**, 1197 (1956).

[26] A. A. Schuppe, R. W. Pidd, and H. R. Crane, *Phys. Rev.* **121**, 1 (1961).

[27] D. T. Wilkinson and H. R. Crane, *Phys. Rev.* **130**, 852 (1963).

[28] G. Gräff, E. Klempt and G. Werth, *Z. Phys.* **222**, 201 (1969).

[29] J. C. Wesley and A. Rich, *Phys. Rev. A* **4**, 1341 (1971).

[30] R. S. Van Dyck, P. B. Schwinberg and H. G. Dehmelt, *Phys. Rev. Lett.* **38**, 310 (1977).

[31] F. J. M. Farley and E. Picasso, The Muon g - 2 Experiments, in [1], p. 479.

[32] J. Bailey et al., *Phys. Lett.* **B28**, 287 (1968).

[33] J. Bailey et al., *Phys. Lett.* **B55**, 420 (1975).

[34] J. Bailey et al. [CERN-Mainz-Daresbury Collaboration], *Nucl. Phys.* **B 150**, 1 (1979).

[35] R. M. Carey et al., *Phys. Rev. Lett.* **82**, 1632 (1999).

[36] H. N. Brown et al. [Muon (g - 2) Collaboration], *Phys. Rev. D* **62**, 091101(R) (2000).

[37] H. N. Brown et al. [Muon (g - 2) Collaboration], *Phys. Rev. Lett.* **86**, 2227 (2001).

[38] G. W. Bennett et al. [Muon (g - 2) Collaboration], *Phys. Rev. Lett.* **89**, 101804 (2002); *Erratum-ibid.* **89**, 129903 (2002).

[39] K. Ackerstaff et al. [OPAL Collaboration], *Phys. Lett.* **B431**, 188 (1998).

[40] M. Acciarri et al. [L3 Collaboration], *Phys. Lett.* **B434**, 169 (1998).

[41] S. Narison, *Phys. Lett.* **B513**, 53 (2001); Erratum-ibid. **B526**, 414 (2002).

[42] J. Schwinger, *Phys. Rev.* **73**, 413 (1948); **76**, 790 (1949).

[43] R. Karplus and N. M. Kroll, *Phys. Rev.* **77**, 536 (1950).

[44] A. Peterman, *Helv. Phys. Acta* **30**, 407 (1957).

[45] C. M. Sommerfield, *Phys. Rev.* **107**, 328 (1957).

[46] C. M. Sommerfield, *Ann. Phys. (N.Y.)* **5**, 26 (1958).

[47] G. S. Adkins, *Phys. Rev. D* **39**, 3798 (1989).

[48] J. Schwinger, Particles, Sources and Fields, Volume III, Addison-Wesley Publishing Company, Inc., 1989.

[49] S. Laporta and E. Remiddi, *Phys. Lett.* **B379**, 283 (1996).

[50] R. Barbieri and E. Remiddi, *Nucl. Phys.* **B 90**, 233 (1975).

[51] M. A. Samuel and G. Li, *Phys. Rev. D* **44**, 3935 (1991).

[52] S. Laporta and E. Remiddi, *Phys. Lett.* **B265**, 181 (1991).

[53] S. Laporta, *Phys. Rev. D* **47**, 4793 (1993).

[54] S. Laporta, *Phys. Lett.* **B343**, 421 (1995).

[55] S. Laporta and E. Remiddi, *Phys. Lett.* **B356**, 390 (1995).

[56] R. Z. Roskies, E. Remiddi and M. J. Levine, Analytic evaluation of sixth-order contributions to the electron's g factor, in [1], p. 162.

[57] T. Kinoshita, Theory of the anomalous magnetic moment of the electron – Numerical Approach, in [1], p. 218.

[58] J. Aldins, S. J. Brodsky, A. Dufner, and T. Kinoshita, *Phys. Rev. Lett.* **23**, 441 (1970); Phys. Rev. D **1**, 2378 (1970).

[59] S. J. Brodsky and T. Kinoshita, *Phys. Rev. D* **3**, 356 (1971).

[60] J. Calmet and M. Perrottet, *Phys. Rev. D* **3**, 3101 (1971).

[61] J. Calmet and A. Peterman, *Phys. Lett.* **B47**, 369 (1973).

[62] M. J. Levine and J. Wright, *Phys. Rev. Lett.* **26**, 1351 (1971); *Phys. Rev. D* **8**, 3171 (1973).

[63] R. Carroll and Y. P. Yao, *Phys. Lett.* **B48**, 125 (1974).

[64] P. Cvitanovic and T. Kinoshita, *Phys. Rev. D* **10**, 3978, 3991, 4007 (1974).

[65] T. Kinoshita and W. B. Lindquist, *Phys. Rev. D* **27**, 867, 877, 886 (1983); D **39**, 2407 (1989); D **42**, 636 (1990).

[66] M. Caffo, S. Turrini, and E. Remiddi, *Phys. Rev. D* **30**, 483 (1984).

[67] E. Remiddi and S. P. Sorella, *Lett. Nuovo Cim.* **44**, 231 (1985).

[68] T. Kinoshita, *IEEE Trans. Instrum. Meas.* **44**, 498 (1995).

[69] H. Suura and E. Wichmann, *Phys. Rev.* **105**, 1930 (1957).

[70] A. Peterman, *Phys. Rev.* **105**, 1931 (1957).

[71] H. H. Elend, *Phys. Lett.* **20**, 682 (1966); *Erratum-ibid.* **21**, 720 (1966).

[72] B. E. Lautrup and E. de Rafael, *Phys. Rev.* **174**, 1835 (1968).

[73] S. Laporta, *Nuovo Cimento* **106A**, 675 (1993).

[74] S. Laporta and E. Remiddi, *Phys. Lett.* **B301**, 440 (1993).

[75] A. Czarnecki, B. Krause and W. J. Marciano, *Phys. Rev. Lett.* **76**, 3267 (1996).

[76] P. J. Mohr and B. N. Taylor, *Rev. Mod. Phys.* **72**, 351 (2000).

[77] B. Krause, *Phys. Lett.* **B390**, 392 (1997).

[78] C. Bouchiat and L. Michel, *J. Phys. Radium* **22**, 121 (1961).

[79] L. Durand III, *Phys. Rev.* **128**, 441 (1962); *Erratum-ibid.* **129**, 2835 (1963).

[80] J. Z. Bai et al. [BES Collaboration], *Phys. Rev. Lett.* **84**, 594 (2000); *Phys. Rev. Lett.* **88**, 101802 (2000).

[81] R. R. Akhmetshin et al. [CMD-2 Collaboration], *Phys. Lett.* **B527**, 161 (2002).

[82] R. Barate et al. [ALEPH Collaboration], *Z. Phys.* **C 2**, 123 (1997).

[83] S. Anderson et al. [CLEO Collaboration], *Phys. Rev. D* **61**, 112002 (2000).

[84] S. Eidelman and F. Jegerlehner, *Z. Phys. C* **67**, 585 (1995).

[85] D. H. Brown and W. A. Worstell, *Phys. Rev. D* **54**, 3237 (1996).

[86] R. Alemany, M. Davier and A. Höcker, *Eur. Phys. J.* **C 2**, 123 (1998).

[87] M. Davier and A. Höcker, *Phys. Lett.* **B419**, 419 (1998).

[88] M. Davier and A. Höcker, *Phys. Lett.* **B435**, 427 (1998).

[89] G. Cvetic, T. Lee and I. Schmidt, *Phys. Lett.* **B513**, 361 (2001).

[90] J. F. de Trocóniz and F. J. Ynduráin, *Phys. Rev. D* **65**, 093001 (2002).

[91] F. Jegerlehner, arXiv:hep-ph/0104304.

[92] M. Davier, S. Eidelman, A. Höcker and Z. Zhang, arXiv:hep-ph/0208177.

[93] M. Perrottet and E. de Rafael, unpublished.

[94] K. Hagiwara, A. D. Martin, D. Nomura and T. Teubner, arXiv:hep-ph/0209187.

[95] S. Peris, M. Perrottet and E. de Rafael, *JHEP* **05**, 011 (1998).

[96] G. 't Hooft, *Nucl. Phys.* **B 72**, 461 (1974).

[97] E. Witten, *Nucl. Phys.* **B 160**, 157 (1979).

[98] M. Perrottet and E. de Rafael, private communication.

[99] J. Calmet, S. Narison, M. Perrottet and E. de Rafael, *Phys. Lett.* **B61**, 283 (1976).

[100] T. Kinoshita, B. Nizic, Y. Okamoto, *Phys. Rev. D* **31**, 2108 (1985).

[101] M. Hayakawa, T. Kinoshita and A. I. Sanda, *Phys. Rev. Lett.* **75**, 790 (1995); *Phys. Rev. D* **54**, 3137 (1996).

[102] J. Bijnens, E. Pallante and J. Prades, *Nucl. Phys.* **B 474**, 379 (1996).

[103] M. Hayakawa and T. Kinoshita, *Phys. Rev. D* **57**, 465 (1998).

[104] M. Knecht and A. Nyffeler, *Phys. Rev. D* **65**, 073034 (2002).

[105] M. Hayakawa and T. Kinoshita, arXiv:hep-ph/0112102, and the erratum to [103] published in *Phys. Rev. D* **66**, 019902(E) (2002).

[106] J. Bijnens, E. Pallante and J. Prades, *Nucl. Phys.* **B 626**, 410 (2002).

[107] J. Wess and B. Zumino, *Phys. Lett.* **B37**, 95 (1971).

[108] E. Witten, *Nucl. Phys.* **B 223**, 422 (1983).

[109] S. L. Adler, *Phys. Rev.* **177**, 2426 (1969).

[110] J. S. Bell and R. Jackiw, *Nuovo Cimento A* **60**, 47 (1969).

[111] J. Bijnens and F. Persson, arXiv:hep-ph/0106130.

[112] M. Knecht and A. Nyffeler, *Eur. Phys. J. C* **21**, 659 (2001).

[113] J. L. Rosner, *Ann. Phys. (N.Y.)* **44**, 11 (1967). Implicitly, the method of Gegenbauer polynomials was already used in the following papers: M. Baker, K. Johnson, and R. Willey, *Phys. Rev.* **136**, B1111 (1964); **163**, 1699 (1967).

[114] M. J. Levine and R. Roskies, *Phys. Rev. D* **9**, 421 (1974); M. J. Levine, E. Remiddi, and R. Roskies, *ibid.* **20**, 2068 (1979).

[115] M. Knecht, A. Nyffeler, M. Perrottet and E. de Rafael, *Phys. Rev. Lett.* **88**, 071802 (2002).

[116] I. Blokland, A. Czarnecki and K. Melnikov, *Phys. Rev. Lett.* **88**, 071803 (2002).

[117] W. J. Bardeen and A. de Gouvea, private communication.

[118] W. A. Bardeen, R. Gastmans and B. E. Lautrup, *Nucl. Phys.* **B 46**, 315 (1972).

[119] G. Altarelli, N. Cabbibo and L. Maiani, *Phys. Lett.* **B40**, 415 (1972).

[120] R. Jackiw and S. Weinberg, *Phys. Rev. D* **5**, 2473 (1972).

[121] I. Bars and M. Yoshimura, *Phys. Rev. D* **6**, 374 (1972).

[122] M. Fujikawa, B. W. Lee and A. I. Sanda, *Phys. Rev. D* **6**, 2923 (1972).

[123] V. A. Smirnov, *Mod. Phys. Lett. A* **10**, 1485 (1995).

[124] S. Peris, M. Perrottet and E. de Rafael, *Phys. Lett.* **B355**, 523 (1995).

[125] A. Czarnecki, B. Krause and W. Marciano, *Phys. Rev. D* **52**, R2619 (1995).

[126] T. V. Kukhto, E. A. Kuraev, A. Schiller and Z. K. Silagadze, *Nucl. Phys.* **B 371**, 567.

[127] S. Peris, M. Perrottet and E. de Rafael, arXiv:hep-ph/0205102.

[128] G. Degrasi and G. F. Giudice, *Phys. Rev. D* **58** (1998) 053007.

Marc Knecht
Centre de Physique Théorique
CNRS-Luminy, Case 907
F-13288 Marseille Cedex 9, France

Poincaré Seminar 2002, 311 – 331

Poincaré Seminar 2002

Physics at the Large Hadron Collider

Bruno Mansoulié

Abstract. The Standard Model of elementary constituants and interactions is a well tested theory, but we clearly see its limitations. In particular, the origin of the particle masses is now a central question. Several new theoretical frameworks are proposed to address it and answer it at least partially. All of them have the mechanism of ElectroWeak symmetry breaking as a cornerstone, and predict new phenomena at its typical energy scale: 1 TeV. The Large Hadron Collider, at Cern, will be the first accelerator to explore this energy scale directly.

Its construction is now in progress, together with the large experiments which will extract the physics out of the particle collisions. All the proposed models have been examined in great detail, and the detectors optimized accordingly. Physicists are confident that indeed LHC will bring crucial new informations and open a path beyond the Standard Model.

1 Introduction

The LHC at Cern is the most ambitious project in particle physics today. The machine and the experiments are huge technical challenges, and the experimental conditions are expected to be difficult in the best case. However, the motivation for this effort is unprecedented: physicists are convinced that LHC will bring key elements to answer the present questions in the field. Starting from the weaknesses of the Standard Model, many larger theories have been put forward, and a large amount of work has been devoted to studying the observable consequences of each of these new theories at LHC. A large part of this work was undertaken by the large collaborations which proposed, and now construct, experiments at LHC[1]. In this talk I will give a global survey of this work. At the same time, I will go into some detail for a few cases, to underline the particular experimental conditions.

1.1 The Standard Model

The Standard Model is the theory which describes all the observations at the microscopic scale today. It assumes a number of input ingredients, namely the nature of the constituants of matter, the type of their interactions, and about 25 arbitrary parameters (mostly particle masses and interaction coupling strengths). Given this, the Standard Model offers a framework which we believe to be essential: quantum mechanics and special relativity, i.e. it is a quantum field theory.

For the experimentalist, the predictive power of quantum field theories comes mostly from the calculations of perturbation series, and renormalizability is the key criterion there. In the Standard Model, renormalizability, and hence an efficient use of perturbation series, is guaranteed by the structure of the model, based on local gauge symmetries.

For a non-insider, it is difficult to imagine how deeply the ideas of gauge invariance and renormalizability have modelled the entire landscape of experimental particle physics. We are now completely used to measuring properties of particles which were never produced in their real, on-shell state, but whose presence is seen through virtual effects. The most spectacular example was the measurement of the top quark mass at LEP ($m_t = 178 \pm 20\,\text{GeV}$) before it was 'discovered' at FNAL in 1994 ($m_t = 174 \pm 5\,\text{GeV}$). Internal and external radiation of real particles (photons, weak bosons, etc.) is routinely observed and taken into account in analyses, with detailed prescriptions for using experimental variables which make sense in the renormalization procedure and avoid divergences in the theoretical calculations.

The constituants are spin 1/2 fermions: leptons and quarks. We know three families of two leptons and two quarks each, with the second and third families replicating the first one in all aspects but the masses of the particles. The interactions are: the electro-weak (E-W) interaction, based on the gauge group SU(2)×U(1), and the strong interaction based on SU(3).

In a given family, the behaviour of the constituant fermions under an interaction is described by their location in the group multiplet representations:

– Leptons and quarks are sensitive to the E-W interaction:

- left-handed states are doublets under SU(2).
- right-handed states are singlets under SU(2).
- One lepton has electric charge -1 (e, μ, τ) , the other is neutral (ν_e,ν_μ,ν_τ), the upper quark in the SU(2) doublet (u, c, t) has charge +2/3, the lower one (d,s,b) has charge $-1/3$.

– Quarks are sensitive to the strong interaction: they are triplets under SU(3) (the strong charge is called 'color', and the strong interaction is referred to as Quantum Chromo Dynamics, in short QCD).

The number of families (3) is unexplained (but it is the minimal number which allows for CP violation, an effect with deep consequences, in particular for cosmology).

Although the particle masses are free parameters in the model, their mere presence is central to the rationale behind the model. Indeed, the structure above would be easily realized if all particles were massless (or in the limit of very high energy where all masses would be negligible). But it is impossible to add masses 'by hand' to the constituants and keep the gauge symmetry structure, and consequently renormalizability. The question is to break gauge symmetry enough to get particle masses, while preserving it in depth. This is achieved by 'spontaneous symmetry breaking'.

In the Standard Model (SM), the electro-weak symmetry is broken down to separate weak and electromagnetic interactions. The standard way to achieve this breakdown is to introduce a scalar field (the Higgs boson) whose energy density is non-zero (positive) in the symmetrical vacuum. The value of this 'vacuum expectation value' determines the scale below which the symmetry appears as broken.

The system breaks the symmetry and choses a new fundamental state with minimum potential energy; then the fundamental fields are determined around this new vacuum. The messengers of the weak interaction (the W^+, W^-, Z^0 bosons) acquire a mass of the order of the Higgs vacuum expectation value, while the photon remains massless. Most importantly, the natural coupling of the constituant fermions with the Higgs provides them with a mass. The value of the masses are still free parameters, but now the theory with these masses is fully gauge invariant and renormalizable.

When the model was set up, the W, Z and top quark had not been yet observed. Thus the experimental detection of the W and Z in 1983, precisely at the mass predicted by other previous measurements (neutrino scattering on nuclei), was a bright confirmation for the model. Since then, millions of Z's have been produced at LEP, and precision measurements have tested the whole scheme in great detail.

All the particles in the SM have now been observed, except the Higgs boson. The model does not predict its mass. For the standard Higgs boson, the LEP experiments have given a lower limit by direct search, $m_H > 113.5\,\text{GeV}$, and an *upper limit* again through virtual effects: $m_H < 193\,\text{GeV}$ at 95% confidence level. There is no real theoretical upper limit to the Higgs mass, but the natural range does not exceed $1\,\text{TeV} = 1000\,\text{GeV}$. For example, the width of a heavy Higgs is $\Gamma_H \sim 0.5 m_H^3$ (Γ_H, m_H in TeV) which shows that the Higgs is no longer a particle beyond $\sim 1\,\text{TeV}$. More precisely, for Higgs masses larger than $\sim 800\,\text{GeV}$, the interactions of W and Z bosons become strong and new structures must appear. We will see that LHC claims to explore completely this mass range. Nevertheless, it is interesting to study carefully the most unfavourable case, with a very heavy Higgs, and a new interaction which would turn on slowly, difficult to see experimentally.

1.2 Beyond the Standard Model

Although the Higgs mechanism is essential in the SM, its simplest implementation by the presence of a single scalar boson is far from satisfactory. The main concern is the 'naturalness' or 'fine-tuning' problem. We think that in the end the SM will be embedded in a more fundamental theory which will include larger mass scales. For example the unification of the strong and E-W interactions is thought to happen around 10^{16} GeV (from measurements of the evolution of their respective coupling strength with energy); even further, ultimately, a quantum gravity theory would bring in its natural scale: the Planck mass (10^{19} GeV). Particles with these large masses would contribute to the Higgs self-energy, driving its mass up to the higher scale, unless a fortuitous cancellation occurs between these contributions. The required accuracy of this cancellation would be given typically by $m_i^2 - m_j^2 \sim m_W^2$, 28 orders of magnitude fine-tuning if $m_i \sim m_j \sim 10^{16}$ GeV, quite an unnatural coincidence.

The candidate theories to go beyond the SM essentially try to solve the fine-tuning problem in their own way.

1.2.1 Composite models/condensate models

In these models the Higgs is not elementary, hence solving the problem. In most implementations, quarks and leptons are also composite. Some of these models also try to explain the number of families as excited states of the same sub-constituants. Although being in the continuation of the 'russian doll' scheme for matter, no good model exists along these lines. Such signals of compositeness could anyway be observed at LHC.

1.2.2 Supersymmetry [2]

Supersymmetry is a symmetry between fermions and bosons. This theory has been developped since a long time, for a number of reasons: first it is the last possible type of symmetry among fields, not yet observed in nature, and up to now we have seen nature using all the symmetries we could think of. Second, it has a deep link with gravity. Our present understanding of gravity is general relativity, a classical field theory, and attempts at a quantum theory have been unsuccessful up to now. The most promising track is string theories, which make use of the connection between gravity and supersymmetry.

Last, supersymmetry solves the fine-tuning problem in an elegant way. The contributions to the Higgs mass, coming from the large mass fermions and bosons, cancel exactly in unbroken supersymmetry. The theory requires superpartners (s-particles) for each of the usual particles. As none of these partners has been observed, supersymmetry has to be broken at some scale. The naturalness argument leads to a supersymmetry breaking scale of the order of the E-W scale. In this scenario, a full spectrum of new particles could be there at masses of order $\sim$ TeV, in the reach of LHC.

Supersymmetry is certainly the favored theory to go beyond the SM, despite the fact that no experimental sign has been found. An enormous amount of work has been devoted to evaluate the potential of LHC experiments on SUSY models. Many models can be constructed, with many free parameters. In order to study well defined cases, the physicists have defined a minimal supersymmetric standard model (MSSM). In this framework the Higgs sector is well defined, as we will see later, but for the other supersymmetric particles there are many variants, essentially in the precise way to implement the breaking of supersymmetry. An effort was brought to selecting the best defined models and exploring their parameter space consistently. The most popular one is the SUGRA model (SUper GRAvity inspired); the connexion to gravity is remote, but technically the model provides a 'reasonable' spectrum of all s-particles and Higgses, with only (!) 5 parameters.

An important aspect of supersymmetry is the link with cosmology, through the dark matter problem. Astrophysical measurements show that a large part of the matter in the universe does not radiate like ordinary matter (for a recent review, see for example [3]). In addition, this dark matter is believed to have a large non-baryonic part, and ordinary neutrinos can only contribute to a small amount. The whole scheme still has uncertainties, but taking it at face value, a large fraction ($\sim 20\%$) of the matter in the universe should be 'cold' dark matter, in the form of new particles, electrically neutral, stable, with a large mass. In many scenarios, the supersymmetric partner of the neutrino, called the neutralino ($\tilde{\chi}^0$), is the lightest supersymmetric particle (LSP), and is a good candidate for this particle. For a given model, one can crunch the usual big-bang scenario and calculate the relic density of neutralinos. In SUGRA models for example, requiring that the neutralino relic density be consistent with the cold dark matter selects a zone in the parameter space [4], which can be explored at LHC.

1.2.3 Extra dimensions

The idea that space-time could have more than 3+1 dimensions goes back to the Kaluza-Klein model, as early as 1919. These authors saw that writing general relativity in 5 dimensions, and 'compactifying' the 5th one on a small radius, one gets the classical theory of electromagnetism. This very appealing remark did not hold its promises, since noone succeeded in building a unified model of gravity and electromagnetism. In the eighties, the idea was revived because string theories, the candidate for a quantum theory of gravity, like to work in a higher-dimensional space-time. In this framework, our usual 4D space-time is what is left after 'compactification' of all other dimensions on a very small scale. It was first thought that this small scale was of the order of Planck's length (10^{-33} cm), or equivalently would be relevant for energies of the order of Planck's mass (10^{19} GeV). Recently, it was realized [5] that this needs not be the case.

In the simplest model[6], only gravity sees the extra dimensions, which could be as large as 1 mm, and the scale for quantum gravity is then $\sim$ 1 TeV. The extreme weakness of gravity at low energies comes from its 'dilution' in the extra dimensional volume, and the large value of Planck's mass is just an illusion: there is no mass scale higher than 1 TeV, which solves the fine-tuning problem. Again, for TeV scale quantum gravity, spectacular effects could be found at LHC.

2 The LHC

2.1 Machine [7] and experimental conditions

As soon as the LEP was approved, and well before its operation, physicists thought about putting a proton-proton collider in its tunnel. In the case of an electron accelerator like LEP, the beam energy is limited by synchrotron radiation losses: the loss must be compensated at each turn by accelerating cavities. The circumference of the LEP tunnel (27 km) was fixed to allow LEP to reach about 50 GeV per beam (100 GeV center of mass energy) with normal cavities, enough to produce on-shell Z^0 bosons, then 100 GeV per beam with superconducting cavities, enough to produce W pairs. In the case of proton beams, the energy is limited by the maximum field available in the bending (dipole) magnets. The design value for the field in the LHC superconducting magnets is 8.4 T, a $\times 1.8$ increase from previous machines (and remember that the magnetic forces go like B^2). The 14 m long magnets operate in superfluid helium at 1.9 K. With this field value, the beam energy is 7 TeV, hence a proton-proton center of mass energy of 14 TeV.

Protons are not elementary: what really counts is the energy available in the collision of the point-like constituants (partons): quarks and gluons. As the quarks and gluons carry a fraction of the momentum of their parent proton, with a statistical distribution (structure function), there is a broad spectrum of collision energies at the constituant level. Of course the most interesting events are those with the highest collision energies: they are also the rarest, since they involve partons which carry an exceptionally large fraction of the proton momentum.

When the US physicists designed a machine to cover the same physics goal, namely explore exhaustively the E-W symmetry breaking mechanism, they chose a center of mass energy of 40 TeV and a circumference of 87 km (the SSC project, unfortunately discontinued in 1993). Limited by the pre-existing tunnel and by the attainable magnetic field, the LHC energy is 'only' of 14 TeV. To increase the discovery reach, the other handle is luminosity, the number of proton-proton encounters per second. The LHC luminosity will be $10^{34} cm^{-2} s^{-1}$, a factor of 10 larger than the SSC design. Typically a factor of 10 in luminosity provides the same rate of rare processes than a factor 1.5 to 2 increase in energy.

This very high luminosity will be achieved by storing a large number of intense proton bunches in each beam. The bunches are only 25 ns apart, and interactions occur at the 4 collision points every 25 ns. The interesting interactions between partons are rare, but the total collision rate between protons is enormous. The total p-p cross-section, from strong interactions, is expected to be about $8 \times 10^{-26} cm^2$, which means about 20 interactions per bunch crossing in average. Each of these interactions is an event with about 60 charged and 60 neutral particles in the acceptance of an experiment around the collision point. The experiment must deal with more than 1000 tracks and 2000 impacts every 25 ns, and yet extract rare signals at a rate of a few events per year. This pattern recognition problem calls for detectors with a very high number of cells or channels, a very fast response, and a large dynamic range.

The other consequence is that the radiation level coming from the interaction point is high, and imposes the use of radiation resistant technologies for most detectors.

2.2 Experiments

Two intersection regions are devoted to high-luminosity p-p collisions,with general purpose experiments: ATLAS and CMS. The other two regions are for the ALICE experiment, which studies ion-ion collisions, and the LHC-B experiment, which studies b-quark physics in medium luminosity p-p collisions. This talk will concentrate on physics at ATLAS and CMS.

When the first ideas of operation at high luminosity appeared (in 1984), the constraints coming from the event rate and radiation environment looked formidable, and it was first thought that the only possible experiment was an 'iron ball' around the interaction point, with only muon detection outside. Through a vigorous R and D program pursued in many labs around the world, it was shown that much more can be done, including precision measurements, detailed particle identification, and inclusive event reconstruction.

The experiments isolate the rare signals against the huge background by selecting processes with good signatures. As the background originates mostly from strong interactions, these signatures may involve the presence in the final state of:

- one or more lepton(s): electrons, muons, and neutrinos (identified by the missing transverse energy).
- photons.
- b-quarks or c-quarks, identified by a displaced vertex.
- hadronic jets of high transverse momentum (from high momentum quarks and gluons).

Although the physics goals and the operation requirements are the same for both experiments, the technical choices for some of the detectors have been rather different, resulting in a real complementarity, as we can illustrate with a few examples:

The magnetic field in CMS is provided by a single, large superconducting solenoid (12 m long, 7 m diameter) with a high field (4T). In ATLAS, the magnet system includes a 'small' solenoid around the central region (7×3m) with a 2 T field, and a large (26m long, 20m diameter) system of 3 toroidal magnets for muon measurements. The CMS solution is conceptually simpler, but the ATLAS sytem should offer a safe measurement of muons in the outer spectrometer alone.

For the electromagnetic calorimeters, which measure the energy of electrons and photons, CMS has chosen scintillating crystals, while ATLAS has chosen a lead/liquid argon sampling technique. The CMS crystals have an excellent intrinsic energy resolution (typically 0.7% at 100 GeV), but it will be difficult to keep the calibration of their light output to the required accuracy (0.4%). On the opposite the ATLAS solution has only a fair intrinsic energy resolution (typ. 1.2% at 100 GeV), but should be very stable in time.

The number of electronic channels amounts to tens of millions in the central track detectors, and hundred of thousands for calorimeters and muons chambers. It is of course impossible to record all the read-outs for every bunch crossing: the trigger system selects interesting events for recording. The selection is made in several (usually 3) levels, the next level up analyzing events in more detail and being more selective. It is very important to establish 'trigger menus' large enough not to miss any new physics processes, but which keep the accepted rate inside the available bandwidth.

ATLAS and CMS are two large international collaborations, each with about 150 participating institutions and more than a thousand physicists. Both were approved in 1996, and are under construction now.

2.3 Simulation

An important part of the preparation work has been devoted to simulations. The collaborations have made exhaustive studies of the LHC physics, starting from available or customized event generators, and going sometimes to the finest detail of the experiment. These simulations have been used to optimize the detectors, design analysis algorithms, and in general evaluate the performance on every physics channel one could think of. Most of the material presented here comes from this work.

3 The Standard Model Higgs

Assuming a mass for the Higgs boson, one can calculate its production cross-section, and the probability for each of its decay modes. As the decay modes change strongly depending on the mass, the search involves different detectors and analyses. Thus the search for the Standard Model Higgs has quickly become the benchmark for detector optimization, and has been studied in great detail.

Several processes contribute to the production of Higgs bosons: $g\overline{g} \to H$ through a heavy quark loop, $qq \to qqH$ ("WWfusion"), $q\overline{q} \to WH$, $gg \to t\bar{t}H$, $gg \to b\bar{b}H$. The relative importance of these processes depends upon the Higgs mass, the first dominates at small mass and the first two become comparable for a Higgs mass of 1 TeV. The Higgs branching ratios are shown in Fig. 1.

3.1 $H \to \gamma\gamma, 115\,\mathrm{GeV} < \mathrm{m_H} < 140\,\mathrm{GeV}$

At low mass ($114\,\mathrm{GeV} < \mathrm{m_H} < 2 \times \mathrm{m_W}$) the main decay modes ($b\bar{b}$, $c\bar{c}$, $\tau^+\tau^-$) cannot be distinguished from the QCD background. One possibility is the decay mode $H \to \gamma\gamma$ which has a tiny branching ratio, but where two photons in the final state offer a rather good signature. This search is very demanding on the detector and has been used as a benchmark for the performance of electromagnetic calorimeters, hence it is interesting to look at it in some detail.

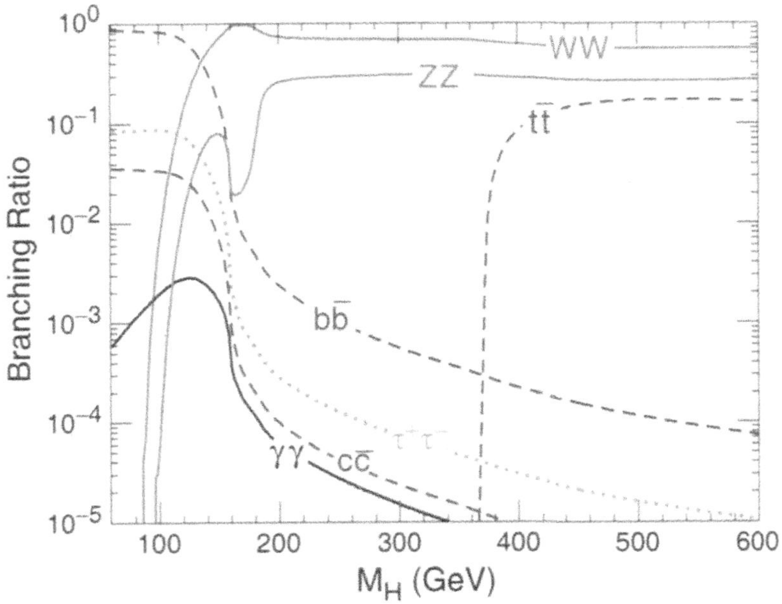

Figure 1: Standard Model Higgs branching ratios as a function of mass

First, one has to identify the two photons. In the same invariant mass range (say, 120 GeV), the rate of jet pairs, coming from QCD processes like $gg \rightarrow gg, qg \rightarrow qg$, etc. is $\sim 10^6$ times larger than the signal; there are also $jet - \gamma$ events at a rate $\sim 10^3\times$ signal. It may seem obvious to discriminate a jet of particles from an isolated photon, but here we need a rejection of more than 1000 against each jet. Jets are made of charged particles (mostly charged pions) and neutral particles, mostly π^0's which decay instantaneously into two photons. Small detector inefficiencies can indeed fake single photons at a very low level. Particular jet configurations are also dangerous: in about 1 case in 1000 a quark hadronizes into a single π^0; with a π^0 momentum of 60 GeV, the two photons from the decay of this π^0 will be only 7 mm apart at the entrance face of the calorimeter, quite difficult to tell from a single photon. The ATLAS and CMS detectors devote 84000 (resp. 140000) read-out channels to a fine-grain section, whose main goal is to gain a factor of 3 rejection against π^0's in this particular search. To reject jets, analyses also require that the energy deposit associated to the photon be isolated, at the expense of a small ($\pm 10\%$) loss in efficiency on the signal.

Then there is a large irreducible background from processes like $q\bar{q} \rightarrow \gamma\gamma$, $gg \rightarrow \gamma\gamma$, $qg \rightarrow q\gamma\gamma$ which produce photon pairs with a continuous mass spectrum. The Higgs would appear as a peak in the photon pair invariant mass distribution, hence the signal to noise ratio depends directly on the mass resolution. The invariant mass is evaluated by $m^2 = 2E_1E_2 \times (1 - cos\theta)$, thus it depends on the energy resolution for each photon, and on the determination of the angle θ between the photons. The energy resolution is given by the performance of the electromagnetic calorimeter. The measurement of the angle poses a challenge quite special to LHC: in an usual experiment, there would be only one interaction vertex, and the direction of a photon would be obtained simply by linking the impact point in the calorimeter to this vertex. At LHC, there are 20 interaction vertices per bunch crossing in average, distributed over 5.6 cm around the nominal crossing point. It is not so easy to associate the right vertex to the photon impact! The solution is to use the calorimeter for measuring not only the energy and position, but also the direction of the photon, and/or to select among all vertices the most probable good one, on other criteria like the multiplicity of tracks above some momentum.

Very detailed simulations have been performed on this channel. The result of such a simulation in CMS is shown in Fig. 2.

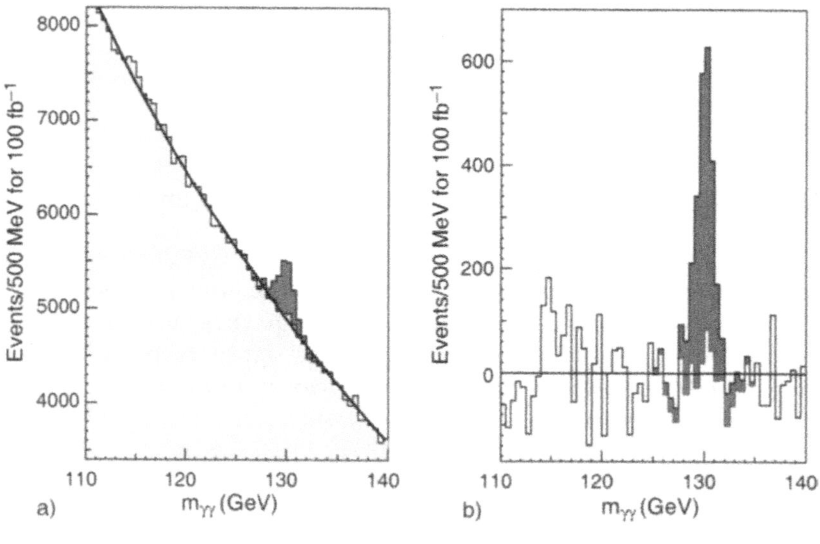

Figure 2: (a) The invariant mass distribution of $\gamma\gamma$ pairs for $M_h = 130\,\text{GeV}$ as simulated by the CMS collaboration. (b) Same, with a smooth background fitted and subtracted. From Ref. [8].

3.2 W(or $t\bar{t}$) + $H \to b\bar{b}$, $115\,\text{GeV} < m_H < 130\,\text{GeV}$

As we said above, it is impossible to extract a signal of a low mass Higgs in the dominant decay mode $H \to b\bar{b}$ if no other signature is present. However, there are processes where the Higgs is produced in association with a W or a $t\bar{t}$ pair.

In this case, one can ask for an electron or a muon from the W (top quark) decay, which reduces the background by a large amount. Then, the capacity of the detector in identifying b-quarks is essential. Mesons and baryons containing b-quarks are known to decay with a typical lifetime of ~ 1.5 picosecond, hence they travel a small distance (hundreds of microns) away from the primary vertex, before decaying. Such displaced vertices can be measured by precision silicon strip track detectors with excellent results as demonstrated at LEP, Tevatron or B-factories. The question was if such precise measurements could be performed in the crowded environment of LHC, and if the silicon detectors, located close to the beam pipe, could survive the radiation.

Building and operating large silicon detectors and their electronics in a radiation environment is a whole field in technology. A lot of progress was done by the LHC experiments, in collaboration with teams interested in other uses, like electronics for space applications. For the pattern recognition problem, LHC vertex detectors have hundreds of times more channels than their predecessors. Again, detailed simulations predict that the b-tagging efficiency will be at least as good as that of the present CDF experiment at FNAL for example, despite the environment. In the end, a signal in this mode would be just visible, and would provide a confirmation of the $\gamma\gamma$ channel.

3.3 $H \to ZZ^* \to 4\,leptons$ (e or μ), $150\,\mathrm{GeV} < \mathrm{m_H} < 600\,\mathrm{GeV}$

If the Higgs mass exceeds $2 \times m_Z$, then the main decay modes are W^+W^- (70%) and ZZ (30%). The Z decays in an e^+e^- or $\mu^+\mu^-$ with a 3% branching ratio (each). $H \to ZZ \to 4\,leptons(e\,or\,\mu)$ are gold plated events, offering excellent signature and mass resolution. This mode allows an easy detection of a Higgs signal for $2 \times m_Z < m_H < 600\,\mathrm{GeV}$; for larger m_H, the Higgs production rate decreases, and at the same time its decay width increases, which spreads the mass peak over the background continuum. The study can be extended to m_H lower than $2 \times m_Z$ down to $\sim 150\,\mathrm{GeV}$: the Higgs can still decay to the same 4-lepton modes, although at least one of the intermediate Z's is off-shell. In this range, the study is more difficult and demands more on the detector resolution. Backgrounds such as $t\bar{t}$ and $Z + b\bar{b}$ contribute, in addition to the ZZ continuum (present at all masses).

3.4 $m_H > 600\,\mathrm{GeV}$

For large Higgs masses, one must search for more frequent decay modes of the W and Z's, at the expense of more difficult signatures. The first mode is $H \to ZZ \to ll\nu\bar{\nu}$, with one Z decaying into an electron or muon pair, and the other into a neutrino pair. Neutrinos are of course not detected individually, but their presence is marked by missing transverse energy when accounting for all the energies measured by the experiment (at a proton machine, the longitudinal momentum balance cannot be used, since the frame of the elementary collision between partons moves along the beam line). The background sources are the physical continuum of ZZ production, but also instrumental effects which can generate fake missing transverse energy, like inefficient areas in the detector. The detectors need to cover the full solid angle around the interaction point, in particular the forward region close to the beam pipe, otherwise the statistical fluctuations of the other events occuring in the same bunch crossing ('pile-up events') would also contribute to the background.

Fig. 3 shows the missing transverse energy spectrum as simulated in ATLAS for a 700 GeV mass Higgs.

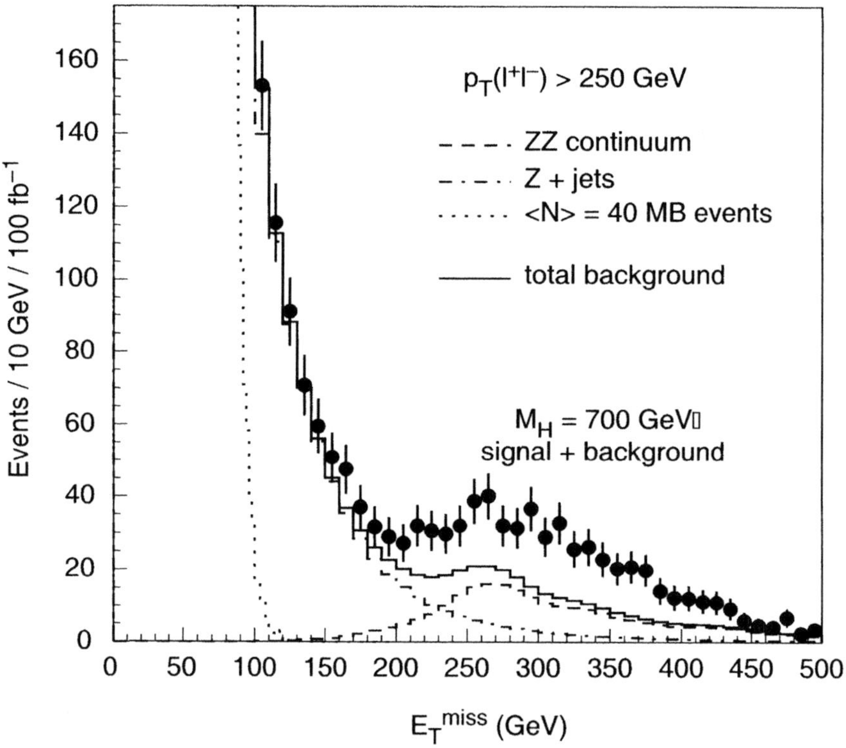

Figure 3: Missing E_T spectrum for the $H \rightarrow ZZ \rightarrow \ell\ell\nu\bar{\nu}$ process. The background contributions are shown separately; $Z + jets$ (dot-dashed); ZZ (dotted) and minimum bias pile up (dashed). The signal is due to a Higgs boson of mass 700 GeV. From Ref. [9].

Then the modes $H \rightarrow WW \rightarrow l\nu + jets$ and $H \rightarrow ZZ \rightarrow l\bar{l} + jets$ have an even larger branching ratio. However, the background from ordinary production of $W + jets$ and $Z + jets$ is very large. In the signal the jet pair invariant mass is m_W or m_Z; the signal to noise ratio depends on the jet pair mass resolution which in turn depends on the performance of the hadronic calorimeter, and on the reconstruction algorithm.

For very high Higgs masses, the dominant production mode is $qq \rightarrow Hqq$, where the Higgs is produced in association with two jets in the forward and backward direction. The detectors have been optimized to measure these jets at a small angle from the beam-line, a difficult region crowded with high-momentum particles and submitted to very high radiation levels. These modes should allow the detection of a Higgs up to a mass of 1 TeV.

3.5 Summary of Standard Model Higgs

Combining the analyses above, the mass range from the LEP limit to 1 TeV is covered. Fig. 4 shows for example in Atlas the statistical significance of a Higgs signal as a function of mass over the whole range.

We should not forget that the LEP results favor the low mass region: 114 GeV to $\sim$ 250 GeV. From 114 to 160 GeV the detection of a Higgs at LHC relies on the mode $H \rightarrow \gamma\gamma$, the mode $W + H \rightarrow b\bar{b}$ and the lowest part of the mode $H \rightarrow ZZ^* \rightarrow 4\,leptons$, and requires all the detector capacity. Above 160 GeV the mode $H \rightarrow ZZ^* \rightarrow 4\,leptons$ allows for an easy detection.

4 The Higgs sector in Supersymmetry

In supersymmetric theories, the Higgs sector is more complex than in the SM. In the MSSM, there are two different Higgs fields with two vacuum expectation values. The analysis of the physical states turn up two charged ($H^{\pm}$) and three neutral (h, H, A) scalar particles. Their masses and couplings are basically determined by two parameters, usually taken as the mass of the A (m_A) and $tan\beta$, the ratio of the two vacuum expectation values. Radiative corrections from loops containing ordinary or supersymmetric particles modify the values of masses and couplings, sometimes substantially. For example, without these corrections, one of the neutral Higgses, the h, would have a mass always *lower* than m_Z, but with corrections, this upper limit can reach 150 GeV for large values of m_A and $tan\beta$.

To limit the parameter space to only these two, we first assume that all supersymmetric partners of usual particles have large masses (TeV); in this case the Higgses can only decay into ordinary particles. The production cross-sections of the 5 Higgses, and their different branching ratios to ordinary particles, vary across the $m_A, tan\beta$ plane. The study of the experiment potential for one particular decay mode of one of the Higgses is expressed as a contour in this plane, inside which a statistically significant signal (5σ) would be observed. Fig. 5 shows the compilation of all these studies in ATLAS. It would be too long to go into the detail of each study, but a few remarks may be made.

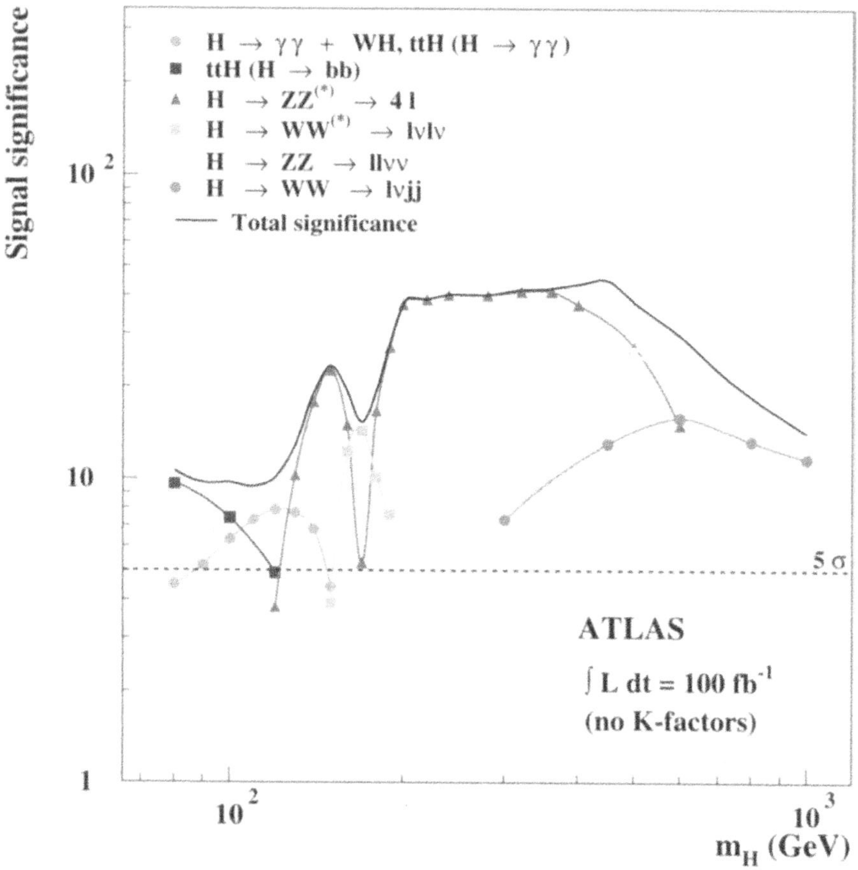

Figure 4: ATLAS experiment sensitivity for the discovery of a Standard Model Higgs boson From Ref. [9].

The first important message is that the entire plane is covered by the reunion of all contours, meaning that in all cases at least one supersymmetric Higgs would be observed. The main features of this coverage go as follows:

- At large m_A, the h behaves like a Standard Model Higgs with a mass lower than 150 GeV. Thus it can be detected in the $h \to \gamma\gamma$ mode as we have seen. However it would be impossible to tell that this is a *supersymmetric* Higgs and not the Standard one.

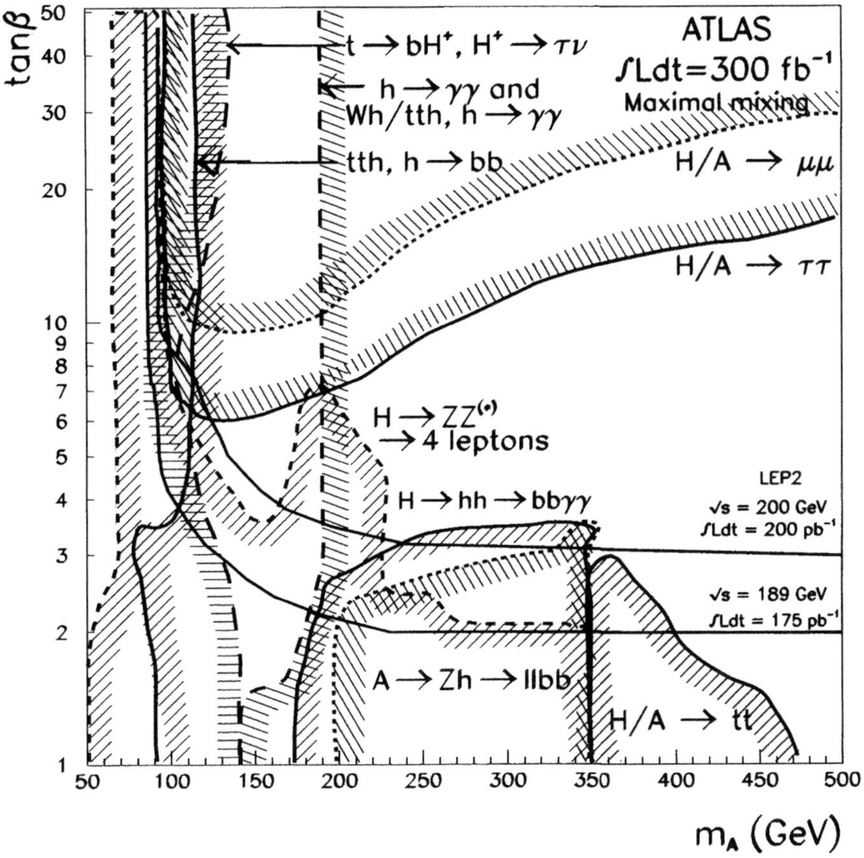

Figure 5: ATLAS experiment sensitivity for the discovery of a Supersymmetric Higgs boson: 5 σ discovery contours in the plane $m_A, tan(\beta)$ From Ref. [9].

- At large $tan\beta$, the branching ratios of H and A into $\tau^+\tau$ (tau lepton pair) is high, and this mode can be detected. This does not have an equivalent in Standard Model studies, and was looked at carefully. The reconstruction of the H or A mass is difficult because the τ decays always contain neutrinos which go undetected. The critical ingredient is the missing transverse energy resolution of the detector.

At lower values of m_A and $tan\beta$, several modes can be observed. The observation of more than one mode would bring redundancy and confirm the supersymmetric nature of the Higgses.

More precise studies must take into account the possibility that the Higgses decay into s-particles or couple to them. There are much too many parameters in the general MSSM, so this can only be attempted in a restricted model as SUGRA. The main conclusions are:

- The overall observability of the h boson through $\gamma\gamma$ or $b\bar{b}$ decays is unaffected.
- In a substantial part of the parameter space, the H boson decays to s-particles (namely neutralinos $\tilde{\chi}^0$ and charginos $\tilde{\chi}^{\pm}$) and this can be detected, although not easily. This would be very important as it would allow to discriminate between a SM Higgs (only seen in $h \to \gamma\gamma$) and a supersymmetric one.
- In a large region of the parameter space, the h can be produced in the cascade decays of s-particles, together with other particles with a very characteristic signature. It can then be detected in its dominant decay mode $b\bar{b}$, which increases the overall sensitivity to the Higgs sector.

5 Supersymmetric particles

In the early searches of supersymmetry at existing machines, or studies for LHC, there were no precise models, and the only signature which people thought of was missing energy. Indeed, if s-particles are produced, their decay products must contain the LSP which would go undetected. The most visible processes would then be of the type $qq \to \tilde{q}\bar{\tilde{q}} \to q + \tilde{\chi}_0 + \bar{q} + \tilde{\chi}_0$. The cross-section for such processes is high because the s-quarks are produced by strong interactions, and the event contains two very hard jets recoiling against nothing, a case with no equivalent in the Standard Model (hence no background).

This simple picture is still valid if we assume very high masses for the s-particles: then the discovery reach is just rate limited. For s-quarks and gluino masses of 2 TeV, we expect a few spectacular events, which would be unambiguous signs of supersymmetry, but would not bring much information beyond this fact.

The studies of the last few years have brought in a different picture, with precise models which give the complete spectrum of s-particles masses and couplings. For a large domain in the parameter space (s-particle masses of the order of, or below, 1 TeV), we now expect a rich phenomenology, with the production of many particle types, complex and beautiful cascade decays, allowing for precision measurements. In fact, the problem would not be to show evidence for supersymmetry as a whole, but to separate the different channels, and discriminate between models. In many cases, the background behind the studied signal comes from other supersymmetric processes!

Let us look at one of these scenarios: a SUGRA model with the parameters chosen to be 'cosmologically' correct. As in the simple case above, the strongest reactions produce squarks and gluinos (which then decay to a squark $\tilde{q}$ and a normal quark q).

Now the decay chain for each squark can be much more complex, and far more interesting:

$$\tilde{q} \to \tilde{\chi}_2^0 q \to \tilde{\ell}^{\pm}\ell^{\mp} q \to \tilde{\chi}_1{}^0 \ell^+ \ell^- q$$

As two squarks were produced, this would give an event with 4 leptons (e or μ), 2 jets, and missing transverse energy. A lot of information can be extracted from such events; in particular the analysis of the event kinematics allows for a determination of the neutralino mass to about 10%, which would be of great importance for cosmology (this was not possible in the early inclusive studies).

Many more studies were performed on SUSY models, which would be too long to report here. Let us mention the GMSB (Gauge mediated symmetry breaking) models, where the LSP is not the neutralino but the gravitino (the s-partner of the graviton). These models have a rather different phenomenology which can be challenging for the detector.

As a summary I would take Fig. 6.

This plot is in the plane of the two most important parameters of SUGRA, for a 'reasonable choice' of the other 3 parameters. The figure shows the 'cosmological' area, (where the LSP relic density is between 10% and 30% of the critical density), the reach of LHC in an inclusive squark or gluino search ($m_{\tilde{q}}, m_{\tilde{g}} < 2\,\mathrm{TeV}$), and the area where the cascade decay above allows for precision measurements and an estimate of the LSP mass. The inclusive search covers all the cosmologically allowed domain, and it is tantalizing that in a large part of it the most interesting studies are possible.

6 Extra dimensions

Since the appearance of the idea that extra-dimensions could be as close as the TeV scale, the number of publications on this topic has exploded: at least 50 papers published each month since year 2000! For the phenomenology at LHC, there are two main classes of models: 'factorizable' and 'non-factorizable' geometries. In factorizable geometries, the extra (compactified) dimensions are just added to the metric, without changing the usual part. Then one can decide which particles have access to all dimensions (the 'bulk') and which remain in our good old world (the 'brane'). In every model, the graviton has access to the bulk, in order to 'dilute' gravity and make it very weak in our world.

In the earliest model [6], only the graviton was allowed to propagate in the bulk. The parameters of the model are the number of extra dimensions n_D and the fundamental mass scale M_D . Planck's mass as it appears to us is related to M_D by the relation: $M^2_{Planck(4D)} = r^{n_D} M_D^{n_D+2}$, where r is the size of the extra dimensions. Taking M_D of order 1 TeV, we see that $n_D = 1$ is obviously excluded as it would make $r \sim 10^{13}$m, and modify gravity in the solar system. $n_D = 2$ and $M_D \sim$ TeV is just allowed, as it would modify gravity at a distance of less than 1 mm. This model appeared because it was realized that we did not have a good measurement of the gravitational force in $1/r^2$ below 1 mm! Since then

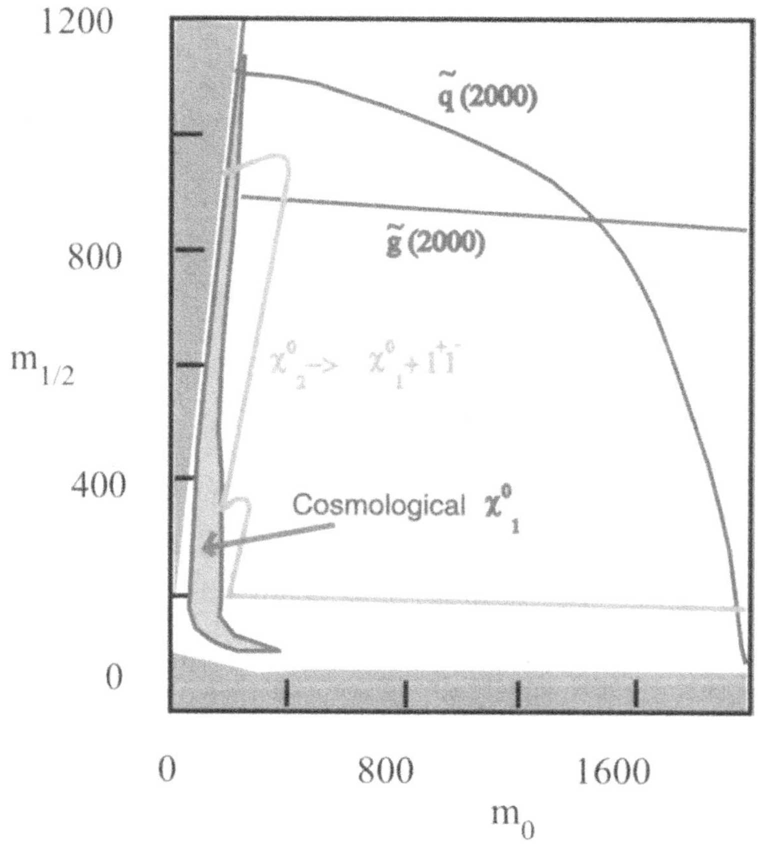

Figure 6: In the plane of the SUGRA parameters, the zone where the neutralino is the cosmologically preferred cold dark matter, with the reach of LHC experiments for $\tilde{q}$ and $\tilde{g}$ discovery (below the curves), and the area where a cascade decay is observable (below the curve, see text).

several laboratory experiments (Cavendish-type) have been started to improve this knowledge, see for example [10]; present limits are $r < 0.2mm$ and $M_D >$ 4 TeV for $n_D = 2$. The graviton has many 'Kaluza–Klein excitations', essentially modes around the extra dimensions compactified on a circle. Take a process like $quark + gluon \rightarrow quark + Graviton$. At low energies, this cross-section would be extremely small: in ordinary terms the right part (graviton/quark coupling) is just the gravitational mass of the quark. In terms of coupling it is suppressed by $1/M_{Planck}$, a very small number indeed. But now, when the energy becomes of the order of M_D, gravity becomes strong and it becomes highly probable to emit a graviton or one of its excitations, which then vanishes into the extra dimensions. Seen in the lab, this appears as an event where an invisible particle has been emitted, and this particle has a continuous spectrum of masses, a very unusual signal. For $n_D = 2$, LHC could see such events for M_D up to 9 TeV.

As an extension of this model, one can allow for example the gauge bosons to propagate in the bulk, a rather natural prescription if m_D is at the weak scale. Then these bosons acquire Kaluza-Klein excitations, with masses given by an harmonic formula such as $m_i^2 = m_0^2 + i^2 m_D^2$. The first states would just look like a W' or a Z', i.e. a heavy W or Z, with the same decay modes as the W and Z. Heavy W' or Z's appear in several other theories, and the potential for their discovery was studied as such. The reach of LHC is about 5 TeV for a Z' and 6 TeV for a W'.

In the other important class of models, non-factorizable geometries, the metric is no longer the simple superposition of extra and normal dimensions; the original model [11] is with 5 dimensions: there is the usual 4D 'brane' of our world, and another similar brane, parallel to the first one and separated from it by some distance in the 5th dimension, and the 4D metric is intricated into the 5D one. Gravity is mainly located on the other brane, and what remains on ours is exponentially weak. All the fields are sensitive to the extra dimension, and have Kaluza-Klein excitations, which appear as new particles. The spacing of these partners is different from the case of factorizable geometry, and would be a strong indication. The graviton also has TeV-scale excitations, which would decay into jets, leptons or photons. Note that the angular distribution of these decays would show the spin-2 nature of the particle, quite an unambiguous sign for a graviton.

As in these models, the Planck mass is at the TeV scale, recent studies have even proposed that black holes could be formed in high-energy collisions [14]. Such microscopic black holes would decay through Hawking radiation. This is "democratic" with respect to particle species; hence these events would be spectacular, with many high momentum leptons and jets with an isotropic distribution.

In summary, extra-dimensions theories are highly speculative. But the same argument is true, that if they have anything to do with electro-weak symmetry breaking, a sign should show up at LHC.

7 And if?

The question is often asked: What if there is no Supersymmetry, no extra-dimensions, and even no Standard Model Higgs below 1 TeV? If the Higgs mass goes beyond 1 TeV, then the interaction between W's would become strong for W momenta of ~ 1 TeV, and ultimately the diffusion process of two W's would violate unitarity (i.e. get an interaction probability greater than 1). So something must happen. One way out is to invoke a strong interaction between W's, which would more or less cancel the problem. There are candidates for such an interaction, like compositeness models or Technicolor models (a kind of new strong force) but as we said above none is really satisfactory. However, one can design phenomenological models without a fundamental basis, just to see what an experiment would detect in such a case. Quite naturally, most phenomenological models involve resonances between W's, which would be seen as large signals at LHC. Now if one really wants to be nasty, it is possible to construct a phenomenological model which removes the unitarity problem 'a minima', without any resonance and with as smooth a behaviour as possible [12]. Then the only possible sign to look at is an abnormal rise of the WW cross-section at the extreme end of the WW mass spectrum. We must admit that this would be very difficult to observe at LHC (a 4σ excess over a large background). Indeed the 40 TeV of the former SSC were chosen to give a clear answer even in this case. Upgrades of the LHC luminosity or energy are being considered to face this very unfavorable situation.

8 Standard Model physics

Besides all the new physics we can dream of discovering, there are many measurements in the Standard Model which will be improved at LHC. As an example the top quark mass can be determined to an accuracy better than ± 2 GeV. Jet and direct photon measurements will be used to test QCD, the theory of strong interactions, into a new domain. A rich program of B-physics will also be possible, with for example a measurement of the CP-violation parameter $sin2\beta$, to ± 0.02.

9 Conclusion

The Standard Model provides a very operative description of what we know about the elementary bricks of nature and their interactions. It is rather frustrating that particle masses (may be the simplest characteristic of a particle) are free parameters in the model. However, we know that there is a deep connection between particle masses and the ElectroWeak symmetry breaking mechanism. This connection was already seen in virtual effects in previous accelerators, like LEP, but LHC will have the potential for studying it at its natural energy scale. It is not surprising that all theories put forward today to subtend the EW breaking mechanism, predict measurable or even spectacular signals at LHC. This is the motivation of hundreds of experimentalists, who devote ten or fifteen years to this very challenging project, and look forward to the first collisions in 2007.

Acknowledgments

I thank all my ATLAS and CMS colleagues who have carried out most of the work presented here. A good guide to this work can be found in the review by J.G. Branson, D.Denegri, I. Hinchliffe, F. Gianotti, F.E. Paige and P. Sphicas [13].

References

[1] Atlas Technical proposal, CERN/LHCC/94-43,
`http://atlas.web.cern.ch/Atlas/`;
CMS Technical proposal, CERN/LHCC/94-38,
`http://cmsinfo.cern.ch/cmsinfo/Welcome.html`

[2] see for example P. Fayet and S. Ferrara, *Phys. Rep.* **C 32**, 249 (1977), H.E. Haber and G.L. Kane, *Phys. Rep.* **C 117**, 75 (1985)

[3] J.R. Primack, astro-ph/0205391.

[4] Ellis et al., *Phys. Lett.* **B474**, 314 (2000).

[5] I. Antoniadis, *Phys. Lett.* **B246**, 377 (1990).

[6] N. Arkani-Hamed, S. Dimopoulos and G. Dvali, *Phys. Lett.* **B429**, 263 (1998).

[7] R. Schmidt, invited paper at Eighth EPAC, Paris 2002, CERN-LHC Project Report 569.

[8] CMS Collaboration, ECAL Project Technical Design Report, CERN/LHCC/97-33,
`http://cmsdoc.cern.ch/cms/TDR/ECAL/ecal.html`.

[9] ATLAS Collaboration, ATLAS Detector and Physics Performance Technical Design Report, CERN/LHCC/99-14,
`http://atlasinfo.cern.ch/Atlas/GROUPS/PHYSICS/TDR/access.html`.

[10] C.D. Hoyle *et al.* (EOTWASH collaboration) *Phys. Rev. Lett.* **86**, 1418 (2001)

[11] L. Randall and R. Sundrum, *Phys. Rev. Lett.* **83**, 3370 (1999).

[12] Barger et al., *Phys. Rev.* **B52**, 3815 (1995) and references within.

[13] J.G. Branson et al. hep-ph/0110021.

[14] S. Dimopoulos and G. Laindsberg, SNOWMASS-2001-P231, June 2001.

Bruno Mansoulié
DAPNIA-SPP
CEA-Saclay
France

GPSR Compliance

The European Union's (EU) General Product Safety Regulation (GPSR) is a set of rules that requires consumer products to be safe and our obligations to ensure this.

If you have any concerns about our products, you can contact us on ProductSafety@springernature.com

In case Publisher is established outside the EU, the EU authorized representative is:

Springer Nature Customer Service Center GmbH
Europaplatz 3
69115 Heidelberg, Germany

Batch number: 10732312

Printed by Printforce, the Netherlands